Casting Alloy Design and Characterization

Casting Alloy Design and Characterization

Guest Editors

Carlos Alexandre Dos Santos
Eleani Maria Da Costa

MDPI

Basel • Beijing • Wuhan • Barcelona • Belgrade • Novi Sad • Cluj • Manchester

Guest Editors
Carlos Alexandre Dos Santos
School of Technology
Pontifical Catholic University
of Rio Grande do Sul
Porto Alegre
Brazil

Eleani Maria Da Costa
School of Technology
Pontifical Catholic University
of Rio Grande do Sul
Porto Alegre
Brazil

Editorial Office
MDPI AG
Grosspeteranlage 5
4052 Basel, Switzerland

This is a reprint of the Special Issue, published open access by the journal *Metals* (ISSN 2075-4701), freely accessible at: www.mdpi.com/journal/metals/special_issues/casting_alloy.

For citation purposes, cite each article independently as indicated on the article page online and using the guide below:

Lastname, A.A.; Lastname, B.B. Article Title. *Journal Name* **Year**, *Volume Number*, Page Range.

ISBN 978-3-7258-3526-3 (Hbk)
ISBN 978-3-7258-3525-6 (PDF)
https://doi.org/10.3390/books978-3-7258-3525-6

Contents

About the Editors

Carlos Alexandre Dos Santos

Carlos A. dos Santos is a full professor at Pontifical Catholic University of Rio Grande do Sul, Brazil. Dr. Santos has a B.Sc. degree in Mechanical Engineering from the School of Engineering of Piracicaba (1994), a Master's degree (1997), and a Ph.D. (2001) degree in Mechanical Engineering from the State University of Campinas, and a postdoctorate from the State University of Campinas (2001–2005). His research is in metallurgical processes and processing microstructure–mechanical properties. He is a member of the Brazilian Association of Metallurgy, Materials and Mining (ABM), the Brazilian Association of Engineering and Mechanical Sciences (ABCM), and the Brazilian Society for the Advancement of Science (SBPC). Dr. Santos has a scholarship in Research Productivity 2—CNPq (The Brazilian National Council for Scientific and Technological Development).

Eleani Maria Da Costa

Eleani Maria da Costa is a full professor at the Pontifical Catholic University of Rio Grande do Sul, Brazil. Dra. Costa graduated with a degree in Chemistry (Federal University of Santa Maria, Brazil) in 1986. She received her M.Sc. (1989) and Ph.D. (1995) in Mining, Metallurgical, and Materials Engineering (Federal University of Rio Grande do Sul, Brazil), with a sandwich period at the Massachusetts Institute of Technology (USA). She has experience in the area of materials and metallurgical engineering, with an emphasis on properties of metals and alloys, working mainly on the following topics: corrosion and degradation of cement by CO_2. Dra. Costa has a scholarship in Research Productivity 2—CNPq (The Brazilian National Council for Scientific and Technological Development).

Preface

Casting processes induce a wide range of solidification conditions, directly affecting the microstructure formation. Grain size and morphology, interdendritic spacing, solute segregation and precipitation, the presence of porosity, and other defects are strongly influenced by the thermal behavior of the metal/mold system during solidification, imposing a close correlation between solidification, the resulting microstructure, and the final properties and applications.

Since control of microstructure during casting processes involves heat and mass transfers, solidification, and phase transformation, numerous studies have focused on the development of correlations to optimize operating parameters as a function of a number of process variables, such as metal/mold chemical compositions, material thermo-physical properties, mold design, cooling conditions, and others.

This reprint is dedicated to studies related to casting alloy design and characterization, especially novel techniques; experimental and theoretical solidification studies; numerical and analytical simulations; heat and mass transfer; processing structure–property relations; and industrial applications.

The guest authors acknowledge all authors for their contributions and the reviewers, editors, and *Metals* editorial team for their time and effort.

Carlos Alexandre Dos Santos and Eleani Maria Da Costa

Guest Editors

metals

Editorial

Editorial for Special Issue "Casting Alloy Design and Characterization"

Eleani Maria da Costa and Carlos Alexandre dos Santos *

Materials Laboratory, School of Technology, Pontifical Catholic University of Rio Grande do Sul, Porto Alegre 90619-900, Brazil; eleani@pucrs.br
* Correspondence: carlos.santos@pucrs.br; Tel.: +55-51-3353-7655

Citation: da Costa, E.M.; dos Santos, C.A. Editorial for Special Issue "Casting Alloy Design and Characterization". *Metals* **2024**, *14*, 1236. https://doi.org/10.3390/met14111236

Received: 6 October 2024
Accepted: 23 October 2024
Published: 29 October 2024

1. Introduction

Solidification, the genesis of metallic materials, is a complex phenomenon encompassing fluid flow, heat transfer, phase transformation, liquid–solid interface, solute redistribution, gas trapping, and others. Thermal solidification parameters, such as the cooling rate, thermal gradient, and solidification rate, directly result from variables such as mold material and design, metal composition, metal–mold interface, molten metal temperature, and pouring conditions. Understanding the interaction of these variables is essential for predicting the performance of cast products, as well as the behavior in subsequent processes, e.g., heat treatment, forming, machining, surface treatment, etc. [1–8].

Although the topic has been extensively investigated in the past sixty years [9–14], many questions remain to be answered in the development of alloys to meet specific characteristics such as biocompatibility, high entropy, shape memory behavior, or high-temperature corrosion and wear resistances, for example. The pursuit for improving casting processes, reducing costs and time, saving energy, and maximizing quality also remains a constant. The successful development of design casting critically depends on understanding the fundamental knowledge of heat and fluid flow during solidification, mainly for processes producing high-quality products and conducting near-net-shape casting. With the evolution of computers since 1980, several analytical and numerical mathematical models have been proposed. The heat and mass transfer conditions during solidification and the consequent simulation of cooling patterns in castings have improved casting processes. The use of simulators contributes to increasing our understanding of these processes; however, some uncertainties must be eliminated before such simulations can be widely accepted as realistic descriptions. In general, these models can be grouped into two categories: solidification under steady-state heat and unsteady-state heat flow conditions. Reliable prediction in unsteady-state heat flow is of fundamental importance since this flow regime covers most industrial solidification processes [15–18].

Consequently, the imposition of a wide range of heat and mass transfer conditions in the casting process generates diverse solidification structures. Hence, structural parameters, such as grain size and morphology, dendritic characteristics, the distribution and number of phases, precipitated particles, the porosity distribution, and others, are highly influenced by the metal/mold thermal behavior during solidification, consequently imposing a close correlation between casting design and the resulting microstructure and subsequent processes [19–23]. The properties of alloys depend on the solidification microstructural arrangement. Under these circumstances, the as-cast microstructure will define the physical, chemical, and mechanical properties of the alloy. Expressions correlating these responses with microstructure parameters are useful, providing insight into the preprogramming of solidification in terms of the desired casting properties [24–28]. Despite the existing knowledge of casting design and alloy characterization, the relationship between the solidification conditions, microstructure evolution, and final casting properties has yet to be fully understood. Therefore, this Special Issue of Metals is dedicated to research related to the design and characterization of cast alloys, especially regarding correlations between processing, properties, and microstructures; numerical and analytical heat and mass transfer simulations; and industrial applications.

2. Published Works

In this Special Issue, ten studies were published based on the following topics: special casting alloys; new trends in industrial applications; interrelationships between material composition, casting conditions, and final properties; and unconventional rheocasting processes for Al-Si alloys.

High-entropy alloys, Fe-based, and Co-based multicomponent alloys were investigated in studies (1) and (6), respectively. Different chemical compositions have been proposed for the design of new alloy systems, including high-entropy and multicomponent doped alloys using experimental approaches. Mahmoud et al. (1) developed six new Fe-Mn-Ni-Cr-Al-Si alloys, exploring relationships between chemical compositions and mechanical properties such as hardness and compression strength. Cabrera-Peña et al. (6) studied the multicomponent CoCrFeMoNi alloy in both pure and Zr-doped conditions for its corrosion resistance response in a simulated seawater environment.

Study (2) reviews the evolution of cast bells, covering the historical evolution of casting bell technology in Asia; the main materials used, such as bronze, brass, cupronickel, and iron alloys; different microstructures; and the technologies currently applied in the casting process. The design of a new gating system to reduce liquid metal turbulence during the sand casting of aluminum alloys was proposed in study (7). Experimental and theoretical analyses were carried out by Bruna et al. (7) to achieve the highest-quality casting considering aspects such as gate design; melting velocity; mold filling conditions; impurity and gas entrapment, and their consequences on mechanical properties; and the porosity of the casting. Similarly, Xie, Lv, and Dong (10) conducted an investigation to improve the casting of a high-power engine exhaust elbow. They used numerical simulations to optimize the operational parameters of the precision casting process. Despite the complex casting, their results demonstrate that the castings exhibit high internal and surface quality, in addition to high dimensional accuracy.

Al-based alloys are widely used in engineering applications due to their enhanced physical–chemical properties—lightweight, high mechanical strength, high thermal/electrical conductivity, corrosion resistance, and manufacturing feasibility. Hence, studies (3), (4), (5), and (8) concern this alloy system. In general, the microstructure of Al alloys is characterized by an α-Al matrix. When alloying elements are added, characteristics such as toughness, hardness, castability, formability, and resistance to elevated temperatures can be improved. Noé et al. (3) investigated the influence of small additions of Be on the formation of the eutectic microconstituent of an Al-33Cu alloy and the effects on the microstructural characteristics and hardness. Under transient solidification, the base alloy presented a microstructure composed of eutectic colonies surrounded by a eutectic mixture of coarser cellular morphology, while the addition of Be promoted dendritic morphology. The effect of adding Y or Er on the formation of the microstructure, tensile property, and heat resistance of an Al-Zn-Mg-Cu alloy was studied by Glavatskikh et al. (4). The alloys were prepared and solidified in different metallic molds, subjected to precipitation hardening heat treatment and hot rolling. Both the as-cast heat-treated and hot-rolled conditions were investigated. With the results, the authors proposed a route to obtain heat-resistant alloys with improved casting characteristics and hot deformation behavior.

The influence of Cr additions to hypoeutectic Al–Cu alloys on the solidification condition, microstructure, hardness, and linear reciprocating sliding wear response was investigated by Lantmann et al. (5). Based on the results, interrelationships between microstructure features and mechanical properties were established considering unsteady-state heat transfer conditions. The study concluded that the addition of Cu and Cr affected solidification transformation temperatures and improved wear response. A study on the impact of different Fe/Mn ratios on the microstructure after the homogenization heat treatment of an Al–Mg–Si alloy was reported by Avalos, Torres, and Valdés (8). They aimed to analyze the transformation of harmful Fe-intermetallic compounds with needle- or plate-shaped morphologies (β-Al_6FeSi) into an α-$Al_{15}(FeMn)_3Si_2$ phase with Chinese-script morphology, which was successfully achieved.

Finally, the unconventional casting process of semisolid rheocasting technology was investigated in study (9) to produce semi-solid slurries from A380 and A356 alloys. Two sets of serpentine channel pouring (SCP) were designed and tested, the first with a vertical configuration and the second with a horizontal design. According to the authors, these designs induce conditions of high heat transfer and nucleation in the channel, obtaining a homogeneous and globular microstructure through an interruption in the nucleus growth by the slurry flow in the channel. The results show that these methods can be applied to modify the primary solid from dendrites to equiaxed grains in the processed alloys. However, the authors advise for further investigation.

3. Conclusions

Since the main properties of casting are directly linked to the microstructure, which is formed depending on the solidification conditions, heat and mass transfer, and fluid flow, among others, it is essential to understand the influence of these characteristics on the final casting properties. This Special Issue overviews state-of-the-art casting designs and alloy characterizations, with different approaches and points of view.

We believe that this subject is extensive and deserves in-depth and constant investigation. Moreover, we hope that these selected contributions are useful for professionals, engineers, and researchers working with casting alloy designs and characterizations.

Acknowledgments: The Guest Authors acknowledge all authors for their contributions and the reviewers, editors, and Metals editorial team for their time and effort.

Conflicts of Interest: The authors declare no conflicts of interest.

List of Contributions :

1. Mahmoud, E.; Shaharoun, A.; Gepreel, M.; Ebied, S. Phase Prediction, Microstructure and Mechanical Properties of Fe–Mn–Ni–Cr–Al–Si High Entropy Alloys. *Metals* **2022**, *12*, 1164. https://doi.org/10.3390/met12071164.
2. Won, C.; Jung, J.; Won, K.; Sharma, A. Technological Insights into the Evolution of Bronze Bell Metal Casting on the Korean Peninsula. *Metals* **2022**, *12*, 1776. https://doi.org/10.3390/met12111776.
3. Rodrigues, A.; Kakitani, R.; Silva, C.; Giovanetti, L.; Dias, M.; Henein, H.; Garcia, A.; Cheung, N. Influence of Minor Additions of Be on the Eutectic Modification of an Al-33wt.%Cu Alloy Solidified under Transient Conditions. *Metals* **2023**, *13*, 94. https://doi.org/10.3390/met13010094.
4. Glavatskikh, M.; Barkov, R.; Gorlov, L.; Khomutov, M.; Pozdniakov, A. Novel Cast and Wrought Al-3Zn-3Mg-3Cu-Zr-Y(Er) Alloys with Improved Heat Resistance. *Metals* **2023**, *13*, 909. https://doi.org/10.3390/met13050909.
5. Lantmann, R.; Mariante, A.; Pinheiro, T.; da Costa, E.; dos Santos, C. Microstructure, Hardness, and Linear Reciprocating Sliding Wear Response of Directionally Solidified Al–(2.5, 3.5, 4.5)Cu–(0.25, 0.50)Cr Alloys. *Metals* **2023**, *13*, 1178. https://doi.org/10.3390/met13071178.
6. Cabrera-Peña, J.; Brito-Garcia, S.; Mirza-Rosca, J.; Callico, G. Electrical Equivalent Circuit Model Prediction of High-Entropy Alloy Behavior in Aggressive Media. *Metals* **2023**, *13*, 1204. https://doi.org/10.3390/met13071204.
7. Brůna, M.; Galcik, M.; Pastircak, R.; Kantorikova, E. Effect of Gating System Design on the Quality of Aluminum Alloy Castings. *Metals* **2024**, *14*, 312. https://doi.org/10.3390/met14030312.
8. Avalos, A.; Torres, J.; Flores Valdés, A. Effect of the Fe/Mn Ratio on the Microstructural Evolution of the AA6063 Alloy with Homogenization Heat Treatment Interruption. *Metals* **2024**, *14*, 373. https://doi.org/10.3390/met14040373.

9. Alfredo, H.; José Federico, C.; Aldo, H.; Miguel Ángel, S. Semi-Solid Slurries for Rheocasting of Hypoeutectic Al-Si-X Alloys Produced by Self-Stirring in Serpentine Channels. *Metals* **2024**, *14*, 413. https://doi.org/10.3390/met14040413.
10. Xie, S.; Lv, Z.; Dong, S. Study on the Optimization of Investment Casting Process of Exhaust Elbow for High-Power Engine. *Metals* **2024**, *14*, 481. https://doi.org/10.3390/met14040481.

References

1. Quaresma, J.M.V.; Santos, C.A.; Garcia, A. Correlation between unsteady-state solidification conditions, dendrite spacings, and mechanical properties of Al-Cu alloys. *Met. Mater. Trans. A* **2000**, *31*, 3167–3178. [CrossRef]
2. Santos, C.A.; Quaresma, J.M.V.; Garcia, A. Determination of transient interfacial heat transfer coefficients in chill mold castings. *J. Alloys Compd.* **2001**, *319*, 174–186. [CrossRef]
3. Jo, Y.H.; Moon, H.-R.; Bae, J.W.; Yoo, J.; Lee, S.G.; Lee, Y.-S. Effects of casting speed on microstructural and tensile properties of Al–Mg–Si alloy fabricated by horizontal and vertical twin-roll casting. *J. Mater. Res. Technol.* **2023**, *26*, 8010–8024. [CrossRef]
4. Barcellos, V.K.; Ferreira, C.R.F.; Spim, J.A.; Santos, C.A.; Garcia, A. The Interrelation between casting size, steel grade, and temperature evolution along the mold length and at the strand surface during continuous casting of steel. *Mater. Manuf. Process.* **2011**, *26*, 113–126. [CrossRef]
5. Çadırlı, E.; Kaya, H.; Büyük, U.; Gunduz, M. Effect of heat treatment on the microstructures and mechanical properties of Al–4Cu–1.5Mg alloy. *Int. J. Met.* **2022**, *16*, 1020–1033. [CrossRef]
6. Khosroaghayani, M.; Khorasanian, M.; Alavi Zaree, S.R.; Eskandari, M. Investigation of microstructure and mechanical properties of AZ91 magnesium alloy produced by directional solidification method in different angles using CAFE simulation. *Int. J. Met.* **2023**, *17*, 195–209. [CrossRef]
7. Goto, I.; Fukuchi, K.; Kurosawa, K. Effects of solidification conditions on the microstructural morphologies and strengths of hypereutectic high-chromium white cast iron castings. *Mater. Sci. Eng. A* **2023**, *886*, 145692. [CrossRef]
8. Machado, P.A.B.; Quaresma, J.M.V.; Garcia, A.; Santos, C.A. Investigation on machinability in turning of as-cast and T6 heat-treated Al-(3, 7, 12%)Si-0.6%Mg alloys. *J. Manuf. Process.* **2022**, *75*, 514–526. [CrossRef]
9. Lipnitsky, A.M. *The Melting of Cast Iron and Non-Ferrous Alloys*; Peace Publishers: Minnetonka, MN, USA, 1960.
10. Chalmers, B. *Principles of Solidification*; John Wiley & Sons: Hoboken, NJ, USA, 1964.
11. Davies, G.J. *Solidification and Casting*; Applied Science Publishers: London, UK, 1973.
12. Flemings, M.C. *Solidification Processing*; McGraw-Hill: New York, NY, USA, 1974.
13. Kurz, W.; Fisher, D.J. *Fundamentals of Solidification*; Trans Tech Publications: Stafa-Zurich, Switzerland, 1984.
14. Campbell, J. *Castings*; Butterworth-Heinemann Ltd.: Oxford, UK, 1993.
15. Thomas, B.G.; Samarasekera, I.V.; Brimacombe, J.K. Comparison of numerical modeling techniques for complex, twodimensional, transient heat-conduction problems. *Metall. Trans. B* **1984**, *15*, 307–318. [CrossRef]
16. Louhenkilpi, S.; Miettinen, J.; Holappa, L. Simulation of microstructure of As-cast steels in continuous casting. *ISIJ Int.* **2006**, *46*, 914–920. [CrossRef]
17. Ferreira, I.L.; Garcia, A. The application of numerical and analytical approaches for the determination of thermophysical properties of Al–Si–Cu–Mg alloys. *Contin. Mech. Thermodyn.* **2020**, *32*, 1231–1244. [CrossRef]
18. Miettinen, J.; Somani, M.; Visuri, V.-V.; Koskenniska, S.; Louhenkilpi, S.; Fabritius, T.; Kömi, J. Simulation of the Solidification and Microstructural Evolution in Steel Casting Processes Using the InterDendritic Solidification Tool. *Steel Res. Int.* **2022**, *93*, 2200120. [CrossRef]
19. Lahbari, A.; Bouchaala, K.; Essoussi, H.; Faqir, M.; Ettaqi, S.; Essadiqi, E.H. Homogenization heat treatment influence on microstructure evolution and mechanical properties of as-cast Al–Li–Cu–Mg–Zr alloy for lightweight aerospace application. *Heliyon* **2024**, *10*, e24426. [CrossRef]
20. Han, J.; Dong, G.; Li, S.; Zheng, J.; Wang, J.; Li, H.; Starostenkov, M.D.; Bi, J. Uneven distribution of cooling rate, microstructure and mechanical properties for A356-T6 wheels fabricated by low pressure die casting. *J. Manuf. Process.* **2024**, *127*, 196–210. [CrossRef]
21. Yang, Q.; Shi, W.; Wang, M.; Jia, L.; Wang, W.; Zhang, H. Influence of cooling rate on the microstructure and mechanical properties of Al–Cu–Li–Mg–Zn alloy. *J. Mater. Res. Technol.* **2023**, *25*, 3151–3166. [CrossRef]
22. Gursoy, O.; Timelli, G. The influence of Gd content on the solidification and microstructure of $AlSi_7Mg_{0.3}$ casting alloy. *J. Therm. Anal. Calorim.* **2024**, *149*, 3125–3139. [CrossRef]
23. Azevedo, H.M.; Mendes, G.H.; Rodrigues, H.C.; Rocha, S.F.; Costa, T.S.; Moreira, A.L.; Rocha, O.L. Effects of solidification, solution and aging processes on an AlCuNb alloy: Thermal and microstructural analysis, microhardness, electrical resistivity and conductivity. *J. Therm. Anal. Calorim.* **2024**, *149*, 953–966. [CrossRef]
24. He, C.; Yu, W.; Li, Y.; Wang, Z.; Wu, D.; Xu, G. Relationship between cooling rate, microstructure evolution, and performance improvement of an Al–Cu alloy prepared using different methods. *Mater. Res. Express* **2020**, *7*, 116501. [CrossRef]
25. Sarkar, J.; Modak, P.; Singh, S.B.; Chakrabarti, D. Effect of cooling-rate during solidification on the structure-property relationship of hot deformed low-carbon steel. *Mater. Chem. Phys.* **2021**, *257*, 123826. [CrossRef]
26. Wang, Y.; Gao, M.; Yang, B.; Zhao, E.; Liu, F.; Guan, R. Microstructural evolution and mechanical property of Al–Mg–Mn alloys with various solidification cooling rates. *Mater. Charact.* **2022**, *184*, 111709. [CrossRef]

27. Obara, T.J.; Minei, R.O.; Kakitani, R.L.; Spinelli, J.E.; Garcia, A.; Cheung, N. Solidification microstructure and mechanical properties of an Al-11wt%Si alloy modified with Nb and Sb. *J. Phys. Chem. Solids* **2023**, *181*, 111521. [CrossRef]
28. Shi, Y.-f.; Guo, C.-j.; Yuan, M.-q.; Jia, Z.-b.; An, G.-h.; Xiao, X.-p.; Yang, B. Effect of solidification rate on dendrite segregation and mechanical properties of Cu–15Ni–8Sn alloy prepared by directional solidification. *J. Iron Steel Res. Int.* **2023**, *30*, 1586–1597. [CrossRef]

metals

Article

Study on the Optimization of Investment Casting Process of Exhaust Elbow for High-Power Engine

Shiyu Xie [1], Zhaozhao Lv [1,*] and Shengquan Dong [1,2]

[1] School of Armament Science and Technology, Xi'an Technological University, Xi'an 710032, China; alittleapple@126.com (S.X.); 13522672868@126.com (S.D.)
[2] School of Materials Science and Chemical Engineering, Xi'an Technological University, Xi'an 710032, China
* Correspondence: lvzhaozhao@xatu.edu.cn

Abstract: The high-power engine exhaust elbow has a complex construction, which makes it susceptible to casting flaws that could negatively impact its functionality. Therefore, the investment casting scheme was established and optimized in this study in order to cast structurally complete exhaust elbows for high-horsepower engines. ProCAST software was used to simulate and optimize the casting and solidification processes. The optimal process parameters were determined as follows: pouring temperature of 1650 °C, pouring speed of 1.5 kg/s, and shell preheating temperature of 1050 °C. The optimization of the primary parameters of the casting process, along with the results of dimensional accuracy analysis, shape and positional deviation, and defect detection, were validated through testing. The results indicated that the optimized castings had no casting defects and complied with the design specifications.

Keywords: investment casting; exhaust elbow; ProCAST software; high-powered engine

Citation: Xie, S.; Lv, Z.; Dong, S. Study on the Optimization of Investment Casting Process of Exhaust Elbow for High-Power Engine. *Metals* **2024**, *14*, 481. https://doi.org/10.3390/met14040481

Academic Editors: Carlos Alexandre Dos Santos and Eleani Maria Da Costa

Received: 12 March 2024
Revised: 10 April 2024
Accepted: 15 April 2024
Published: 20 April 2024

1. Introduction

The exhaust pipe is a crucial component of the engine and is necessary for it to operate properly [1–3]. Exhaust pipes for high-horsepower engines, like those found in specialty vehicles, maritime vessels, and huge internal combustion engines, must be structurally strong and perform exceptionally well for the engine to run properly [4,5]. However, the exhaust pipes for high-horsepower engines that are currently being manufactured have some defects, particularly in the case of the exhaust pipes with a curved shape, which has an impact on the exhaust pipes' performance and the advancement of high-horsepower engines [6–9]. These defects are caused by problems in the process of preparing exhaust pipes for high-horsepower engines. Therefore, it is crucial to research and optimize the exhaust elbow manufacturing process for high-horsepower engines to produce high-quality exhaust elbows.

Currently, the manufacturing methods for exhaust elbows mainly include castings, machining, welding–forming, and other technologies [10–14]. Setiyorini Yuli et al. [15] fabricated an acetabular prosthesis using the investment casting technique. In order to reduce the production cost, the casting scheme was optimized using process simulation. According to Kumar P's research on the investment casting process, shrinkage-related casting flaws can be minimized by applying a thin layer of wax coating to the plastic model [16]. Smruti Ranjan Pradhan [17] compared the mechanical, dimensional, and biocompatibility features of canine dental crowns fabricated by direct metal laser sintering (DMLS) and DMLS-waste-assisted investment casting using two different alloys. The study found that investment casting is cost-effective and offers better mechanical properties compared to DMLS parts. Chander Prakash [18] contrasted the technology of additive manufacturing with a few other methods, and the findings indicated that additive manufacturing is a quick, affordable, and environmentally friendly production process. The majority of parts are now made utilizing casting and welding–machining procedures, even though additive manufacturing is a relatively new technology that is utilized to make some sophisticated parts. It has been discovered that the investment casting technique offers several noteworthy benefits when it comes to creating intricate castings with thin walls [19–21]. Complex castings can be produced in a single process thanks to investment casting, which lowers manufacturing costs and speeds up production [22–25]. However, casting flaws including shrinkage and shrinkage holes, as well as air bubble holes, are frequently present in castings made using the investment casting process [26–29].

In this work, the investment casting process has been refined using the numerical modeling technique to prepare exhaust elbows for high-horsepower engines with structural integrity that are free of flaws. Through simulation optimization, an ideal set of process parameters was discovered and used in the manufacturing of exhaust elbows for high-horsepower engines. The manufactured exhaust elbows were subjected to industrial CT testing in order to confirm the outcomes of the improved process.

2. Program Design

2.1. Structural Design of Exhaust Elbow Casting

A three-dimensional model of the exhaust elbow is shown in Figure 1. The exhaust elbow construction consists of a straight pipe, flange, and elbow, with external dimensions of approximately 217 × 150 × 120 mm. Figure 1a shows the straight section of the exhaust elbow which has a length of 93 mm, an outer diameter of 90 mm, an inner diameter of 87 mm, and an average wall thickness of 1.5 mm. Figure 1b shows a flange between the two end faces which is 16 mm thick with an annular groove of 13 mm depth and less than 6 mm width on the inside. This flange has an outside diameter of 91 mm and its annular groove wall thickness is not more than 2.5 mm. Figure 1c shows the bent portion of the exhaust elbow, which has an average wall thickness of less than 2 mm. The bend is approximately 90°, and the end of the bend is 20° outward-flared. It can be seen by analyzing the structure of the exhaust elbow that the exhaust elbow has a thin wall and a large curvature. Therefore, shrinkage holes, hot cracks, and other casting defects can easily occur during the casting process.

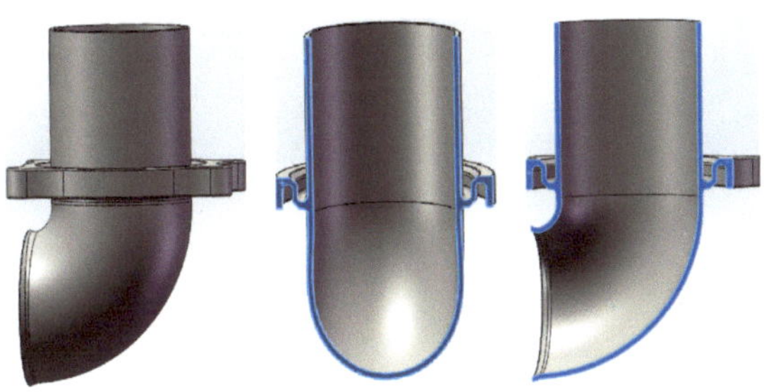

(a) (b) (c)

Figure 1. Three-dimensional diagram of the exhaust elbow and its cross-section: (**a**) three-dimensional drawing of exhaust elbow; (**b**) side view; (**c**) front view.

2.2. Exhaust Elbow Materials

Table 1 shows the main components of the materials used for casting exhaust elbows. High-chromium–nickel austenitic stainless steel (1Cr20Ni14Si2) alloy is used for the exhaust elbows.

Table 1. The main components of 1Cr20Ni14Si2 alloy (adapted from Ref. [30]).

Elemental	C	Cr	Ni	Si	Mn	P	S	Fe
Standard Range Value	≤0.2	19–22	12–15	1.5–2.5	≤1.5	≤0.035	≤0.03	Remainder
Actual Value	0.0629	21.66	12.44	1.7	1.02	0.0296	0.0055	Remainder

2.3. Pouring System Design

ProCAST casting simulation software (version 2021.5) is used to simulate the casting process. Ease of filling and avoiding casting flaws should be taken into account while designing the casting system [31–34]. There are high dimensional accuracy requirements for the exhaust elbow's straight pipe and elbow at the thin-walled construction, so avoid setting up in these inner-runner sections. Nonetheless, an inner runner with a sizable cross-sectional area at the flange is achievable due to the comparatively thick flange. For two primary reasons, the inner runner is arranged with a large cross-sectional area: first, it guarantees a smooth flow of the molten metal liquid; second, it provides sufficient pressure to fill the areas of the casting with metal liquid, preventing casting flaws. The inner runner is positioned on the flange of the casting, and a side-injection type is utilized to establish four inner runners on the four lugs of the flange. The three-dimensional diagram and the mesh model are illustrated in Figure 2 below. The mesh-edge length of the casting system part and other non-focused parts is 3 mm, while the mesh-edge length of the exhaust pipe casting is 1 mm. The number of casting face meshes is 658,572, the number of casting system face meshes is 104,176, and the total number of body meshes is 2,055,150. The mesh division of the 3D group tree model of the pouring system is shown in Figure 3.

Figure 2. Schematic diagram of the pouring system model.

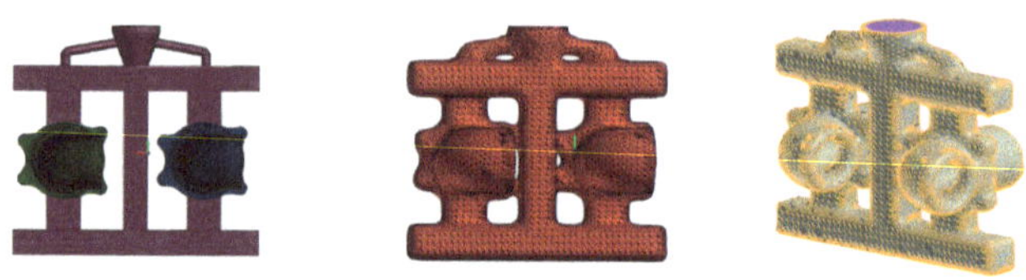

(**a**) (**b**) (**c**)

Figure 3. Face and body meshing of the 3D group tree model of the pouring system: (**a**) Meshing of castings; (**b**) Shell face meshing; (**c**) Mesh division of castings and molded shells.

In the simulation process, the castings are built up with an alloy casting system type and a 1550 °C temperature. The exhaust casting's stress type is configured to an elastoplastic model, while the casting system part's default stress type is a linear–elastic model.

The type of mold shell is Mould Shell, the material is Chemical Mold Shell Sand, the filling is 100%, and the temperature of the mold shell is 800 °C. The rest of the parts are set to rigid.

3. Casting Process Simulation

3.1. Analysis of the Simulation Results of the Filling Process

Figure 4 shows the exhaust pipe casting pouring system filling process. The liquid metal is injected from the riser and flows into the inner gate at the bottom after filling the cross gate below from the middle straight gate when the mold is filled to a quarter of its capacity. As the liquid metal moves from the gate through the straight gate to the bottom of the cross gate during liquid splashing, gas may be produced; however, this only happens in the pouring system and has no bearing on the casting. Because of its wider cross-section, the inner gate makes it easier to fill with metal liquid and prevents flaws. As shown in Figure 4c, when 75% of the mold is filled, the liquid metal has completely filled the two castings, and the top-two inner sprues are also filled. Cooling and solidification begin from the thin-walled part of the casting. When 98% of the mold is filled, the entire 3D model is filled, and cooling and solidification start at the thinnest part of the casting wall. Until the entire cavity was filled, the temperature of the alloy liquid continuously decreased throughout the filling process, and there was no underpouring of the casting. The total filling time was 9.81 s.

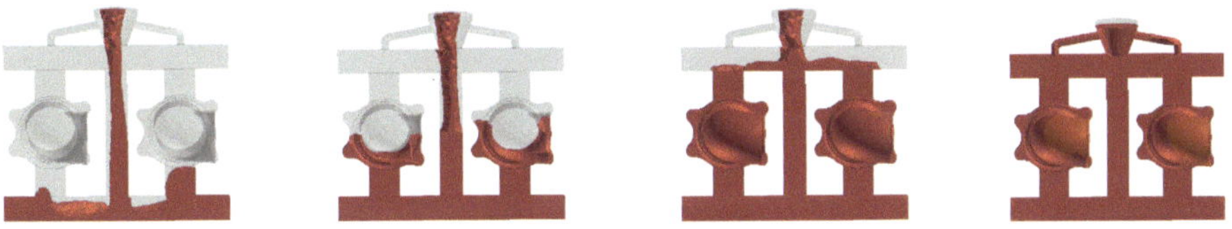

(a) (b) (c) (d)

Figure 4. Simulation of the filling process of the exhaust elbow: (**a**) 25% of the charge; (**b**) 50% of the charge; (**c**) 75% of the charge; (**d**) 98% of the charge.

3.2. Analysis of Simulation Results of Solidification Process

Figure 5 shows the solidification process of the exhaust elbow casting system. When the solidification volume fraction reaches 25%, the casting is essentially solidified. At this point, only the sprue and casting-connection position have a solid-phase rate of 0.6, while the rest of the casting shows a solid-phase rate between 0.9 and 1. Throughout the casting process from the position distant from the gate, the solid-phase rate hits 1 when the solidification volume fraction reaches 50%. The casting is nearly entirely solidified when the volume percentage of solidification reaches 75%, with the exception of the solid-phase ratio of 0.5 at the upper-gate intersection. The total time required for complete solidification is calculated to be 1743 s.

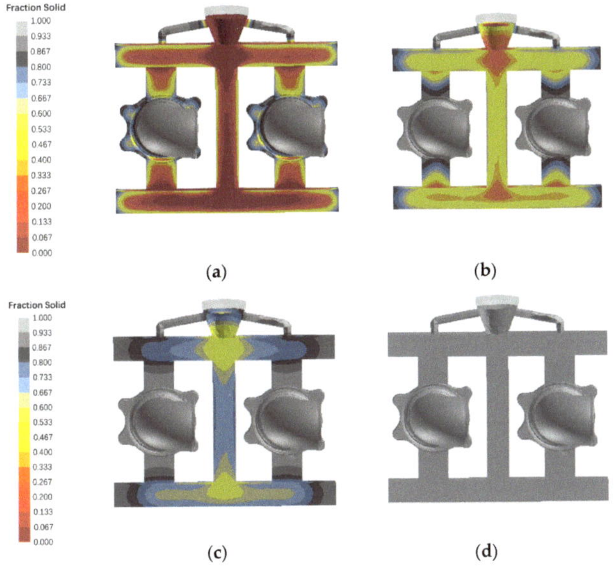

Figure 5. Simulation of the solidification process of the exhaust elbow: (**a**) 25% of the charge; (**b**) 50% of the charge; (**c**) 75% of the charge; (**d**) 100% of the charge.

3.3. Location Distribution of Casting Defects

Figure 6 shows the simulation results of casting flaws. The figure indicates that shrinkage holes are the primary flaws at the corners of the casting system and at the riser–cross-gate connections. There is comparable shrinkage in both castings, mostly on the flanges, in terms of both place and proportion. However, a small amount of shrinkage was observed at the straight pipes and elbows. Calculations reveal that the total volume of shrinkage inside both castings amounts to approximately 2.12 cc, with each casting having around 1.06 cc.

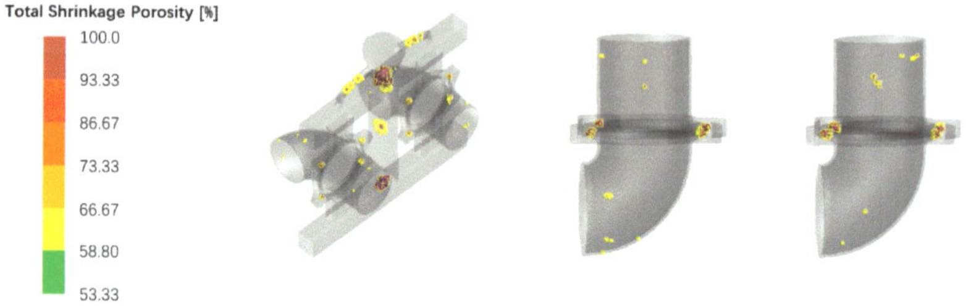

(a) (b)

Figure 6. Exhaust elbow shrinkage distribution results: (**a**) Shrinkage distribution with casting system; (**b**) Distribution of shrinkage in castings without pouring system.

A diagram illustrating the tendency of exhaust elbow castings to crack thermally is shown in Figure 7. In the figure, (a) and (b) indicate two specific locations where thermal cracking defects may occur, respectively. Area A is the section with the highest curvature on the exhaust pipe (the vulnerable point on the exhaust pipe casting), exhibiting an average effective stress of 439.5 MPa and an HTI (Horizontal Transverse Isotropy) value of 0.000163, whereas area B shows an average effective stress of 258 MPa and an HTI value of 0.000760.

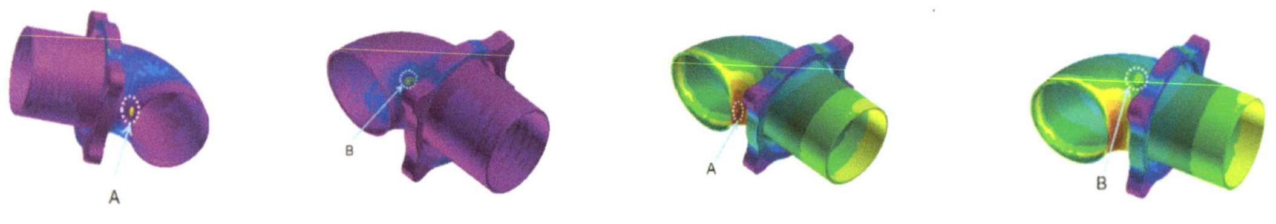

(a) (b) (c) (d)

Figure 7. Thermal cracking tendency diagram and effective stress of exhaust pipe casting: (**a**) Exhaust pipe bend maximum position HTI value: 0.000163; (**b**) Maximum position of exhaust elbow HTI: 0.000760; (**c**) Effective stress at area A: 439.5 MPa; (**d**) Effective stress at area B: 258 MPa.

3.4. Programme Optimization

3.4.1. Optimization of Casting Parameters

The casting technique was adjusted to address casting problems that showed up in the first simulations, including shrinkage, shrinkage holes, and thermal cracks. Seven groups with different pouring temperatures were utilized for simulation experiments, and the results are presented in the Table 2. At a pouring temperature of 1650 °C, the shrinkage volume of the exhaust elbow casting is minimized. Particularly, the shrinkage defects in the thin-walled part of the casting are eliminated. Shrinkage on the flange can be ignored as the flange needs to be machined later. At 1650 °C, the stress at the maximum HTI value is 272 MPa, while the ultimate yield strength of the alloy is not less than 295 MPa, and no thermal cracking defects are produced at points A and B.

Table 2. The effect of different pouring temperature on the volume of internal shrinkage of exhaust pipe castings, HTI values, and effective stresses in A and B regions of exhaust pipe castings [1].

Pouring Temperature (°C)	Crude Volume (cc)	HTI (Region A)	HTI (Region B)	Effective Stress A (Mpa)	Effective Stress B (Mpa)
1530	2.75	0.000190	0.000837	416	264.9
1550	2.12	0.000240	0.000751	435.7	275.6
1570	2.54	0.000441	0.000640	447.8	276.9
1590	1.73	0.000495	0.000554	454.6	283.6
1610	1.97	0.000270	0.000608	371.5	265.6
1630	1.49	0.000398	0.000543	477.1	337.8
1650	1.45	0.000339	0.000486	379.4	272

[1] Pouring temperature 1550 °C, Shell preheating temperature 800 °C.

Seven different casting speeds are set for simulation experiments. The casting temperature and shell preheating temperature are initialized with 1550 °C and 800 °C parameters, respectively. The casting speed is simulated by taking one parameter every 0.5 s in the range of 1.5–4.5 kg/s and the results are shown in Table 3.

Table 3. The effect of different pouring speed on the volume of internal shrinkage of exhaust pipe castings.

Pouring Temperature (°C)	Shell Preheating Temperature (°C)	Pouring Speed (kg/s)	Crude Volume (cc)
1550	800	1.5	1.77
		2.0	1.93
		2.5	2.37
		3.0	2.12
		3.5	3.42
		4.0	3.51
		4.5	3.91

Seven groups with different shell preheating temperatures are established for the simulation. The results are presented in Table 4, indicating that elevating the preheating temperature of the shell can effectively minimize the shrinkage defects within the casting.

Table 4. The size of internal shrinkage volume of exhaust pipe castings under different shell preheating temperatures.

Pouring Temperature (°C)	Pouring Speed (kg/s)	Shell Preheating Temperature (°C)	Crude Volume (cc)
1550	3	750	1.77
		800	1.93
		850	2.37
		900	2.12
		950	3.42
		1000	3.51
		1050	3.91

When the preheating temperature of the shell reaches 1050 °C, it is observed that the shrinkage defects in all thin-walled parts disappear, and only a small amount of shrinkage defects are concentrated in the flange lugs. This indicates that an increase in the preheating temperature of the mold will significantly slow down the cooling and solidification rate of the alloy liquid. This can effectively guarantee continuous shrinkage of the alloy liquid in thin-wall and flange-wall sections during casting, thereby reducing the likelihood of shrinkage defects occurring.

3.4.2. Simulation of Optimal Casting Solutions

Three characteristics were combined to model the exhaust casting process and reduce faults. As shown in Figure 8, the distribution of internal shrinkage in the exhaust pipe casting is illustrated. Under these process parameters, shrinkage and shrinkage holes inside the casting are primarily concentrated in the thickest part of the flange. No shrinkage or shrinkage hole defects occur in the thin-walled sections of the casting, such as straight pipes and elbows.

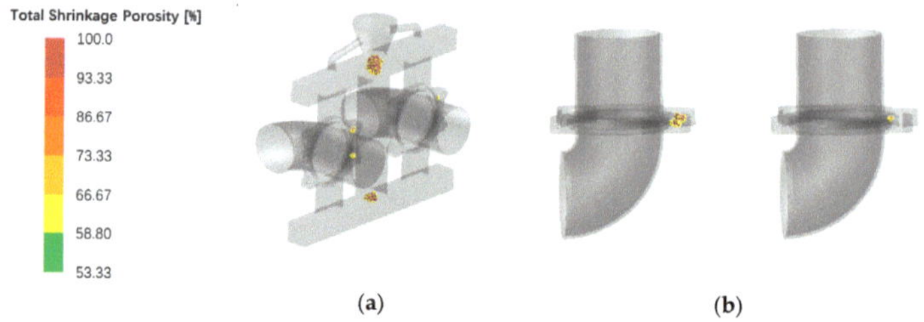

Figure 8. Shrinkage distribution inside the exhaust pipe casting: (**a**) Shrinkage distribution with casting system; (**b**) Shrinkage distribution without casting system.

Under ideal process conditions, the dynamics of the mechanical behavior of particular exhaust bend sections A and B are depicted in Figure 9. As the temperature increases, the effective stress in the corresponding region decreases. At the end of solidification, the effective stress in regions A and B is less than 100 MPa. The corresponding elastic strain fluctuates in the range of 0.4–1%, which is small and does not show any obvious change. The mechanical behavior for both locations suggests a low probability of flaws resulting from thermal cracking at the end of solidification.

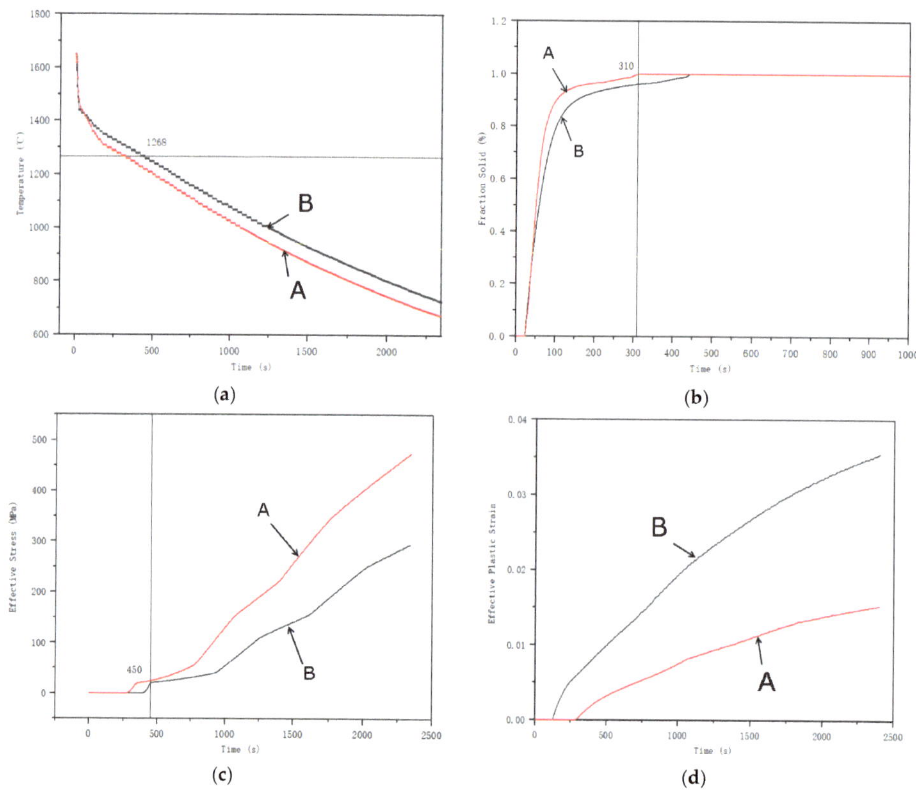

Figure 9. Dynamic changes in mechanical behavior at specific regions A and B on the exhaust bend with optimal process parameters: (**a**) Temperature–Time; (**b**) Fraction Solid Rate–Time; (**c**) Effective stress–Time; (**d**) Effective Plastic Stress–time.

In summary, the process parameters are as follows: alloy temperature at casting is 1650 °C, casting filling speed is 1.5 kg/s, and shell temperature is 1050 °C.

4. Casting Trial Production and Process Inspection

4.1. Preparation and Production for Investment Casting

The wax molds, shells [35], and castings with a pouring system used in the trial production are shown in Figure 10. Investment casting mold shells are constructed in a continuous fashion, with the final layer sealed with slurry, for a total of four and a half layers with a wall thickness of 9–10 mm. The viscosity of the slurry used for sealing the layers is presented in Table 5.

Table 5. Slurry viscosity and drying process parameters for each layer.

Storey	Slurry Composition	Stickiness	Type of Sanding	Grits	Drying Time (H)
Top layer	Zirconium Powder Silica Sol	40 ± 5 s	Zircon Sand	120	15
Second layer	Mullite Powder Silica Sol	35 ± 5 s	Mullite Sand	60	24
Third layer	Mullite Powder Silica Sol	25 ± 2 s	Mullite Sand	30	24
Fourth layer	Mullite Powder Silica Sol	20 ± 2 s	Mullite Sand	30	24
Seal coat	Mullite Powder Silica Sol	12 ± 1 s	-	-	24

Ambient temperature: 22–28 °C Humidity: 40–60%.

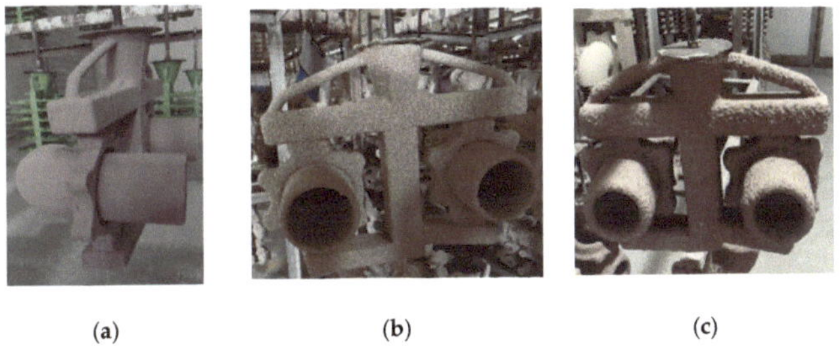

Figure 10. Exhaust Bend Module Shell Making: (**a**) Surface sanding; (**b**) Sanding of the transition layer; (**c**) Backing slurry.

The castings are poured separately using the original and optimized parameters. After the casting is molded, insulation is spread on the sprue location, and then it is air-cooled. After cooling, the sand shell outside the casting was removed by vibration using a vertical shell-vibration machine with a vibration impact frequency of not more than 1200 times per minute. After that, the casting was cut, polished at the sprue area, and subjected to shot blasting and pickling passivation.

As can be clearly seen in Figure 11, the connection parts of the integrated molding process of investment casting are significantly better than those of the traditional machining–welding process. The appearance of castings before and after optimizing process parameters is essentially the same.

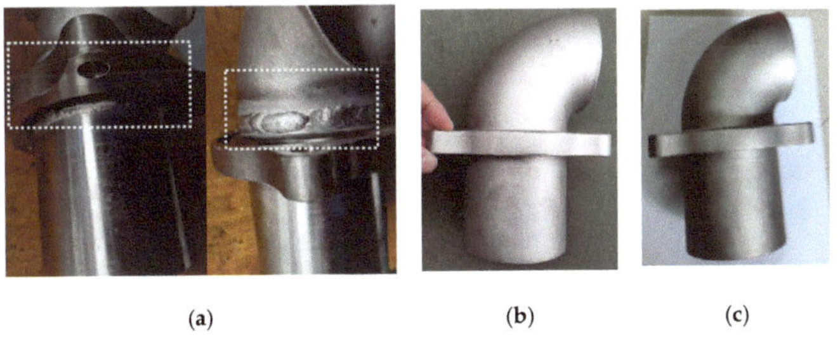

Figure 11. Comparison of different process parameters: (**a**) Traditional machining–welding process; (**b**) Initial casting process parameters castings; (**c**) Optimized process parameters castings.

4.2. *Quality Verification of Exhaust Pipe Castings*

In the trial production stage of the product, a total of 12 trial production pieces were made, resulting in the successful production of 10 exhaust bend castings, with a yield rate of 83.3%. The remaining two pieces have obvious casting defects on the surface. Two factors contribute to this: first, shape damage from workers' unavoidable operational errors during the production process; second, holes from the mold shell formed in the flange's two annular grooves during the shell formation process: iron clamping happens when alloy liquid enters the holes during pouring. The latter was significantly improved by extending the drying time of the prepared shell.

4.2.1. Exterior Inspection of Exhaust Elbows

The dimensions of the exhaust elbow castings were measured using a Coordinate Measuring Machine (CMM) and vernier calipers, as illustrated in Table 6. Following testing, the dimensions of the produced castings align with the design specifications, and the tolerances fall within the acceptable range.

Table 6. Exhaust elbow casting dimensional inspection results.

Dimensions	Required Size (mm)	Maximum Deviation Detection Value	Number of Tests (Pieces)	Test Results
Straight tube inner diameter	$\varphi 86^{+0.3}_{0}$	φ86.2	10	Qualified
Outer diameter of straight tube	$\varphi 91^{0}_{-0.2}$	φ91	10	Qualified
Flange thickness (plus allowance)	$22^{0}_{-0.3}$	22	10	Qualified
Ring groove size	$\varphi 120^{+0.2}_{0}$	φ120.18	10	Qualified
Flange outer diameter	$\varphi 128^{0}_{-0.2}$	φ127.9	10	Qualified
Elbow Size	$61^{0}_{-0.5}$	60.8	10	Qualified
Wall thickness of exhaust pipe	2	2	10	Qualified
Exhaust elbow centre distance	$70^{+0.5}_{-0.5}$	70.2	10	Qualified
Straight pipe + flange length	$99.5^{0}_{-0.5}$	99.5	10	Qualified

Visual inspection of the casting surface quality reveals a relatively high standard, with no signs of oxidized skin, mottled scars, wrinkled skin, or other defects.

4.2.2. Internal Quality Inspection of Exhaust Pipe Castings

X-ray inspection equipment used: ICT-450 for industrial CT inspection [36], following inspection standards as per GB/T 29070-2012 [37]. Voltage: 430 kV, current: 1.5 mA, focal size: 0.4 mm, integration time: 1000 ms, ray source-detection distance: 1100 mm, ray source-sample distance: 700 mm. The acquisition method is a standard cone beam, with an amplitude of 1600 and a voxel size of 0.127 mm. The test method is shown in Figure 12.

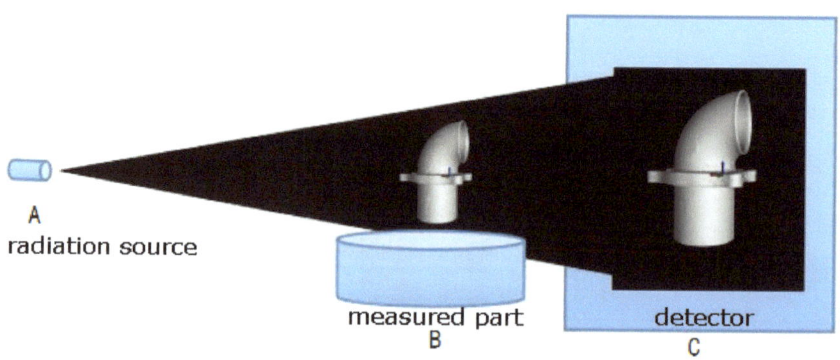

Figure 12. Schematic diagram of industrial CT inspection.

After X-ray inspection, the exhaust elbow castings produced under the initial process parameters (casting temperature 1550 °C, casting speed 3 kg/s, shell preheating temperature 800 °C) were found to have shrinkage, shrinkage holes, and porosity defects, as shown in Figure 13. The casting shrinkage in the elbow part is generally consistent with the simulation results, and no cracks were detected.

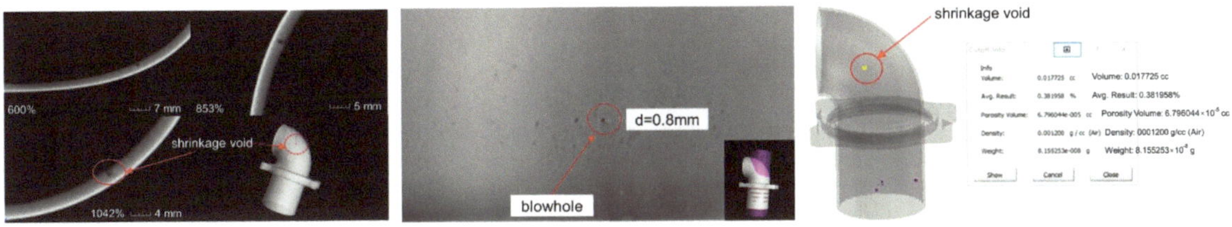

(a) (b) (c)

Figure 13. CT inspection results of exhaust elbow castings under initial pouring process parameters: (**a**) Shrinkage of bent pipe sections; (**b**) Dimensions of blowhole defects in straight sections; (**c**) (Analogue) Bend section corresponding to shrinkage position.

The CT section of the optimized exhaust pipe is illustrated in Figure 14. Figure 14a represents a cross-section along the center of the ring for the bend section. Figure 14b depicts a transverse section for the straight section. Figure 14c displays DR (Digital Radiography) result, and Figure 14d shows a cross-section at the flange. No shrinkage hole defects are observed in the exhaust bend castings within the bend section, straight section, and flange area. Additionally, no shrinkage phenomenon is detected in thin-walled parts or casting system hot joints, and there are no signs of thermal cracks present. The internal quality of these castings meets all requirements.

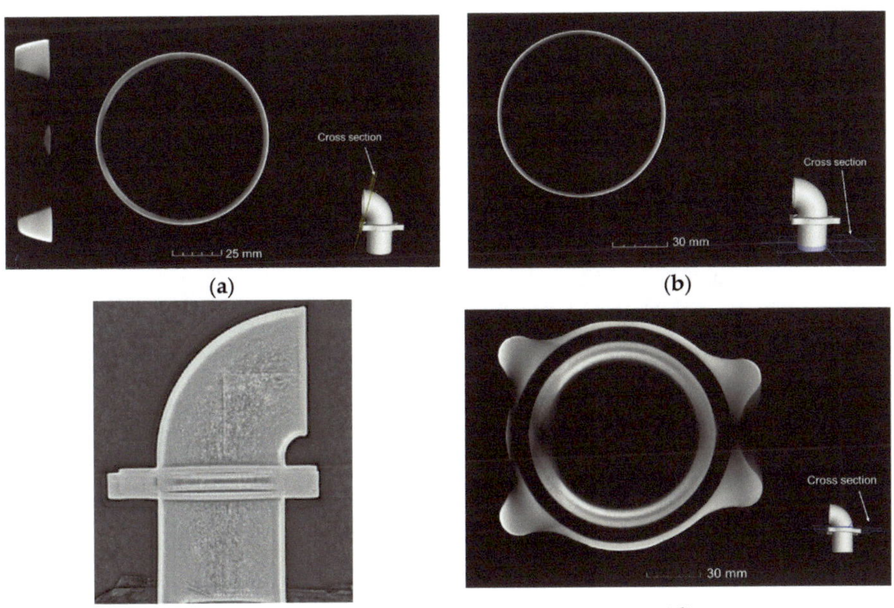

(c) (d)

Figure 14. CT section of exhaust elbow for casting scheme with optimized process parameters: (**a**) Transverse section of a bend section of pipe; (**b**) Transverse section of a straight section of pipe; (**c**) DR result; (**d**) Transverse section of flange.

5. Conclusions

In this paper, the precision casting process design and optimization of a high-horsepower engine exhaust pipe bend through an investment casting program are discussed.

1. Determine the optimal process parameters: pouring temperature of 1650 °C, pouring speed of 1.5 kg/s, and shell preheating temperature of 1050 °C.
2. Analysis of individual process parameter changes on the casting of the internal size of the volume of shrinkage: it is indicated that an increase in the preheating temperature of the mold will significantly slow down the cooling and solidification rate of the alloy liquid, thereby reducing the likelihood of shrinkage defects occurring.
3. The process flow was summarized and analyzed, along with the castings manufactured before and after parameter optimization. Dimensional testing, visual inspection, and industrial CT testing were conducted to demonstrate that optimization can effectively produce castings with high internal quality, surface quality, and dimensional accuracy to achieve the desired outcomes for whole casting forming.
4. The research resulted in a reduction in material and labor costs associated with machining and welding, while significantly improving product pass rates.
5. The casting system has some limitations: its structure is complex, there are many redundant parts, the casting requirements are high, and maintaining a high pass rate is difficult. Future research can enhance the pouring system to decrease production costs and complexity, consequently boosting the product qualification rate.

Author Contributions: Conceptualization, methodology, software, validation, S.X.; formal analysis, investigation, resources, data curation, writing—original draft preparation, Z.L.; writing—review and editing, visualization, supervision, project administration, funding acquisition, S.D. All authors have read and agreed to the published version of the manuscript.

Funding: This work was supported by the Natural Science Basic Research Program of Shaanxi Province (Grant No.: 2023-JC-QN-0612).

Data Availability Statement: The data presented in this study are available on request from the corresponding author. The data are not publicly available due to privacy.

Conflicts of Interest: The authors declare that the research was conducted in the absence of any commercial or financial relationships that could be construed as a potential conflict of interest.

References

1. Mendes, A.; Tomoyose, R. *Exhaust Manifold Structural Analysis Focusing Mass Reduction for Vehicular Diesel Engine Application*; SAE International: Warrendale, PA, USA, 2010. [CrossRef]
2. Meda, L.; Shu, Y.; Romzek, M. *Exhaust System Manifold Development*; SAE Technical Paper 2012-01-0643; SAE International: Warrendale, PA, USA, 2012; p. 10. [CrossRef]
3. Guo, S.C.; Zhu, Y.; Chen, D.; Gao, X.R.; Cheng, T.; Lin, Z.H. Numerical simulation of the flow performance of a diesel engine exhaust system and its structural optimization. *Chin. Intern. Combust. Engine Eng.* **2017**, *38*, 113–119.
4. Galindo, J.; Luján, J.M.; Serrano, J.R.; Dolz, V.; Guilain, S. Design of an exhaust manifold to improve transient performance of a high-speed turbocharged diesel engine. *Exp. Therm. Fluid Sci.* **2004**, *28*, 863–875. [CrossRef]
5. Lorenzini, M.; Barbieri, S.G.; Mangeruga, V.; Giacopini, M. Development of an experimental/numerical validation methodology for the design of exhaust manifolds of high performance internal combustion engines. *Eng. Fail. Anal.* **2023**, *152*, 1350–6307.
6. Xu, Y.; Huang, S.; Su, L.; Zhang, X.; Li, Y. Elbow flow analysis and optimal exhaust port profile design. *J. Mech. Sci. Technol.* **2022**, *2*, 1251–1262. [CrossRef]
7. Liu, F.; Xu, Y.; Rui, J.; Shi, Z.; Li, Y. Flow Analysis and Theoretical Design of Diesel Engine Exhaust Port. *Int. J. Automot. Technol.* **2020**, *21*, 189–196. [CrossRef]
8. Bohac, S.; Landfahrer, K. Effects of pulsating flow on exhaust port flow coefficients. *SAE Int. J. Engines* **1999**, *108*, 299–309.
9. Zu, B.F.; Kang, X.L.; Fu, G.Q.; Xu, Y.L.; Liu, J. Structural design and estimation on intake and exhaust ports of multi-valve engines. *Trans. Chin. Soc. Agric. Mach.* **2005**, *36*, 28–35.
10. Pattnaik, S.; Karunakar, D.B.; Jha, P.K. Developments in investment casting process—A review. *J. Mater. Process Technol.* **2012**, *212*, 2332–2348. [CrossRef]
11. China Foundry Association. *Investment Casting Manual*; China Machine Press: Beijing, China, 2000.

12. Morsiya, C.V.; Pandya, S.N. Recent Advancements in Hybrid Investment Casting Process—A Review. In Proceedings of the 2nd International Conference on Recent Advances in Manufacturing (RAM 2021), Surat, India, 10–12 June 2021; pp. 817–831. [CrossRef]
13. Dave, H.K.; Dixit, U.S.; Nedelcu, D. *Lecture Notes in Mechanical Engineering*; Springer: Singapore, 2022; pp. 817–831.
14. Santos, E.C.; Shiomi, M.; Osakada, K.; Laoui, T. Rapid manufacturing of metal components by laser forming. *Int. J. Mach. Tools Manuf.* **2006**, *46*, 1459–1468. [CrossRef]
15. Yuli, S.; Sungging, P. Investment Casting Simulation of ASTM F75 Acetabular Implant with Mold Pattern Orientation and Gating System Geometry Variation. *Appl. Mech. Mater.* **2024**, *7052*, 157–168.
16. Kumar, P.; Singh, R.; Ahuja, I.P.S. Investigations on dimensional accuracy of the components prepared by hybrid investment casting. *J. Manuf. Process* **2015**, *20*, 525–533. [CrossRef]
17. Pradhan, S.R.; Singh, R.; Banwait, S.S. Comparison of DMLS and DMLS-waste assisted investment casting. *Mater. Lett.* **2022**, *324*, 132782. [CrossRef]
18. Prakash, C.; Singh, S.; Mohan, V. Comparative job production based life cycle assessment of conventional and additive manufacturing assisted investment casting of aluminium: A case study. *J. Clean. Prod.* **2021**, *289*, 125164. [CrossRef]
19. Zhou, L.; Wang, D.; Cui, J.; Zhao, D.; Liu, S.; Shuai, S.; Yin, Y.; Shu, D. Modeling and experimental study on deformation prediction of thin-walled turbine blades during investment casting process. *Proc. Inst. Mech. Eng. Part B J. Eng. Manuf.* **2024**, *238*, 223–234. [CrossRef]
20. Dejene, N.; Lemu, H.; Gutema, E. Critical Review of Comparative Study of Selective Laser Melting and Investment Casting for Thin-Walled Parts. *Materials* **2023**, *16*, 7346. [CrossRef] [PubMed]
21. Brotzu, A.; Felli, F.; Mondal, A.; Pilone, D. Production issues in the manufacturing of TiAl turbine blades by investment casting. *Procedia Struct. Integr.* **2020**, *25*, 79–87. [CrossRef]
22. Cheah, C.M.; Chua, C.K.; Lee, C.W.; Feng, C.; Totong, K. Rapid prototyping and tooling techniques: A review of applications for rapid investment casting. *Int. J. Adv. Manuf. Technol.* **2005**, *25*, 308–320. [CrossRef]
23. Vaghela, J.; Valaki, J.; Thanki, S.; Pandey, A. Sustainability Analysis of Rapid Tooling-Based Investment Casting: A Comprehensive Review. *Smart Sustain. Manuf. Syst.* **2023**, *7*, 54. [CrossRef]
24. Jones, S.; Yuan, C. Advances in shell moulding for investment casting. *J. Mater. Process Technol.* **2003**, *135*, 258–265. [CrossRef]
25. Beeley, P.R.; Smart, R.F. *Investment Casting*; CRC Press: Boca Raton, FL, USA, 2023.
26. Singh, S.; Kumar, R.; Kumar, R. Aluminum metal composites primed by fused deposition modeling-assisted investment casting: Hardness, surface, wear, and dimensional properties. *Proc. Inst. Mech. Eng. Part L J. Mater. Des. Appl.* **2022**, *236*, 674–691. [CrossRef]
27. Kumar, P.; Ahuja, I.; Singh, R. Effect of process parameters on surface roughness of hybrid investment casting. *Prog. Addit. Manuf.* **2016**, *1*, 45. [CrossRef]
28. Tewo, R.K.; Rutto, H.L.; Focke, W.; Seodigeng, T.; Koech, L.K. Formulations, development and characterization techniques of investment casting patterns. *Rev. Chem. Eng.* **2019**, *35*, 335–349. [CrossRef]
29. Kumar, P.; Ahuja, I.S.; Singh, R. Experimental investigations on hardness of the biomedical implants prepared by hybrid investment casting. *J. Manuf. Process* **2016**, *21*, 160–171. [CrossRef]
30. *GB/T 4238-2015*; Heat-Resisting Steel Plate, Sheet and Strip. China ISC S38240. Standardization Administration of China: Beijing, China, 2015.
31. Liao, Q.; Ge, P.; Lu, G.; Song, Y.; Ye, W.; Gao, J.; Luo, X. Simulation Study on the Investment Casting Process of a Low-Cost Titanium Alloy Gearbox based on ProCAST. *Adv. Mater. Sci. Eng.* **2022**, *2022*, 4484762. [CrossRef]
32. Zhang, X.; Li, S.; Lu, H. Simulation of Ni_3Al-based Alloy and Investment Casting Process of its Thin Wall Castings. *Isr. J. Chem.* **2010**, *47*, 363–368. [CrossRef]
33. Miao, L.; Tang, Z.; Zhang, D. Investment Casting Simulation and Analysis of Shell AlSi11 Alloy Part Journal. *J. Phys. Conf. Ser.* **2021**, *2002*, 12–40. [CrossRef]
34. Gebelin, J.-C.; Jolly, M.R. Modelling of the investment casting process. *J. Mater. Process Technol.* **2003**, *135*, 291–300. [CrossRef]
35. Yarlagadda, P.K.D.V.; Hock, T.S. Statistical analysis on accuracy of wax patterns used in investment casting process. *J. Mater. Process Technol.* **2003**, *138*, 75–81. [CrossRef]
36. Silva, W.; Lopes, R.; Zscherpel, U.; Meinel, D.; Ewert, U. X-ray imaging techniques for inspection of composite pipelines. *Micron* **2021**, *145*, 103033. [CrossRef]
37. *GB/T 29070-2012*; Non-Destructive Testing—Industrial Computed Tomography (CT) Testing—General Requirements. China ISC 19.100. Standardization Administration of China: Beijing, China, 2012.

metals MDPI

Article

Semi-Solid Slurries for Rheocasting of Hypoeutectic Al-Si-X Alloys Produced by Self-Stirring in Serpentine Channels

Hernández Alfredo [1,2,*], Chávez José Federico [1], Hernández Aldo [3] and Suárez Miguel Ángel [4]

1 Department of Engineering in Metallurgy and Materials, ESIQIE, Instituto Politécnico Nacional, UPALM, Mexico City 07738, Mexico; jfchavez@ipn.mx
2 Department of Materials Engineering, Tecnológico de Estudios Superiores de Coacalco, Coacalco 55710, Mexico
3 Department of Mechatronics Engineering, Universidad Politécnica del Valle de México, Tultitlán 54910, Mexico; aldo.hernandez.ramirez@upvm.edu.mx
4 Department of Materials, Azcapotzalco, Universidad Autónoma Metropolitana, Mexico City 02128, Mexico; masuarez@azc.uam.mx
* Correspondence: ahernandezh@ipn.mx

Abstract: Nowadays it is common to see the production of complex and critical automotive and aeronautical components reduced in weight for energy efficiency using light alloys with improved microstructural and mechanical properties. The casting processes involved in this trend are strong; in this study, an optimized design of a vertical serpentine channel and a novel design of a horizontal serpentine channel to produce semi-solid slurry (S2S) with thixotropic behavior by self-stirring for rheocasting of A380 and A356 alloys are tested. Simultaneously, chilling during solidification, flow development, and shearing on the alloys to improve the performance of solid fractions and self-stirring at high shear rate are applied. The effects of these conditions on the modification of the morphology transition of the α(Al) phase from dendrite to equiaxed grain are discussed. The results suggest the ability of the mentioned processes to promote the morphological transition of the primary solid due to the produced equiaxed grains of α(Al) phase having sizes between 25–50 µm from A380 alloy processed by vertical self-stirring. On the other hand, the treatment of the A356 alloy using the new horizontal serpentine channel produces equiaxed grains with an average size of 39 µm. Unexpected Si crystals, trapped in the α(Al) phase using both methods with both alloys, are detected. The applied operation parameters were aided by gravity-pouring close to the liquidus temperature, and the obtained microstructural results show the ability for S2S to form alongside thixotropic behavior and non-dendritic solidification by mean of self-stirring in the serpentine channels, suggesting the potential for further experiments under die-casting conditions.

Keywords: casting; rheocasting; solidification; microstructure; characterization

Citation: Alfredo, H.; José Federico, C.; Aldo, H.; Miguel Ángel, S. Semi-Solid Slurries for Rheocasting of Hypoeutectic Al-Si-X Alloys Produced by Self-Stirring in Serpentine Channels. *Metals* **2024**, *14*, 413. https://doi.org/10.3390/met14040413

Academic Editors: Carlos Alexandre Dos Santos and Eleani Maria Da Costa

Received: 29 February 2024
Revised: 25 March 2024
Accepted: 25 March 2024
Published: 30 March 2024

1. Introduction

Semi-solid processing (S2P) was introduced 50 years ago by MIT researchers; the rheological behavior, including the decrease in viscosity of Pb15Sn alloy, was published by Spencer et al. [1]. The concept of dendritic structure transformation during solidification was proposed by Flemings; since then, understanding of the microstructural change to the shape of the primary solid from dendrite to sphere by shearing, which produces a significant decrease in shear stress and viscosity, has been attained. This behavior, identified as thixotropy, is common in ceramics and polymers, but until now unknown in metallic systems [2,3].

The S2P technology made available in the last 35 years has created a new trend in near-net shape manufacturing, in the form of casting light alloys with enhanced microstructural and mechanical properties. Most critical and complex aeronautical and automotive components are conventionally processed by die-casting and further heat treatments to produce dendritic solidification followed by quenching, aging, and spheroidization to obtain the final microstructure at a low cost.

In conventional die-casting of hypoeutectic Al-Si-X alloys, the melt is injected into the mold, then solidification begins, since the filled mold, the stationary liquid alloy, and the as-cast microstructure are in a dendritic α(Al) phase surrounded by an Al-Si eutectic; in rheocasting, on the other hand, a controllable slurry or mush state containing the non-dendritic primary solid within the liquid phase matrix is produced from the melt for further die-injection into the mold, and the equiaxed morphology of the solid α(Al) phase directly in its as-cast condition is finally obtained due to dendrite trimming during the growth of the solid phase.

Strong agitation breaks up the dendrites, promoting the formation of new grains, since inserting the arms into the bulk improves the distribution of new globular particles; under this condition, the slurry viscosity decreases while shear rate increases because of the non-Newtonian behavior of the semi-solid alloy.

Recently, efforts have been focused on adding a rheocasting process to the die-casting industry [3]. Rheocasting is an S2P method that enables the alloy to transition from the liquid to a semi-solid state with thixotropy due to the increase of the solid phase with non-dendritic morphology by simultaneously solidifying it at controlled cooling rate and stirring it at high shear rate [4–9].

Due to the interest in this field, several rheocasting process are in the R&D stages, with researchers seeking a less complex formation route for slurries directly from melt with minimal investment in equipment and minimal modifications to the process. The processing energy saving, the reduction in cycle time, the decreased damage to dies, and minimal casting defects, such as porosity and shrinkage, allows for the most cost-effective process [2,10,11].

Unlike thixoforming, rheocasting does not requires special billets. The feedstock is the liquid alloy produced by one or another conventional melting method. The removed gating and risering systems, rejected parts, and scrap can be simply recycled in-plant, key reasons to choose S2S casting over conventional casting as a high quality and cost-effective process.

In comparison to high-pressure die-casting (HPDC), rheo-die-casting (RHDC) comes with advantages including improving microstructures, increasing mechanical properties, and decreasing defects in a cost-effective manner. The presence of pinhole/porosity in injected parts produced by HPDC is widely known, and the difficulty of the control process due to the cold chamber design and the operation parameters such as injection pressure and flow rate are responsible for flow turbulence and air entrapment to produce these typical defects. The sound production of parts, due to the better control of RHDC, is possible because the aspiration effect disappears and the flow turbulence is decreased.

The non-dendritic microstructure produced by S2P, which would increase the mechanical properties, the comparison of UTS, and strain for hypoeutectic Al-Si alloys produced by conventional casting and rheocasting, is reported on in Table 1. It can be noted that non-dendritic conditions increase the ductility of the alloy without being harmful to the tensile strength. The heat treatment T6 is defined by quenching and aging together. There are no data available for the non-dendritic A380 alloy.

Table 1. Mechanical properties of Al-Si alloys according to the casting method.

Alloy	Casting Process	Tensile Strength, (MPa)	Strain, (%)
A356	HPDC + T6	227	3
A356	RHDC	243	4
A356	RHDC + T6	277	14
A380	HPDC + T6	317	3

The performance of RHDC depends on the method used for S2S forming. In this study, two different melted light alloys were gravity-poured into two serpentine pouring channels to produce thixotropic S2S as potential feedstock for HPDC.

2. Recent Developments in Slurry Forming

According to Flemings [2], the slurry casting route is preferable via immediate industrial applications and potentially a route for further R&D. Initial attempts to bring a rheocasting process into the die-casting industry involved modifying the process, machines, and dies, including in some cases, significant investment in new equipment and shifts in technology; however, these are unacceptable ideas.

Those initial proposals have been refocused to highlight two ideas: the objective must be process improvement, not process change; and the concept must be shifted from how to where to produce slurry on demand. Thus, a different approach to S2S formation has been applied to produce the mushy state at a low formation temperature close to the liquidus, with control of the solid fraction for simplified pouring into a shot sleeve with minimal modifications to the process, die system, and equipment.

Several methods have been developed to prepare S2S directly from the melt; these include mechanical stirring, electro-magnetic stirring, gas bubble induction, and self-stirring by channel flow. Most of them involve high nucleation for enough solid particles to form by forced convection during solidification to cause dendrite trimming in the expected equiaxed structure in the α(Al) phase [2]. A complex process window in each slurry formation must be expected due to a wide combination of process parameters such as viscosity; shear rate; solid fraction; cooling rate; shear time; forming temperature; alloy composition; solute distribution; and history of the melt treatment including degassing, grain refinement, and eutectic modification [12–16].

By mechanical stirring, the dynamics of the rheological treatment are commonly generated by mean of impellers or rods immersed inside the melt. The process has advanced from a batch method to a continuous slurry supply; however, this technique has found limited application on the industrial scale due to its drawbacks, such as erosion of the stirrer, contamination of the slurry by oxides and dross, gas entrapment, low productivity, and difficulty in process control [17–19]. This kind of method produces a typical grain size between 100 and 400 μm during the α(Al) phase.

Electromagnetic stirring was developed to overcome the technical complications associated with mechanical stirring, including advantages in cost-effectiveness. In this method, a ceramic crucible contains the melt for refining treatment to produce abundant nuclei; in this state, the melt, which is close to the liquidus in temperature, is charged into an induction furnace to produce vigorous stirring for further formation of the product inside the die-casting machine. The shear is induced to the charged melt by an electromagnetic field acting on the increase of the solid fraction by the temperature decrease to obtain new fragmented and equiaxed crystals with a size normally about 30 μm.

The UBE process, also known as new rheocasting (NRC), includes electromagnetic stirring to produce slurry on demand. The combined stirring–cooling close to the liquidus temperature of an Al-Si-X alloy results in abundant nucleation of the primary α(Al) particles in the slurry. Furthermore, these methods could have favorable effects on the formation of intermetallic particles, promoting precipitation hardness [20].

The gas-induced superheated slurry (GISS) method was introduced by Wannasin et al. [18,19]. A graphite diffuser is immersed into the melt just above its liquidus temperature to inject micro-sized inert gas bubbles under suitable conditions; chill and vigorous convection are achieved by the flow of the fine inert gas bubbles around the cold diffuser to produce grain refinement of a large number of fine solid particles with a non-dendritic structure surrounded by the remains of the melt. The slurry, with significantly lower heat content with respect to the melt and the controlled solid fraction, is then ready for die-casting, squeeze-casting, or gravity-casting [19]. More than 100 pieces of GISS equipment have been acquired by top companies around the world.

Considering the basic ideas of controlling nucleation and inhibiting dendrite growth, serpentine channel pouring (SCP) has been developed as a different method. The serpentine channel is contained within a heat conductor block. The melt is poured into the block, and simultaneously, solidification–stirring is performed inside the serpentine channel, modifying the solid phase shape while the slurry, with its increasing solid fraction, is obtained. A slurry with the desired controlled solid fraction is the block output.

The SCP process has two advantages with respect to the above-mentioned processes. On one hand, the serpentine channels have strong heat transfer capacity enabling them to produce a chilling effect, which creates a heterogeneous nucleation substrate. The growth of the dendrites is effectively inhibited by the increase of the primary α(Al) nuclei. On the other hand, the stirring occurs when the alloy slurry flows through the closed and curved serpentine channel, and the direction of the alloy slurry changes several times with the effects of gravity; in this way, the movement at decreased viscosity of the formed slurry has the function of self-stirring [21–23].

Zhi-Young Liu et al. recommended an increase to the curve number and decrease of the curve diameter to improve microstructural results; furthermore, the pouring temperature has an effect on grain size. The temperature gradient and the concentration gradient of the slurry state decrease in the serpentine channel by means of self-stirring [21]. Similar simplified methods for slurry formation have been developed, such as the cooling slope [24–26], the cooling channel [27], and the ultrasonic treatment (UST) [28].

3. Experimental Procedure

In this study, two SCP assemblies were designed and tested to produce S2S; two arrangements, optimized vertical SCP and a new horizontal form of SCP, with enough contact surface were created to produce chilling for abundantly nucleation and many bends for self-stirring at high shear rate to obtain a change from a dendrite to a globular shape in the primary solid, free of the agglomeration effect on the grains. The vertical design includes a rectangular cross-section on the channel, while circular transverse section was designed for the horizontal channel. In both cases, the treated S2S was discharged into a cast iron mold to produce a flat slab.

The solid fraction $f_{\alpha(Al)}$ was calculated by the Scheil model [29–34] according to the Equations (1) and (2) in terms of the process temperature (T), and the solute distribution coefficient between the solid and the liquid phases (k), where T_m is the melting temperature of Al, T_l is the liquidus temperature of the alloy, and T_e is the eutectic temperature of the Al-Si system; the solute distribution coefficient for hypoeutectic Al-Si alloys is equal to 0.1309 just at the eutectic temperature. The boundary conditions for hypoeutectic Al-Si alloys are $T_e \leq T \leq T_l$ for the process temperature and $0 \leq X_0 \leq 0.126$ for the composition field, where X_0 is the alloy composition and X_L is the solute content at T into the liquid. Both data are reported in fractions by weight.

$$f_{\alpha(Al)} = 1 - \left[\frac{T_m - T_l}{T_m - T}\right]^{\left(\frac{1}{1-k}\right)} = 1 - \left(\frac{X_0}{X_L}\right)^{\left(\frac{1}{1-k}\right)} \quad (1)$$

$$k = 0.1309\frac{\left[1 - exp\left(\frac{T_m - T}{T}\right)\right]\left[1 - exp\left(\frac{T_e - T_m}{T}\right)\right]}{\left[1 - exp\left(\frac{T - T_m}{T}\right)\right]\left[1 - exp\left(\frac{T_m - T_e}{T}\right)\right]} \quad (2)$$

Two common hypoeutectic alloys of the Al-Si-X system were selected for the experimental tests for SCP self-stirring: A380 alloy (Al-Si-Cu) (Form Technologies, Charlotte, NC, USA), which is in wide use for block engines, and A356 alloy (Al-Si-Mg), extensively applied in wheels, two automotive components produced at high scale. The chemical compositions of the cited alloys determined by optical emission spectrometry are reported in Table 2.

Table 2. Chemical composition of alloys, % by weight.

	Si	Mg	Cu	Fe	Mn	Zn	Ni	Ti	Sr
A380	9.00	0.012	4.050	0.361	0.012	0.051	0.015	0.014	0.0001
A356	7.12	0.338	0.003	0.369	0.011	0.032	0.015	0.011	0.0001

The liquidus temperature of both alloys was determined to define the pouring temperature for rheocasting experiments close to the appearance of the α(Al) phase. The samples were obtained from the ingots used as feedstock for melting. The liquidus temperatures tested by differential scanning calorimetry (DSC) are 585–595 °C for A380 alloy and 605–620 °C for A356 alloy, respectively, according to the tests run shown in Figure 1.

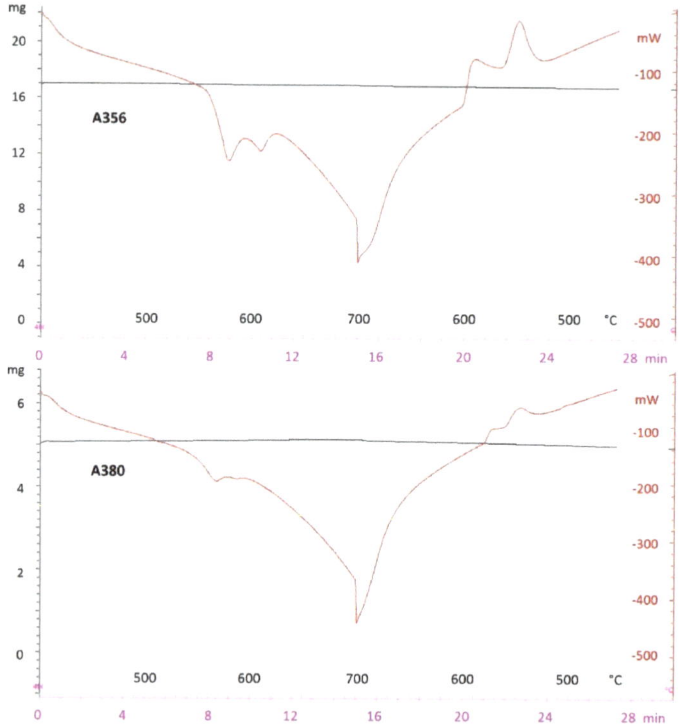

Figure 1. DSC tests carried out for samples of A380 and A356 alloys obtained from the feedstock (ingots).

The melts in weight of 420–460 g were placed in an electric furnace at 720–730 °C, degassed with C_2Cl_6, refined with Al-5Ti-B, and modified with Al-10Sr to obtain 100 ppm of Sr. The master alloy Al-5Ti-B was added into the melt to promote the heterogeneous nucleation, refining in this way the size of the primary solid. The crystallographic structures of TiB_2 and Al_3Ti formed by the inoculant made it easier to increasing the number of nuclei in the poured liquid due to the coincidence of favorable crystallographic planes which were, on one hand, perpendicular to the [111] growth direction and on the other hand, parallel to the $[\bar{1}10]$ and $[\bar{1}12]$ directions of the α(Al) phase. The quantity of inner refining by the master alloy in addition to the provided outward by the channel for the chilling effect were the support for abundant nuclei, the experimental set-up was showed in Figure 2 and the mechanism for the morphological transition of the primary solid by chill and trimmed dendrites was proposed in Figure 3.

The Sr added by mean of the Al-10Sr master alloy is a modifying agent commonly used in Al-Si-X alloys to refine the eutectic structure because it depresses the needed temperature to 15 °C and then stabilizes the eutectic transformation temperature; under these conditions, the morphology of the eutectic silicon is changed from a fine flake shape to a refined fibrous shape. Additions of 100–150 ppm are suitable for this eutectic modification.

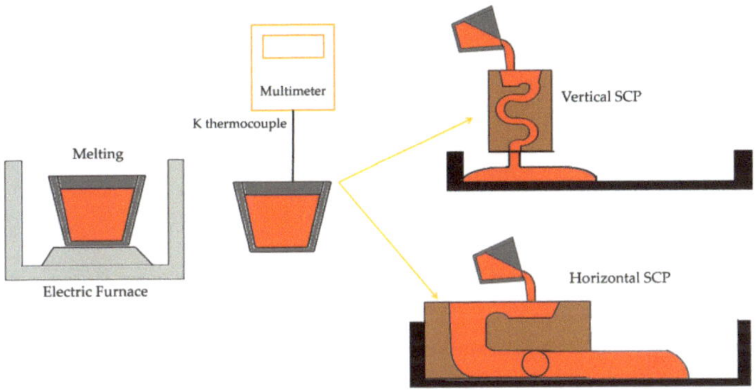

Figure 2. Schematic diagram of the vertical and the horizontal SCP processes.

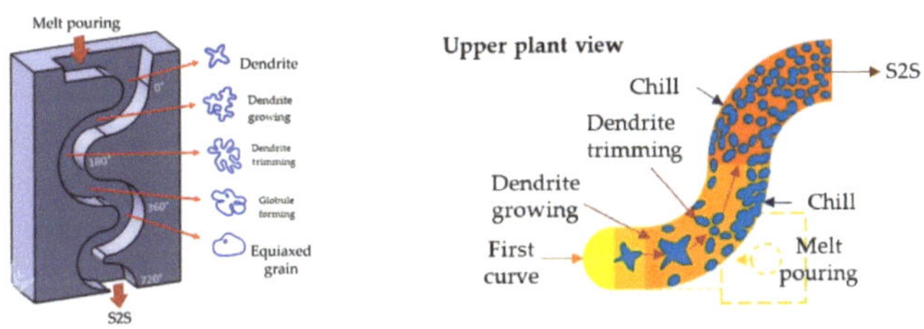

Figure 3. Schematic diagram of the microstructural evolution mechanism from dendrite to S2S with equiaxed grains in the α(Al) phase in both vertical SCP and horizontal SCP.

The A380 and A356 alloys were processed by vertical SCP and horizontal SCP respectively, and in both cases the preheated at 200–220 °C serpentine channels were created in cooper blocks to obtain significant nucleation by chilling and self-stirring from the curve number and the curve diameter.

On one hand, the A380 alloy was gravity-poured into the vertical SCP shape at temperature between 605–610 °C. The measured cooling rate is 5.7–5.8 °C/s. The design of the vertical serpentine channel included four curves equivalent to 720° in rotation with a progressive reduction in the curve radii from 14 to 9 to 4.5 mm; the cross-section of 300 mm^2 remained constant along the channel. Solidified samples between 225–275 g in weight were produced.

Meanwhile, the A356 alloy was gravity-poured via horizontal SCP in a different design including three curves to produce 270° in rotation with 283.5 mm^2 of constant cross-section. For these experiments, the pouring temperature varied between 615 and 690 °C to modify the solid fraction. Solidified samples 205–240 g in weight were obtained.

The vertical and the horizontal serpentine channels, including their respective runner castings, are shown in Figure 4; the samples produced by the second method can be observed in Figure 5.

In this study, horizontal SCP is introduced as a new technique for S2S formation by self-stirring with advantages over vertical SCP. The flow trajectory changes from vertical to horizontal, thus decreasing the height of the liquid column to control the pressure and velocity during gravity-pouring; furthermore, the capacity for channel filling is improved, thus reducing the entrapment of air bubbles.

Other key benefits are the improvement of flow development and heat transfer by the use of a circular cross-section due to the rectangular cross-area along the channel, a shorter route with fewer curves has been included to increase yield by minimal consumption of the alloy in the channel; under these conditions, it is possible to obtain a semi-solid alloy further from the unloading gate despite the low pouring temperature.

Figure 4. A380 and A356 alloys rheocast by vertical SCP and horizontal SCP respectively.

Figure 5. Runner and samples of rheocast A356 from left to right at 615 °C, 630 °C, 660 °C and 690 °C, respectively, made using horizontal SCP.

The chilling process is not affected due to the adequate availability of contact surface area between the slurry and the wall of the channel; moreover, dendrite formation is inhibited by the large number of nuclei associated with the space reduction in the fluid.

The experimental assembly for tests by horizontal SCP is a simplified and compact device allowing for greater ease of execution and control of experiments, and the kit can be installed close to the furnace for fast testing. The container with the serpentine channel is a $125 \times 75 \times 75$ mm cooper block put on top of the mold.

Microstructural characterization of the A380 alloy was carried out by OM and SEM on samples selected from two positions, at rotations of 360° (the middle of the route) and 720° (end of the vertical channel); on the other hand, the A356 samples were chosen at 270°, corresponding to the end of the horizontal channel. The samples for metallographic analysis were etched for 10 s by hydrofluoric acid at a concentration of 1 % by volume.

The globule size (G_{LS}) was determined using the direct measurement method, which consists of measuring the diameter of each of the globules in the micrograph. The shape factor ($C = 1 \ \therefore \ sphere\ ; C \rightarrow 1 \ \therefore \ needle$) was calculated using the following expression: $C = \frac{(4\pi A)}{P2}$, where A is the area of the globule and P is the perimeter [35].

These microstructural parameters were measured with the software program Image J (https://imagej.net/ij/, accessed on 24 March 2024), and at least 30 measurements were taken to obtain average values.

4. Results and Discussion

4.1. A380 Alloy Processed by Vertical SCP

The metallographic analysis by optical microscopy and SEM in Figure 6 shows the morphological transition to equiaxed grain in the α(Al) phase. The selected regions of the vertical serpentine channel to be analyzed were at the middle and at the end of the route; halfway along the channel, globular α(Al) phase in the 25–100 μm size range was produced, while at the end of the channel, particles in order of 25–50 μm were obtained. In both cases it is evident the absence of dendrites; however, some clusters or agglomerated globules were detected.

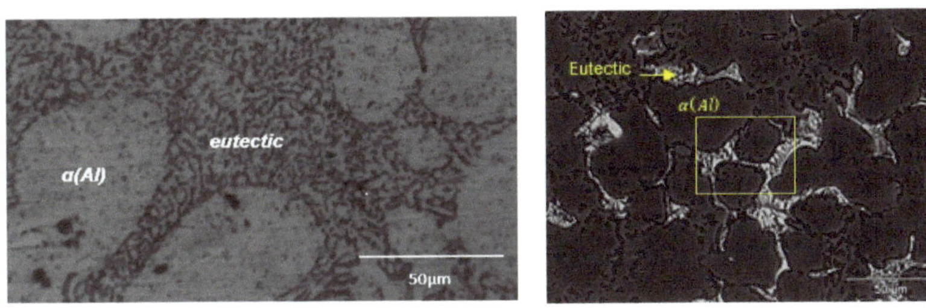

Figure 6. Non-dendritic A380 alloy produced by self-stirring through vertical SCP at rotations of 360° (middle of the route) and 720° (end of the route).

Several stains inside the α(Al) phase were observed by optical microscopy; a detailed analysis of the globules by SEM showed the presence of entrapped Si crystals within the primary solid due to the saturation of solute in the solid phase by its limited solubility. Furthermore, the rejection and diffusion of Si outside of the grain was inhibited by rapid solidification.

Linear EDS was carried out to identify the compositional changes in the primary solid and the eutectic structure. The enlarged region in Figure 6 was analyzed according to the trajectory showed in Figure 7; in this case, the primary solid is the spheroidal α(Al) phase with a rich content of Al containing dissolved Si and Cu. The detected stain at the middle of this grain is a retained Si crystal supported by the decrease in Al and the increase in Si content; from this point, the EDS test reports symmetrical behavior in the composition profile. The eutectic structure around the equiaxed crystals of the α(Al) phase was formed from the remains of the liquid enriched in Si and Cu and alternating fibers of eutectic Al-Si, and some Cu_2Al precipitates are observed in this zone.

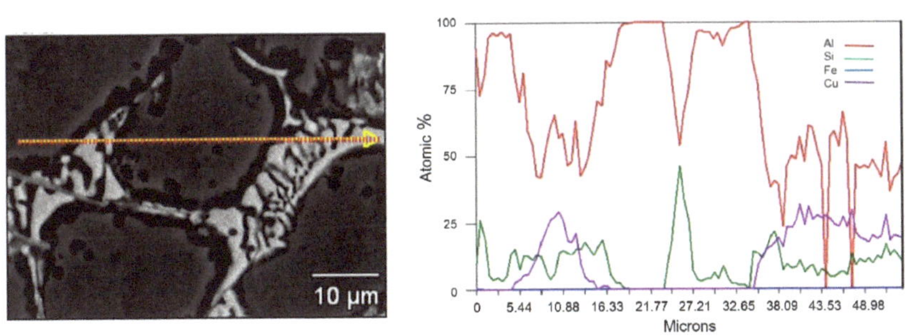

Figure 7. EDS linear scanning of the highlighted region in the previous micrograph including the eutectic structure around the equiaxial grain of the α(Al) phase in A380 alloy.

The EDS mapping in Figure 8 shows the compositional field on the equiaxed particles of the α(Al) phase with average size on 30 μm surrounded by a Si-enriched eutectic structure. The results of the SEM and EDS tests coincide in the presence of Si crystals entrapped in the α(Al) phase due to the saturation of solute of the primary solid, low diffusion, and the chilling effect.

The A380 alloy is difficult to treat by semi-solid processing due to its short temperature interval in the mushy state and the fact that the top of solid fraction available for the treatment is 0.33 at eutectic temperature; however, vertical SCP showed the capacity for S2S formation and non-dendritic solidification of this type of alloy with high Si content close to eutectic composition.

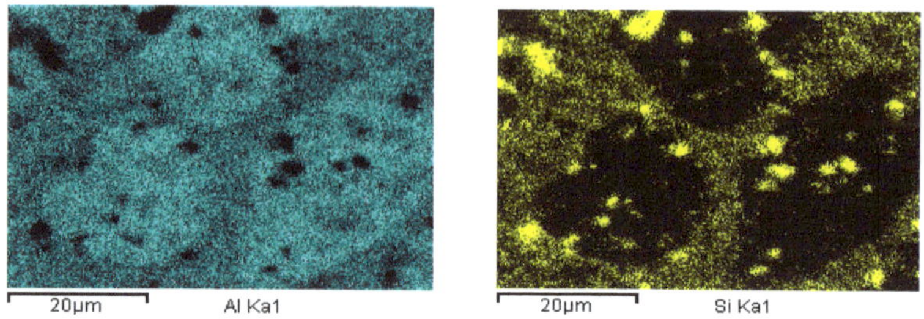

Figure 8. EDS mapping of the non-dendritic α(Al) phase and the eutectic structure in A380 alloy.

4.2. A356 Alloy Processed by Horizontal SCP

The microstructural analysis of the A356 alloy showed the presence of α(Al) phase in a globular shape surrounded by a eutectic structure, as shown in Figure 9. The effect of the pouring temperature on the morphology transition was analyzed. The liquidus temperature for this alloy is 614 °C at 7wt% Si. The solid fraction ($f_{\alpha(Al)}$) is affected by the pouring temperature: $f_{\alpha(Al)} = 0$ when the melt was gravity-poured into the serpentine channel, $0 \leq f_{\alpha(Al)} \leq 0.46$ during slurry formation and the defined limit $f_{\alpha(Al)} \approx 0.46$ at 577 °C. The solid fraction was calculated at the end of the channel according to the measured temperature at that position.

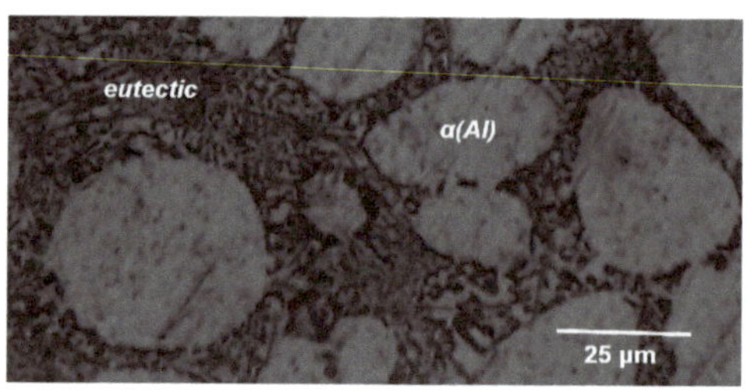

Figure 9. Non-dendritic A356 alloy produced by horizontal SCP.

The solid fraction was calculated by means of the Scheil model according to Equations (1) and (2) for A380 and A356 alloys; the calculations are only valid for hypoeutectic compositions and for the temperature interval between the liquidus and the eutectic transformation. The drawing line for the calculated solid fraction of the A356 alloy showed in Figure 10 is very close to that reported by Mei-Lau Hu et al. by means of the Pandat database [34]. Moreover, Figure 10 contains the traced *k* coefficient; this calculation is based on the solute distribution, and the solute distribution on both phases does not remain constant and depends on *T*.

The micrographs in Figure 11 show the microstructural evolution according to the operation parameters. The melt poured and unloaded at 690 °C and 620 °C, respectively, is above the liquidus temperature and the solid appearance; in this case, a small number of globules by chilling and a high retention of dendrites were produced. The measured cooling rate is 5.83 °C/s.

At 660 °C for pouring and 602 °C for unloading, irregular globules at low concentrations of agglomeration were produced due to the insufficient number of nuclei. The cooling rate was 4.83 °C/s, and the solid fraction calculated by Scheil equation at the horizontal SCP unload was $f_{\alpha(Al)} \approx 0.17$.

At 630 °C and 592 °C for pouring and unloading, respectively, smaller globules at an increasing concentration were produced due to the increase of the solid fraction calculated by the Scheil equation at the channel unload $f_{\alpha(Al)} \approx 0.31$; however, the agglomeration effect persists, probably due to insufficient shear rate during the slurry formation. The cooling rate decreased to 3.16 °C/s compared to the previous tests.

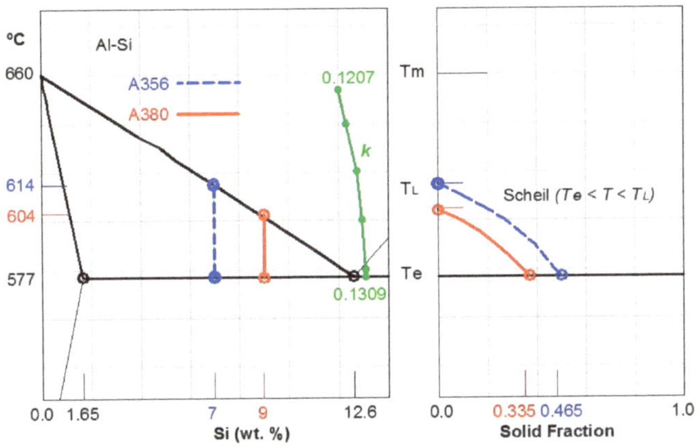

Figure 10. Solid fraction calculated by the Scheil equation for the A380 and A356 alloys.

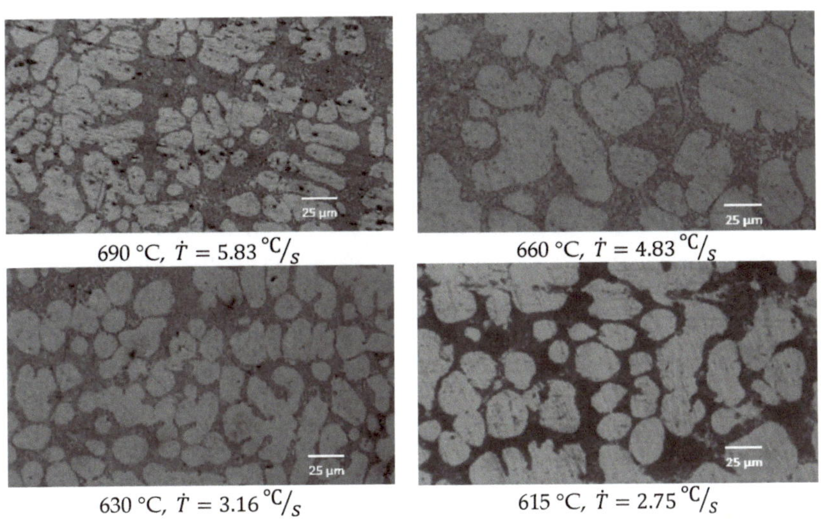

Figure 11. Effect of the pouring temperature and the cooling rate on the solid fraction and shape morphology on A356 alloy produced by horizontal SCP.

Finally, more regular globules were produced at 615 °C for pouring, 582 °C for unloading, the lowest cooling rate $\dot{T} = 2.75\ \frac{°C}{s}$, and solid fraction $f_{\alpha(Al)} \approx 0.41$ (Scheil). The melt was poured close to the solid appearance, the slurry is formed into the horizontal SCP with flow capacity further to the unload position.

Table 3 shows the shape factor (circularity, C) of the globules and their average size. The lowest form factor was obtained at the highest pouring temperature, and this value is consistent with the elongated morphology of the α(Al) phase globules; on the other hand, the highest form factor was obtained from the alloy poured at the lowest temperature (615 °C), where the α(Al) phase globules presented greater circularity.

Table 3. Microstructural parameters of alloy A356 at different pouring temperatures.

Pouring Temperature, (°C)	Globule Size G_{LS}, (μm)	Shape Factor, C
690	48 ± 4	0.41 ± 0.03
660	51 ± 2	0.52 ± 0.03
630	43 ± 2	0.59 ± 0.03
615	39 ± 2	0.68 ± 0.03

5. Conclusions

Equiaxed grains in the α(Al) phase uniform in size and distribution are produced by vertical SCP due to the chilling effect associate to the increase on curve number and decrease on curve diameter; however, the alloy was totally solidified into the channel at the end of the route. The non-dendritic A380 alloy produced by vertical SCP is a notable mention even though the technical difficulties such as the low semi-solid temperature interval, the top on availability of solid fraction and the gravity-pouring.

The horizontal SCP concept is introduced as a new design with some advantages over vertical SCP, notably the improvement on flow development and heat transfer; the results suggest that the chilling effect and the shearing by self-stirring produce the capacity for the change of morphology of the primary solid; moreover, a short route including a smaller number of turns increases the yield with the minimal consumption of the alloy in the channel.

The effect of pouring temperature on the solid fraction produced in the slurry and the shape conversion of the α(Al) phase on A356 alloy was studied. The top of the solid fraction was produced by pouring close to the liquidus temperature, and in this condition, enough solid alloy was available to modify its morphology. The increase in pouring temperature decreased the solid fraction, causing insufficient nuclei, irregular globules, and agglomeration. At a high pouring temperature, the limited availability of the solid phase produced a low number of globules, and the dendritic growth was dominant.

The two SCP methods described in this study for S2S formation have shown the ability to modify the primary solid from dendrites to equiaxed grains on the processed alloys. The morphological transition of the α(Al) phase is irreversible; the probable mechanisms for the shape evolution of the primary solid are exposed and proposed in the schematic diagram in Figure 3. The experiments were carried out to obtain on one hand, a chill of the constructed runner in addition to grain refinement as adequate nucleation agents; on the other hand, self-stirring at high shear rate to produce the slurry with thixotropic behavior by the morphology evolution of the α(Al) phase.

The vertical and horizontal SCP tests were assisted by gravity-pouring with its own flow characteristics such as pressure, velocity, kinetic energy, etc. A suitable design of the serpentine channel contained within a chilling device installed inside a mold for die-casting could be a study of interest to modify those flow conditions. Two ideas for further experiments are suggested with respect to the location of the SCP block: on one hand, it could be placed inside the mold between the injection plunger and the gate to the castings; and on the other hand, it could be located in the charge gate of the cold chamber of the die-casting machine. In both cases, the melt should be poured into the inlet of the SCP channel close to the liquidus temperature of the alloy.

Author Contributions: Conceptualization, formal analysis, writing-original draft preparation, H.A. (Hernández Alfredo); supervision, resources, funding acquisition, project administration, writing-review and editing, C.J.F.; investigation, methodology, H.A. (Hernández Aldo); validation, writing-review and editing, S.M.Á. All authors have read and agreed to the published version of the manuscript.

Funding: The authors acknowledge financial support from National Council for Humanity, Science and Technology (CONAHCyT), SIP-IPN and ESIQIE.

Data Availability Statement: The raw data supporting the conclusions of this article will be made available by the authors on request.

Conflicts of Interest: The authors declare no conflicts of interest.

References

1. Spencer, D.B.; Mehrabian, R.; Flemings, M.C. Rheological behavior of Sn-15 pct Pb in the crystallization range. *Metall. Mater. Trans. B* **1972**, *3*, 1925–1932. [CrossRef]
2. Wannasin, J. *Semi-Solid Slurry Forming of Alloys*; North American Die Casting Association NADCA: Arlington Heights, IL, USA, 2013.
3. Kirkwood, D.; Suéry, M.; Kapranos, P.; Atkinson, H.; Young, K. *Semi-solid Processing of Alloys*; Springer: Berlin/Heidelberg, Germany, 2010. [CrossRef]
4. Li, G.; Lu, H.; Hu, X.; Lin, F.; Li, X.; Zhu, Q. Current Progress in Rheoforming of Wrought Aluminum Alloys: A Review. *Metals* **2020**, *10*, 238. [CrossRef]
5. Modigell, M.; Pola, A.; Tocci, M. Rheological Characterization of Semi-Solid Metals: A Review. *Metals* **2018**, *8*, 245. [CrossRef]
6. Pola, A.; Tocci, M.; Kapranos, P. Microstructure and properties of semi-solid alloys: A literature review. *Metals* **2018**, *8*, 181. [CrossRef]
7. Menargues, S.; Baile, M.T.; Martin, E.; Raya, O.M.; Picas, J. New mechanical stirring system to study the rheological behavior of A380 second melting aluminium alloy. *Res. Sq.* **2023**. [CrossRef]
8. Rosso, M. Thixocasting and rheocasting technologies, improvements going on. *J. Achiev. Mater. Manuf. Eng.* **2012**, *54*, 110–119.
9. Martinez, R.A.; Flemings, M.C. Evolution on particle morphology in semisolid processing. *Metall. Mater. Trans. A* **2005**, *36*, 2205–2210. [CrossRef]
10. Abdulrahman, K.; Gonda, V.; Réger, M.; Varga, P. Comparison of Microstructure, Density and Shrinkage Porosity for Casting and Rheocasting of AlSi7Mg Alloy. *Res. Sq.* **2023**.
11. Hu, X.G.; Zhu, Q.; Midson, S.P.; Atkinson, H.V.; Doug, H.B.; Zhang, F. Blistering in semi-solid die casting of aluminium alloys and its avoidance. *Acta Mater.* **2017**, *124*, 446–455. [CrossRef]
12. Salleh, M.S.; Omar, M.Z.; Syarif, J.; Mohammed, M.N. An Overview of Semisolid Processing of Aluminium Alloys. *Int. Sch. Res. Not.* **2013**, *2013*, 679820. [CrossRef]
13. Kotadia, H.R.; Hari, B.N.; Zhang, H.; Arumuganathar, S.; Fan, Z. Solidification Behavior of Intensively Sheared Hypoeutectic Al-Si Alloy Liquid. *Metall. Mater. Trans. A* **2010**, *42*, 1117–1126. [CrossRef]
14. Nafisi, S.; Ghomashchi, R. *Semi-Solid Processing of Aluminum Alloys*; Springer International Publishing: Cham, Switzerland, 2016. [CrossRef]
15. Boluri Gashti, A.H.; Abedi, H.R.; Salehi, M.T. Microstructure evolution and constitutive modeling of as-cast A356 aluminum alloy in semi-solid regime. *J. Mater. Res. Technol.* **2023**, *24*, 7720–7731. [CrossRef]
16. Brabazon, D.; Browne, D.J.; Carr, A.J. Experimental investigation of the transient and steady state rheological behavior of Al-Si alloys in the mushy state. *Mater. Sci. Eng.* **2003**, *A326*, 69–80. [CrossRef]
17. Honarmand, M.; Salehi, M.; Shabestari, S.G.; Saghafian, H. Impact strength and structural refinement of A380 aluminum alloy produced through gas-induced semi-solid process and Sr addition. *Trans. Nonferrous Met. Soc. China* **2022**, *32*, 1405–1415. [CrossRef]
18. Wannasin, J.; Flemings, M. Development of the Gas Induced Semi-Solid Metal Process for Aluminum Diecasting Applications. In Proceedings of the 10th International Conference on Semi-Solid Processing of Alloys and Composites, Aachen, Germany, 16–18 September 2008; pp. 97–102.
19. Wannasin, J.; Janudom, S.; Rattanochaikul, T.; Canyook, R.; Burapa, R.; Chucheep, T.; Thanabumrungkul, S. Research and development on the gas induced semi solid process for industrial applications. In Proceedings of the 11th International Conference on Semi-Solid Processing of Alloys and Composites, Busan, Republic of Korea, 11–13 September 2010; pp. 544–548.
20. Liu, Y.; Gao, M.; Meng, S.; Fu, Y.; Li, W.; Li, C.; Guan, R. Solidification behavior and enhanced properties of semi-solid Al-8.5Si-0.5Fe alloys fabricated by rheo-diecasting. *J. Mater. Res. Technol.* **2022**, *19*, 3160–3171. [CrossRef]
21. Li, N.-Y.; Mao, W.-M.; Geng, X.-X. Preparation of semi-solid 6061 aluminum alloy slurry by serpentine channel pouring. *Trans. Nonferrous Met. Soc. China* **2022**, *32*, 739–749. [CrossRef]
22. Liu, Z.-Y.; Mao, W.-M.; Wang, W.-P.; Zheng, Z.-K. Preparation of semi-solid A380 aluminum alloy slurry by serpentine channel. *Trans. Nonferrous Met. Soc. China* **2015**, *25*, 1419–1426. [CrossRef]
23. Liu, Z.-Y.; Mao, W.-M.; Wang, W.-P.; Zheng, Z.-K.; Yue, R. Investigation of rheo-diecasting mold filling of semi-solid A380 aluminum alloy slurry. *Int. J. Miner. Metall. Mater.* **2017**, *24*, 691–700. [CrossRef]

24. Das, P.; Kumar, M.; Samanta, S.; Dutta, P.; Ghosh, D. Semisolid Processing of A380 Al Alloy Using Cooling Slope. *Mater. Manuf. Process.* **2014**, *29*, 422–428. [CrossRef]
25. Das, P.; Samanta, S.K.; Das, R.; Dutta, P. Optimization of degree of sphericity of primary phase during cooling slope casting of A356 Al alloy: Taguchi method and regression análisis. *Measurement* **2014**, *55*, 605–615. [CrossRef]
26. Das, P.; Samanta, S.K.; Kumar, P.; Dutta, P. Phase Field Simulation of Equiaxed Microstructure Formation during Semi-solid Processing of A380 Alloy. *ISIJ Int.* **2014**, *54*, 1601–1610. [CrossRef]
27. Hirt, G.; Kopp, R. (Eds.) *Thixoforming, Semi-Solid Metal Processing*; Wiley-VCH: Weihnheim, Germany, 2009; pp. 344–345.
28. Khalifa, W.; El-Hadad, S. Ultrasonication effects on the microstructure characteristics of the A380 die cast alloy. *Int. J. Met.* **2019**, *13*, 865–879. [CrossRef]
29. Kurz, A.W.; Fisher, D.J. *Fundamentals of Solidification*; Trans Tech Publications Ltd.: Wollerau, Switzerland, 1992; ISBN 0-87849-804-4.
30. Bonatti, R.S.; Bortolozo, A.D.; Baldo, R.F.G.; Poloni, E.; Osório, W.R. Anisotropic Tensile and Compressive Strengths of Al–4 wt.%Cu Alloy Powder: Part 1—Effects of Compaction Loads and Heat Treatments. *Metals* **2023**, *13*, 1710. [CrossRef]
31. Dantzig, J.A.; Rappaz, M. *Solidification*; CRC Press: Cham, Switzerland, 2009.
32. Hernández, A.; Chávez, F.; Hernández, R. Simplified rheocasting of Al-Si alloys sheared by mechanical stirring. *Key Eng. Mater.* **2015**, *651–653*, 1545–1550.
33. Hernández, A.; Chávez, J.F. Stir Casting and Atmospheric Pouring of A380 Alloy with Sr Addition. *AIP Conf. Proc.* **2019**, *2113*, 140003. [CrossRef]
34. Hu, M.-L.; Guo, H.-M.; Ma, T.-F.; Jin, H.-L. Microstructure evolution in semi-solid A356 aluminium alloy treated by ICTE process. *Mater. Sci. Technol.* **2023**, *39*, 1050–1058. [CrossRef]
35. Dantas, A.V.R.; Brollo, G.L.; Tamayo, D.V.; Zoqui, E.J. Thixoforming of an Al-Si-Zn-Mg Alloy–Thermodynamic Characterization, Microstructural Evolution and Rheological Behavior. *Mater. Res.* **2021**, *24*, e20200313. [CrossRef]

metals MDPI

Article

Effect of the Fe/Mn Ratio on the Microstructural Evolution of the AA6063 Alloy with Homogenization Heat Treatment Interruption

Adolfo Galván Avalos *, Jesús Torres Torres * and Alfredo Flores Valdés

Cinvestav Saltillo, Industria Metalúrgica 1062, Ramos Arizpe 25900, Mexico; alfredo.flores@cinvestav.edu.mx
* Correspondence: adolfo.galvan@cinvestav.mx (A.G.A.); jesus.torres@cinvestav.edu.mx (J.T.T.); Tel.: +52-844-243-4744 (A.G.A.); +52-844-438-9600 (J.T.T.)

Abstract: The casting structure of the AA6063 alloy contains intermetallic particles of β-Al_5FeSi, which can result in the fragility of the cast pieces. However, with heat treatment, the β phase transforms from a needle or plate form into an intermetallic phase known as α, which resembles Chinese-script in its morphology. To analyze the effect of the ratio of Fe/Mn with different ratios of 0.5, 0.75, and 1, a heat treatment process is used with intermittent interruptions. The alloy is subjected to a temperature of 575 °C for 12 h to determine the microstructural evolution of the β-Al_6FeMn and α-$Al_{15}(FeMn)_3Si_2$ phases. This study used scanning electron microscopy to conduct point analyses and elemental mappings of the intermetallics found in the casting and heat treatment samples. Additionally, X-ray diffraction was employed to determine the stoichiometry of the present phases. The results indicated that the cast structure contains β-Al_6FeMn and α-$Al_{15}(FeMn)_3Si_2$ phases and that the β-Al_6FeMn phase transforms into the α-$Al_{15}(FeMn)_3Si_2$ phase upon completion of the heat treatment process. By using specific Fe/Mn ratios, the formation of the needle-shaped Al_5FeSi phase in the casting structure of the alloy can be inhibited, leading to the precipitation of phases such as β-Al_6FeMn and α-$Al_{15}(FeMn)_3Si_2$ instead.

Keywords: heat treatment; AA6063 alloy; homogenization; mapping; Chinese-script morphology

Citation: Avalos, A.G.; Torres, J.T.; Flores Valdés, A. Effect of the Fe/Mn Ratio on the Microstructural Evolution of the AA6063 Alloy with Homogenization Heat Treatment Interruption. *Metals* **2024**, *14*, 373. https://doi.org/10.3390/met14040373

Academic Editors: Carlos Alexandre Dos Santos and Eleani Maria Da Costa

Received: 21 February 2024
Revised: 19 March 2024
Accepted: 20 March 2024
Published: 23 March 2024

1. Introduction

The aluminum alloy Al–Mg–Si is a combination of medium mechanical strength, good corrosion resistance, excellent formability, weldability, and easy recycling. The main alloying elements are Mg and Si, which provide heat treatment capabilities [1].

As a result, it is widely used to make large casts with sophisticated structures for the transportation industry, including chassis frames and battery module housings for electric cars [2]. The Al 6063 is known as a ternary Al–Mg–Si alloy; also, it can be considered a quaternary alloy since Fe is a controlled impurity. In the casting of primary or recycled aluminum alloys, Fe, Si, and Mg migrate to grain boundaries and interdendritic regions of the α-Al matrix during solidification. This promotes the formation of iron-enriched intermetallic compounds [3].

Iron levels ranging from 0.4 to 0.8% in recycled aluminum scrap can have a detrimental effect on the alloy's properties. This is particularly true for the formation of needle-shaped β-Al_5FeSi intermetallics during solidification, which can compromise mechanical properties, including ductility and fatigue strength. As such, careful attention is necessary to ensure that these high iron levels are appropriately managed to maintain the desired quality of the alloy [4,5].

Alloys containing Al–Mg–Si are not recommended for hot forming while in their cast state. This is due to the presence of particles, mainly consisting of β-Al_5FeSi, α-Al_8Fe_2Si, and Mg_2Si during solidification. Various factors such as alloy composition, cooling rate, and Fe content influence the morphology of Fe-rich phases. As for the solidification process, α-Al_8Fe_2Si typically forms at higher cooling rates than β-Al_5FeSi [6].

The large size needle/platelet β-Al_5FeSi phase can induce severe casting defects and declined properties. The other kind of Fe-rich phase in Al–Mg–Si alloy is the α-AlFeSi phase, whose chemical formula is α-Al_8Fe_2Si (or α-$Al_{15}Fe_3Si_2$). Depending on the alloying elements and cooling conditions, the α-AlFeSi phase appears as hexagonal, star-like, dendritic, Chinese script, or globular shape. Hereinto, the needle-like β-AlFeSi phase deteriorates the mechanical properties since its sharp interfaces act as the location of high stress concentration. Conversely, the Chinese script-like α-AlFeSi phase is relatively harmless [7,8].

The homogenization process is critical for low-cost mechanical deformation without defects. It involves heat treatment to improve the quality of billets [9,10]. The heat treatment aims to dissolve particles of Mg_2Si and low-melting silicon, which cause cracking during the process. It also aims to obtain the phase transformation of β-Al_5FeSi to α-$Al_{12}(FeMn)_3Si$ particles and spheroidize α particles, improving extrusion properties [11,12].

The process of homogenization treatment involves the formation of globular phases, reduction of segregation, and precipitation of dispersoids. In Al–Mn–Fe–Si (–Mg) alloys, the term dispersoid refers to intermetallic particles usually on a nanoscale level. The type of dispersoids formed in Al–Mn–Fe–Si alloys depends on the total quantity of Si present. For alloys with low Si content, the dispersoids are generally in the form of the plate-like Al_6FeMn phase, whereas, for alloys with high Si content, the dispersoids are in the cubic α-Al(FeMn)Si phase [13].

To reduce the fragility of the intermetallic compound β-AlFeSi, researchers have focused on changing the morphology of this phase. One method adds elements like Be, Mn, Cr, or V that promote dispersoid formation. Among these elements, manganese is the most efficient. Manganese acts as a catalyst and increases the phase transformation rate from β-AlFeSi to α-Al(FeMn)Si at high holding temperatures. The resulting α-Al(MnFe)Si phase has a body-centered cubic structure and is known for its Chinese-script morphology [14,15].

A study suggest that the stability zones of the αH-AlFeSi (hexagonal) and αc-AlFeMnSi (cubic) phases are limited to a concentration range where the weight of Mn is less than 0.5% [16].

Adding Mn to aluminum–silicon alloys increases the total number of intermetallic compounds containing (Fe + Mn), such as Al_6FeMn and AlFeMnSi. To promote the precipitation of the α-Fe phase, it is recommended to maintain an Fe/Mn ratio greater than 0.5 [17]. As a common alloying element, Mn has a relatively low cost and is generally used for alloy strengthening. Moreover, the addition of Mn can eliminate the β-AlFeSi phase and promote the formation of the α-Al(FeMn)Si phase, which is beneficial for the mechanical properties [18].

The solidification conditions and the presence of other elements such as Cu and Mg lessen the efficiency of Mn in aluminum–silicon alloys. However, the phase transformation from the β plate-like form to α with the Chinese-script form is not completely inhibited, even with a Mn/Fe ratio of up to 2. When the Mn/Fe ratio is around 1.2, the intermetallic phase β-Al_5FeSi is wholly converted to the α-$Al_{13}(FeMn)_4Si_2$ phase [19]. Table 1 shows the phases present in the 6XXX series [20].

A recent work reports the analysis of variables like that of this study, which are the effect of variations in the Mn/Cr ratios (0.5–8) on the precipitation behavior of dispersoids in 6XXX series aluminum alloys, with homogenization thermal treatment at different temperatures 500, 530, and 560 °C and residence times of 0–24 h [21]. Another study analyzed the individual and synergistic effects of Mn and Cu on the microstructure evolution and mechanical properties [22].

Table 1. Intermetallic phases in the 6xxx series.

Phase	Structure	Stoichiometry
$\alpha_c(\alpha)$AlFeSi	Cubic	$Al_{12}Fe_3Si$, $Al_{12\text{-}15}Fe_3Si_{1\text{-}2}$
	Cubic	$Al_{12}(FeMn)_3Si$, $Al_{15}(FeMn)_3Si_2$
	Cubic	$Al_{12}Mn_3Si$, $Al_{15}Mn_3Si_2$, Al_9Mn_2Si
$\alpha_h(\alpha')$-AlFeSi	hexagonal	Al_8Fe_2Si
β-AlFeSi	monoclinic	$Al_{4.5}FeSi$

The objective of this research is to analyze the impact of the Fe/Mn ratio on microstructural alterations that occur during homogenization heat treatment, specifically how this ratio influences the β→α transformation and its behavior during mechanical forming. To evaluate the microstructural changes in the β and α phases, a homogenization heat treatment on the alloy using varying Fe/Mn ratios (0.5, 0.75, and 1) was performed and halted at a temperature of 575 °C for 12 h; the samples were cooled in water at room temperature every 30 min until we reached a total time of 660 min. Subsequently, we mechanically roughed the samples with SiC paper from #80 to #2400, polished them with 3 and 1 μm diamond paste, and finished them with 0.06 μm colloidal silica to attain a mirror-like finish. Lastly, we prepared the samples for microstructural analysis.

Scanning electron microscopy was used to conduct point analyses and elemental mappings of the intermetallic presents in the casting and heat treatment samples. Also, X-ray diffraction was utilized to determine the stoichiometry of the phases present. The results showed that the β-Al_6FeMn and α-$Al_{15}(FeMn)_3Si_2$ phases are in the cast structure and the phase β-Al_6FeMn is transformed into phase α-$Al_{15}(FeMn)_3Si_2$ when the heat treatment ends.

2. Materials and Methods

To manufacture preforms, the raw material (aluminum profile) is melted. The raw material needed to obtain the AA6063 alloy was prepared using aluminum profile scrap in a 50 kg silicon carbide crucible gas furnace. This process results in ingots with the same chemical composition. Table 2 shows the results obtained.

Table 2. Chemical composition of the ingots (%wt).

Si	Fe	Mn	Mg	Cu	Zn	Others	Al
0.55	0.237	0.014	0.23	0.03	0.02	0.019	Bal

An induction furnace with a capacity of 14 kg was used to obtain the preforms. The process consisted of melting the raw material to obtain liquid metal. Once this step was complete, the required amount of Fe determined for each of the Fe/Mn ratios was added in the form of a wire at 850 °C. A time of 1 h was provided to ensure its dissolution in the metal bath with mechanical agitation for 15 min. The next stage involved adding the determined amount of Mn flakes at 850 °C without mechanical agitation. Finally, argon gas was injected into the metal bath to degas for 10 min. The final stage involves the removal of the liquid metal. This was conducted at a temperature of 750 °C, using a steel mold preheated to 450 °C with a rectangular shape in a vertical position. The process was repeated for each Fe/Mn ratio used. The chemical composition of the alloys was determined by atomic emission spectroscopy. Table 3 shows the alloys' chemical composition and Fe/Mn ratios.

Table 3. The chemical composition (%wt) and Fe/Mn ratios of the alloys used in this study.

Fe/Mn	Si	Fe	Mn	Mg	Cu	Zn	Others	Al
1	0.55	0.7	0.7	0.23	0.3	0.02	0.019	Bal
0.75	0.55	0.6	0.8	0.23	0.3	0.02	0.019	Bal
0.5	0.55	0.55	1.1	0.23	0.3	0.02	0.019	Bal

The obtained preforms have dimensions of 381 mm length, 102 mm width, and 12.7 mm thickness. These preforms had a rectangular shape and underwent a homogenization heat treatment at a temperature of 575 °C for 11 h. A sample was obtained every 30 min until reaching 11 h. After the heat treatment, they were cooled in water at room temperature. This heat treatment aimed to analyze its effect on the precipitation of the phases β and α for each of the different Fe/Mn ratios.

The samples were sectioned and prepared for microstructural analysis by being first mechanically roughed with SiC paper from #80 to #2400. Then, they were polished with 3 and 1 μm diamond paste, and, finally, finished with a mirror-like finish using 0.06 μm colloidal silica.

The specimens underwent chemical exposure to the Keller reagent, which comprises 2 mL HF, 3 mL HCl, 5 mL HNO_3, and 190 mL H_2O. This reagent made it possible to expose the β and α phases, which are distinguished by distinct coloration. To assess the percentages of each phase present in each condition, for the Fe/Mn ratio and heat treatment time, Image-Pro 6.0 software was utilized. To capture measurements and quantify the phases, 10 field micrographs were taken of each sample using a Keyence optical microscope VHX-6000 (Keyence, Osaka, Japan) and Image-Pro Plus 6.0 software.

A JEOL model JSM-7800F PRIME scanning electron microscope (JEOL Ltd., Akijima, Japan) with a BRUKER QUANTAX EDS microanalysis system was used to obtain micrographs of cast and heat-treated samples after 660 min. The microstructure was analyzed to determine the presence and quantification of each element present in the intermetallic compounds of each sample. Elemental mappings were carried out to carry out this analysis. Additionally, BRUKER model D8 ADVANCE ECO X-ray diffraction equipment (Bruker, Billerica, MA, USA) was used to examine the samples. The results obtained were analyzed with Match! 3 software version 3.16 Build 288 to determine which phase corresponds to the intermetallics present in the microstructure. The Inorganic Crystal Structure Database (ICSD) and the Match! 3 software were used for the analysis of the diffraction patterns obtained from each of the samples.

3. Results and Discussion

3.1. Analysis of the Microstructure of the β and α Phases Using SEM

To understand how variables such as the duration of the homogenization heat treatment and the Fe/Mn (0.5, 0.75, and 1) ratios in the chemical composition affect microstructural changes in the samples, we conducted a punctual analysis and elemental mapping using scanning electron microscopy. Figure 1 shows micrographs taken at 500× magnification of the casts and homogenized samples that were treated for 660 min.

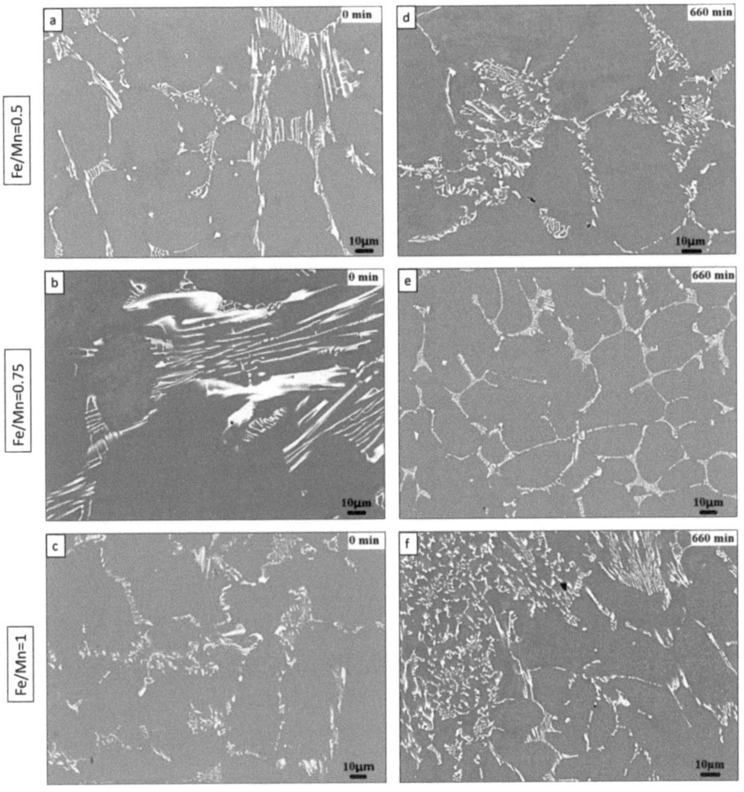

Figure 1. Micrographs of cast samples (**a–c**) and homogenized samples (**d–f**) taken at 500× magnification.

3.1.1. Fe/Mn = 0.5 Ratio

Figure 2 displays a micrograph at 5000× of the cast sample, comprising a porous light gray phase and a dark gray phase. Figure 3 illustrates the mapping of these phases at 5000×, revealing that the light phase contains Al, Fe, Mn, and Si, while the dark phase contains Al, Fe, and Mn only. The morphology of the microstructure exhibits irregular polygonal holes for the light phase and irregular polygonal shapes for the dark phase. Further analysis, quantification, and mapping confirm that the intermetallics present are Al_6FeMn and $Al_{15}(FeMn)_3Si_2$. As in the previous ratios, these intermetallics are Al_6FeMn and $Al_{15}(FeMn)_3Si_2$. This ratio has the highest amount of manganese among the three used, which means the Al_6FeMn phase for this ratio has the highest amount of manganese in the quantitative analyses.

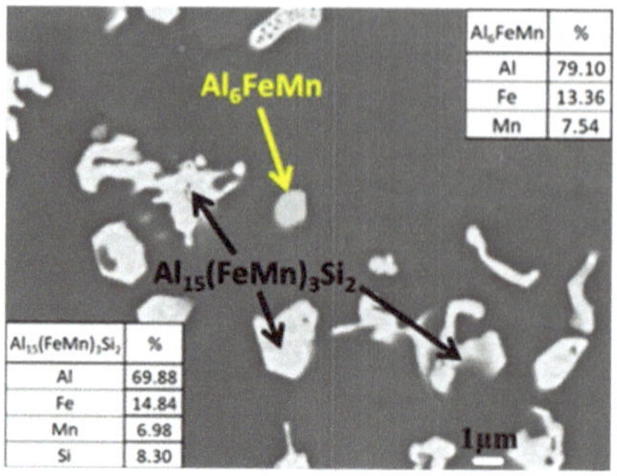

Figure 2. SEM micrograph of a cast sample with an Fe/Mn ratio of 0.5.

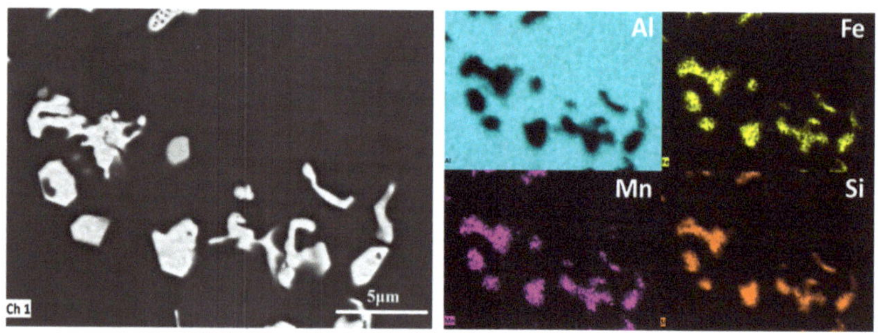

Figure 3. Elemental mapping using EDS on a cast sample with an Fe/Mn ratio of 0.5.

Figures 4 and 5 display a micrograph and mapping at 2500× of the sample that underwent 660 min of homogenization heat treatment from the Fe/Mn = 0.5 ratio. The mapping carried out shows only one intermetallic in the analyzed area. The light gray phase observed in the mapping analysis contains the following elements: Al, Fe, Mn, and Si. Based on the mapping, punctual analysis, and quantification results, it can be concluded that this intermetallic is $Al_{15}(FeMn)_3Si_2$.

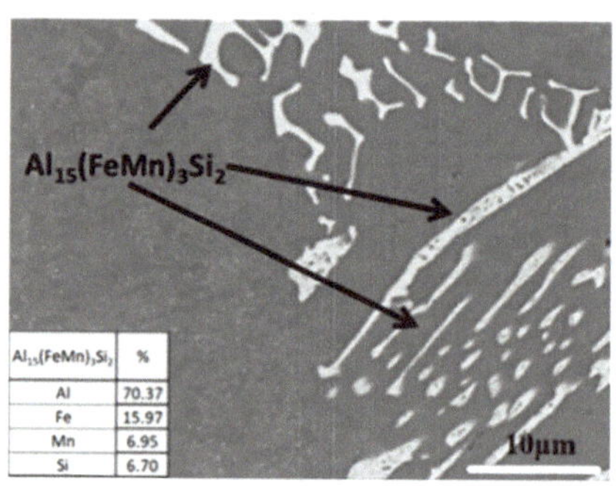

Figure 4. SEM micrograph of a sample homogenized for 660 min, with an Fe/Mn ratio of 0.5.

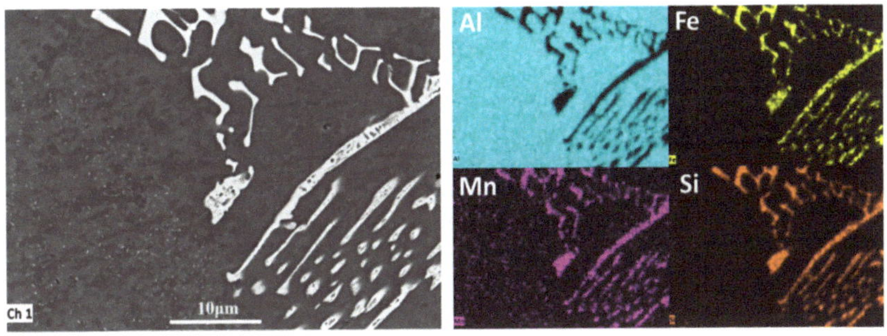

Figure 5. Elemental mapping using EDS on a sample homogenized for 660 min, with an Fe/Mn ratio of 0.5.

3.1.2. Fe/Mn = 0.75 Ratio

The micrograph in Figure 6, taken at 2500× magnification, displays two intermetallic compounds. One appears light gray and porous, while the other appears shadow gray and contains only Al, Fe, and Mn. The elemental mapping in Figure 7, also at 2500×, reveals that the light phase includes the following elements: Al, Fe, Mn, and Si. On the other hand, the dark gray phase shows Al, Fe, and Mn. These intermetallic compounds are identified as Al_6FeMn and $Al_{15}(FeMn)_3Si_2$.

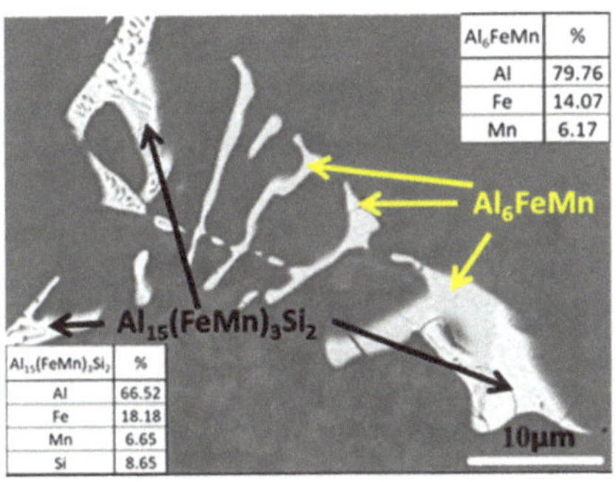

Figure 6. SEM micrograph of a cast sample with an Fe/Mn ratio of 0.75.

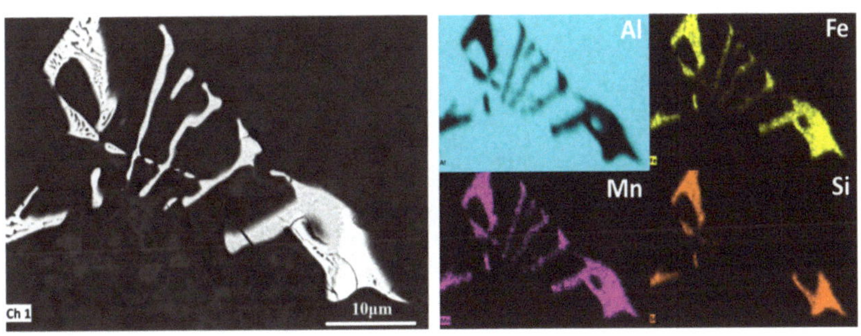

Figure 7. Elemental mapping using EDS on a cast sample with an Fe/Mn ratio of 0.75.

It should be noted that these phases were also observed in an Fe/Mn ratio of 1, but the amount of Mn in the Al_6FeMn compound was found to be 2% higher compared to the Fe/Mn = 1 ratio.

Figures 8 and 9 show a micrograph and map at 2500× of a sample that underwent 660 min of homogenization heat treatment. The map indicates the presence of a single intermetallic in the analyzed area, with a light phase showcasing the elements Al, Fe, Mn, and Si. Further analysis revealed that this phase is $Al_{15}(FeMn)_3Si_2$. However, due to the homogenization heat treatment, the shape of the phase changed from an irregular polygon with holes to a Chinese-script morphology. A light gray phase is visible in both figures, with elemental mapping confirming the presence of Al, Fe, Mn, and Si. Based on the point analysis and quantification, this phase is identified as $Al_{15}(FeMn)_3Si_2$. The heat treatment resulted in changes in the casting structure's morphology, transforming it from an irregular polygon with holes to a Chinese-script morphology.

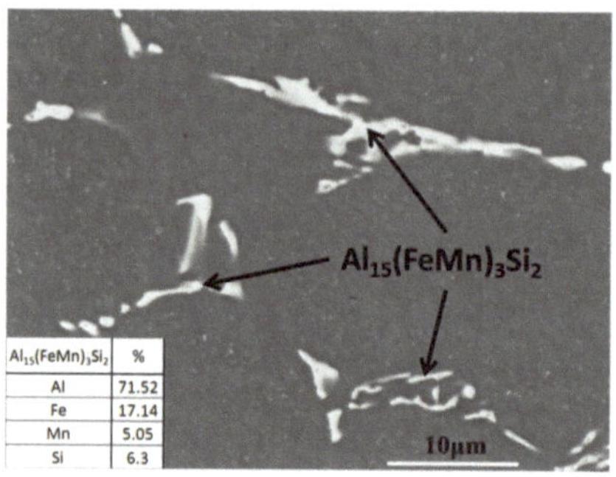

Figure 8. SEM micrograph of a sample homogenized for 660 min with an Fe/Mn ratio of 0.75.

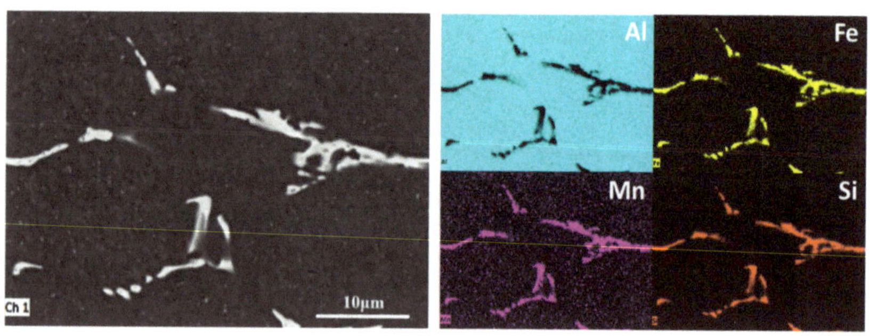

Figure 9. Elemental mapping using EDS on a sample homogenized for 660 min with an Fe/Mn ratio of 0.75.

3.1.3. Fe/Mn = 1 Ratio

Figures 10 and 11 display mappings and micrographs of cast samples obtained from Fe/Mn = 1 ratio. The figures show two different phases of particles. The mapping shows two phases present in the analyzed zone, represented with different colors. The light gray phase contains elements such as Al, Fe, Mn, and Si, while the dark gray phase contains Al, Fe, and Mn. Both phases are irregular polygons in morphology. The punctual analysis, quantification, and mapping concluded that the phases correspond to the Al_6FeMn and $Al_{15}(FeMn)_3Si_2$ intermetallics.

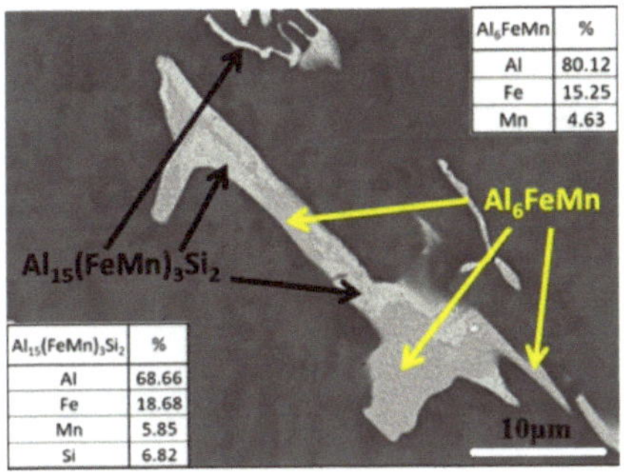

Figure 10. SEM micrograph of a cast sample with an Fe/Mn ratio of 1.

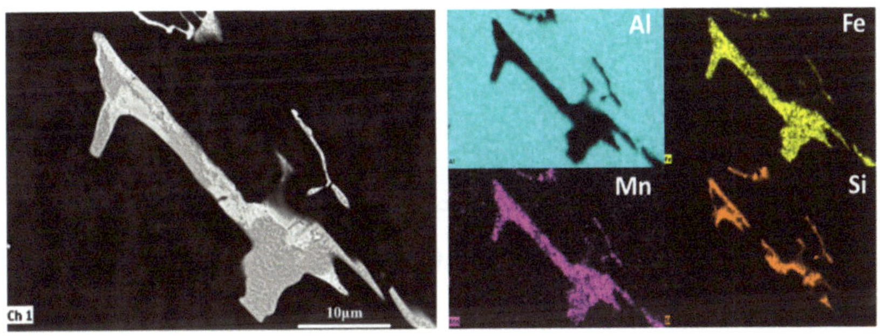

Figure 11. Elemental mapping using EDS on a cast sample with an Fe/Mn ratio of 1.

Figures 12 and 13 show the heat treatment results on Fe/Mn = 1 ratio alloy samples for 660 min. The micrograph at 2500× magnification depicts a light gray phase that contains Al, Fe, Mn, and Si elements. Based on the mapping results, analysis, and quantification, it was determined that the phase was $Al_{15}(FeMn)_3Si_2$. The sample's morphology changed to irregular polygonal shapes with holes compared to the cast sample. The main objective of the heat treatment was to transform the Al_6FeMn phase into the $Al_{15}(FeMn)_3Si_2$ cubic phase.

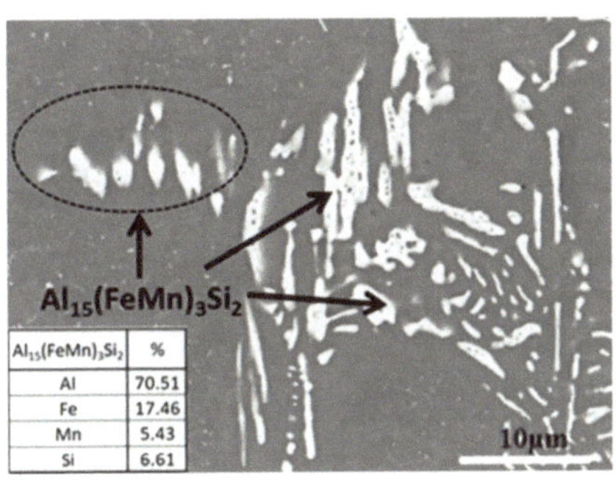

Figure 12. SEM micrograph of a sample homogenized for 660 min with an Fe/Mn ratio of 1.

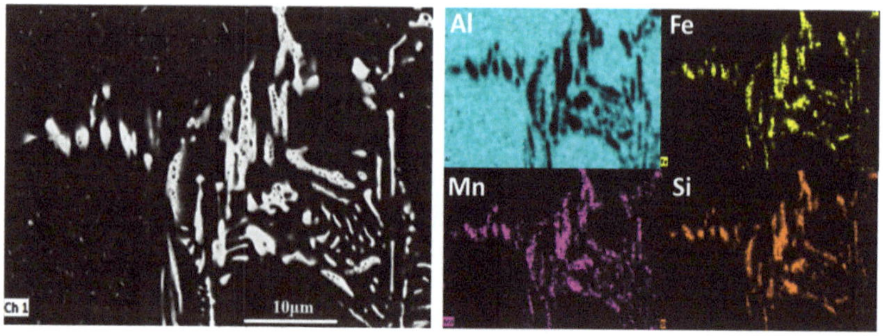

Figure 13. Elemental mapping using EDS on a sample homogenized for 660 min with an Fe/Mn ratio of 1.

In each of the micrographs presented for the Fe/Mn ratios of 0.5, 0.75, and 1, the changes in morphology and transformation of phases are a result of the homogenization heat treatment. The presence of the $Al_{15}(FeMn)_3Si_2$ phase in the casting microstructure can be attributed to manganese, which acts as a catalyst and accelerates the reaction to promote the formation of the $Al_{15}(FeMn)_3Si_2$ phase.

The microstructures show an increased volume fraction of the phase with Chinese-script and irregular polygon morphology in the Fe/Mn = 0.5 ratio after 660 min of heat treatment compared to the other ratios. This phase also occurs at shorter times, which can be attributed to the fact that the amount of manganese in this ratio is twice as much as that of iron.

It has been observed that for the Fe/Mn ratios of 1 and 0.75, the same pattern is seen in the casting structure containing both the β and α phases, while an increase in heat treatment time leads to a rise in the α phase and a decrease in the β phase.

Based on the results obtained, it can be concluded that the casting microstructure contains β-Al_6FeMn and α-$Al_{15}(FeMn)_3Si_2$ phases with an irregular polygon morphology. However, after subjecting the samples to homogenization heat treatment for 660 min, only the α-$Al_{15}(FeMn)_3Si_2$ phase with Chinese-script and holes of irregular polygon morphology was observed. This suggests that homogenization heat treatment transformed the β-Al_6FeMn phase obtained from casting into the α-$Al_{15}(FeMn)_3Si_2$ phase. This transformation occurred through the diffusion of silicon in the β phase and, subsequently, nucleation and growth of the α phase.

The literature reports the mechanism of α phase transformation, which starts with nucleation of α-$Al_{15}(FeMn)_3Si_2$ at particle–matrix interfaces, grows through the particles, and invades the Al_6FeMn phase [23,24].

The growth process can be identified as a eutectoid process, where the Al_6FeMn phase decomposes into a mixture of α-phase and an aluminum solid solution. The nucleation process controls the rate of the overall transformation, as the growth of the α-phase is rapid. The transformation from β to α is dependent on the diffusion of silicon from the matrix into the particles, thus making silicon an important factor in this process [25,26].

The quaternary intermetallic compound AlFeMnSi has a stoichiometry that is defined based on its chemical composition. For an Fe/Mn ratio of 1, the stoichiometry is $Al_8FeMnSi_2$. For an Fe/Mn ratio of 0.99, the stoichiometry is $Al_{0.66}Fe_{0.081}Mn0_{.082}Si_{0.175}$. For an Fe/Mn ratio of 0.63, the stoichiometry is $Al_{11.8}FeMn_{1}._{6}Si_{1.6}$. Finally, for an Fe/Mn ratio of 0.25, the stoichiometry is $Al_{16}(FeMn)_4Si_3$ [27]. A previous study found an acicular morphology α-$Al_{12}(FeMn)_3Si$ phase with an fcc structure. In another study, with alloy 6016, the chemical composition was modified by increasing the Fe and Mn contents (1.4% Fe and 1.1% Mn), and the morphology reported was the Chinese-script with a cubic structure, as discovered by EDS analyses, and thermodynamic simulations indicated that it could be the α-AlFeSi or the α-Al(FeMn)Si phase [28].

3.2. *Analysis of the β and α Phases by X-ray Diffraction*

The results of the analysis to identify the phases present in the casting samples were corroborated through X-ray diffraction, as scanning electron microscopy had been previously used. The analysis was performed on samples with different Fe/Mn ratios before and after a 660 min homogenization heat treatment. Figure 14 displays the diffraction patterns for each of the ratios.

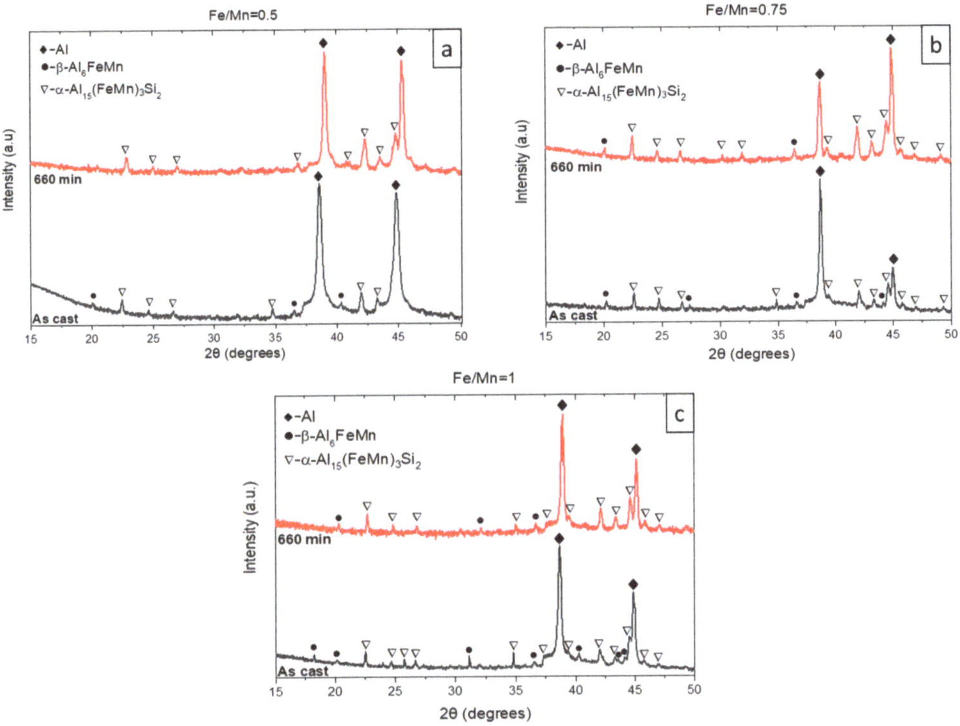

Figure 14. X-ray diffraction patterns with different Fe/Mn ratios: (**a**) 0.5, (**b**) 0.75, and (**c**) 1.

The analysis revealed the presence of an aluminum matrix and two phases: β-Al_6FeMn and α-$Al_{15}(FeMn)_3Si_2$. These phases were observed in the casting samples, but only the α-$Al_{15}(FeMn)_3Si_2$ phase was present after the homogenization heat treatment. This was because the β-Al_6FeMn phase had precipitated in the casting structure and was fully converted into the α-$Al_{15}(FeMn)_3Si_2$ phase after the heat treatment.

For the Fe/Mn = 0.5 ratio, only the α-$Al_{15}(FeMn)_3Si_2$ phase was present, which can be attributed to the higher manganese content in this ratio. The addition of manganese promoted the formation of this phase. In contrast, for the Fe/Mn = 1 and 0.75 ratios, the β-Al_6FeMn phase was present even after the heat treatment but in reduced proportions. The objective of the heat treatment was to transform the β-Al_6FeMn phase into the α-$Al_{15}(FeMn)_3Si_2$ phase, and this was achieved to a greater extent in the Fe/Mn = 0.5 ratio due to the higher amount of manganese.

Overall, this study concludes that the presence of the β-Al_6FeMn and α-$Al_{15}(FeMn)_3Si_2$ phases can be observed in the casting samples, but the homogenization heat treatment can transform the former into the latter phase, especially when the manganese content is high.

The X-ray diffraction results identified the presence of the α-$Al_{15}(FeMn)_3Si_2$ phase, this validates the information obtained in the SEM analyses, in which the morphology of this phase was observed to be divided into massive and regular polyhedrons, hollow and regular polyhedrons, and multi-branched polyhedrons. The results of X-ray diffraction, and the morphology for the α-$Al_{15}(FeMn)_3Si_2$ phase of this study, are consistent with a previous report, where similar morphologies and regular polyhedrons, hollow and regular polyhedrons, and multi-branched polyhedrons for the α-$Al_{15}(FeMn)_3Si_2$ phase were found [29]. Additionally, the α-$Al_{15}(FeMn)_3Si_2$ phase was found to have several morphologies, including polygon + thick Chinese-script, dense Chinese-script + polygon, and polygonal + dense Chinese-script [30]. Table 4 shows the crystallographic information of the analyzed phases.

Table 4. Crystallographic information of the analyzed phases.

Phase	Crystal System	Stoichiometry	Space Group	Cell Parameters
β-AlFeMn	Orthorhombic	Al_6FeMn	C m c m (63)	a = 7.49800 Å b = 6.49500 Å c = 8.83700 Å
α-Al(FeMn)Si	Cubic	$Al_{15}(FeMn)_3Si_2$	P m-3 (200)	a = 12.5600 Å

3.3. *Quantification of β-Al_6FeMn and α-$Al_{15}(FeMn)_3Si_2$ Phases*

During phase quantification, thirteen samples underwent analysis for each Fe/Mn ratio. To examine the microstructure, optical microscopy images were captured at 1000× magnification, revealing both β and α phases in all samples with Fe/Mn ratios of 0.5, 0.75, and 1. The casting structure displayed consistent microstructural characteristics across all three ratios, with the phases being evident.

Through the analysis of the β and α phases, an area of approximately 5% was determined. To ensure utmost precision, the percentages of each phase were evaluated in ten different fields. The micrographs of the casting samples at Fe/Mn ratios of 0.5, 0.75, and 1 confirm the presence of the α phase, which was then named casting alpha. The corresponding casting α phase percentages for Fe/Mn ratios of 0.5, 0.75, and 1 were obtained from the casting sample and found to be 2.31%, 2.08%, and 2.10% of the area for this phase and the area of the β phase was 2.69%, 2.92% and 2.90% for Fe/Mn ratios of 0.5, 0.75, and 1, respectively; these values are considered as the initial amount of this phase, from which the percentage of transformation to α phase is determined.

The analysis of the β→α transformation involved setting the percentage of the casting α phase to zero, which allowed for the determination of the percentage of the β phase. Micrographs of samples attacked with Keller reagent for the quantification of the β and α phases are presented in Figure 15. The distinct color tones of the phases in the micrographs are a result of a chemical attack with the Keller reagent. This approach enabled the quantification of each phase in the analyzed sample using Image-Pro 6.0 software.

The transformed fraction (θ) changes as the reaction takes place, until the reagents are transformed into products. In a reaction, the value of (θ) can be calculated from the reduction of the area of the compound intermetallic β to time (t), related to the total decrease in area, which corresponds to the end of the reaction. The transformed fraction (θ) is determined using Equation (1):

$$(\theta) = \frac{\theta_1}{\theta_0 + \beta_0} * 100 \quad (1)$$

where θ_0 is the initial transformed fraction of phase α, θ_1 is the percentage of phase α transformed at time t, and β_0 is the percentage of phase β [31].

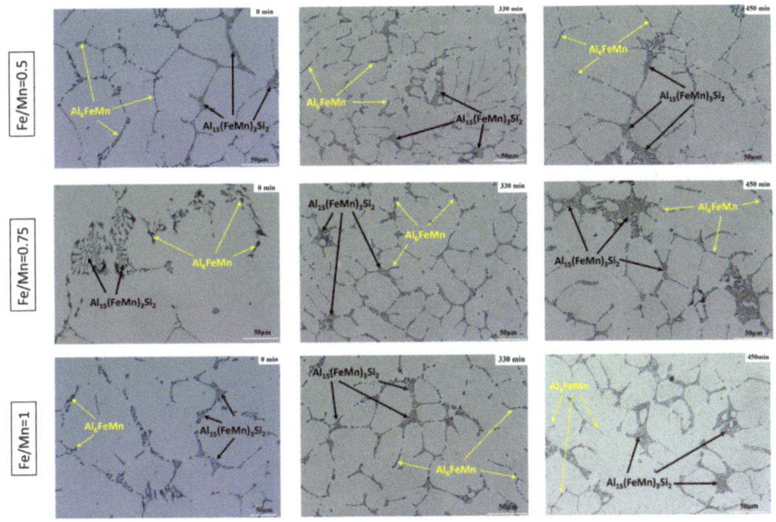

Figure 15. Micrographs of samples attacked with Keller reagent for the quantification of the β and α phases.

Table 5 displays the percentage of the transformed fraction obtained for both intermetallics β and α in each heat treatment condition.

Table 5. Results of the quantification for the percentages of the transformed fraction β and α phases.

Time	Ratio					
	Fe/Mn = 0.5		Fe/Mn = 0.75		Fe/Mn = 1	
Min	β (%)	α (%)	β (%)	α (%)	β (%)	α (%)
As cast	2.69	0	2.92	0	2.90	0
120	2.19	0.50	2.32	0.60	2.58	0.32
210	1.99	0.70	2.17	0.75	2.37	0.53
240	1.83	0.86	2.10	0.82	2.31	0.59
270	1.75	0.94	2.02	0.90	2.23	0.67
300	1.64	1.05	1.89	1.03	2.10	0.80
330	1.53	1.16	1.79	1.13	2.09	0.81
360	1.3	1.39	1.71	1.21	1.92	0.98
390	1.15	1.54	1.65	1.27	1.81	1.09
420	0.98	1.71	1.4	1.52	1.70	1.20
450	0.76	1.93	1.12	1.80	1.62	1.28
540	0	2.69	0.72	2.20	1.32	1.58
660	0	2.69	0.4	2.52	0.81	2.09
Total (%)	0	100	14	86	28	72

According to the findings, the longer the heat treatment time, the more the β phase converts to the α phase. This is due to an increase in the α phase fraction and a decrease in the β phase fraction, which begins at the 120 min mark for all three ratios. Initially, the β phase fraction is comparable among the three ratios, ranging from 2.69 to 2.92. Notably, the Fe/Mn = 0.5 ratio achieved a complete transformation from β to α phase in 540 min, which was faster than the other ratios. Figure 16 displays the percentages of β and α phases for different ratios with homogenization heat treatment at 575 °C.

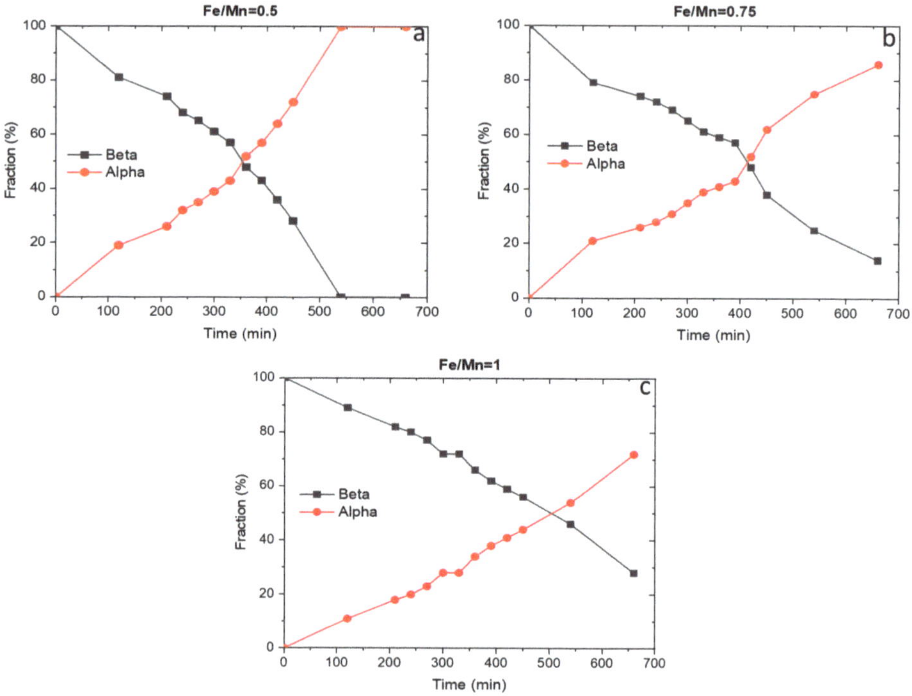

Figure 16. Percentage of β and α phase for Fe/Mn ratios: (**a**) 0.5, (**b**) 0.75, and (**c**) 1, with homogenization heat treatment at 575 °C.

The results of the quantification of the α and β phases indicate that the variations in the chemical composition of the elements, in this case, Fe and Mn for each of the analyzed ratios, result in both β and α phases being present in the casting microstructure; the formation of an α phase from casting has a great impact so that the β→α transformation is carried out in very short times when subjecting them to the homogenization thermal treatment. This transformation occurred through the diffusion of silicon in the β phase and, subsequently, nucleation of α-$Al_{15}(FeMn)_3Si_2$ at particle–matrix interfaces, growing through the particles and invading the Al_6FeMn phase.

4. Conclusions

The use of Fe/Mn ratios has been found to prevent the formation of needle-shaped β-Al_5FeSi in the casting structure of the alloy. Instead, it encourages the precipitation of β-Al_6FeMn and α-$Al_{15}(FeMn)_3Si_2$ phases. The presence of these phases improves the mechanical properties of the alloy as the α-type intermetallic is characterized by ductility, which provides the material with easy deformation. The casting structure's formation of these β-Al_6FeMn and α-$Al_{15}(FeMn)3Si_2$ phases is attributed to the manganese, which acts as a catalyst.

The presence of manganese as a catalyst speeds up the formation of the $Al_{15}(FeMn)_3Si_2$ phase reaction during the heat treatment process of homogenization. Manganese diffuses through the alloy, reducing the diffusion time and leading to the nucleation and growth of the β-Al_6FeMn phase. This phase is already present in the casting structure due to the quantity of iron and manganese used.

During the heat treatment process of homogenization, the β-Al_6FeMn phase obtained during casting is transformed to the α-$Al_{15}(FeMn)_3Si_2$ phase. This transformation happens due to silicon diffusion in the β phase, which results in the nucleation of the α phase at the particle–matrix interfaces. The growth of the α phase then occurs through the particles, replacing the Al_6FeMn phase.

The mapping-based elemental analysis revealed that the cast and heat-treated samples have α phases that contain Al, Fe, Si, and Mn, and β phases that contain Al, Fe, and Mn elements. X-ray diffraction confirmed that these phases correspond to the β-Al_6FeMn and α-$Al_{15}(FeMn)_3Si_2$ phases.

Author Contributions: Conceptualization J.T.T. and A.F.V.; methodology, A.G.A., J.T.T. and A.F.V.; software A.G.A.; validation, A.G.A., J.T.T. and A.F.V.; formal analysis, A.G.A., J.T.T. and A.F.V.; investigation, A.G.A., J.T.T. and A.F.V.; resources, J.T.T. and A.F.V.; data curation, A.G.A. and J.T.T.; writing—original draft preparation, A.G.A.; writing—review and editing, J.T.T. and A.F.V.; visualization, J.T.T. and A.F.V.; supervision, J.T.T. and A.F.V.; project administration, J.T.T. and A.F.V.; funding acquisition, J.T.T. and A.F.V. All authors have read and agreed to the published version of the manuscript.

Funding: The research work was carried out with economic resources and in Cinvestav-Saltillo. laboratories. A.F.V., J.T.T., and A.G.A. would like to thank CONAHCYT (Consejo Nacional de Humanidades, Ciencias y Tecnologías: 20450002) for the financial support.

Data Availability Statement: The raw data supporting the conclusions of this article will be made available by the authors on request.

Acknowledgments: Marco Antonio Leyva Ramirez and Cinvestav-Zacatenco for technical support.

Conflicts of Interest: The authors declare no conflicts of interest.

References

1. Yan, L.Z.; Zhang, Y.G.; Li, X.W.; Li, Z.H.; Wang, F.; Liu, H.W.; Xiong, B.Q. Investigation of the microstructure evolution of 6016 aluminum alloy during homogenization. *Adv. Mater. Res.* **2013**, *668*, 784–788. [CrossRef]
2. Singh, P.; Ramacharyulu, D.A.; Kumar, N.; Saxena, K.K.; Eldin, S.M. Change in the structure and mechanical properties of Al-Mg-Si alloys caused by the addition of other elements: A comprehensive review. *J. Mater. Res. Technol.* **2023**, *27*, 1764–1796. [CrossRef]
3. Whalen, S.; Overman, N.; Taysom, B.S.; Bowden, M.; Reza-E-Rabby, M.; Skszek, T.; DiCiano, M. Effect of high iron content on direct recycling of unhomogenized aluminum 6063 scrap by Shear Assisted Processing and Extrusion. *J. Manuf. Process.* **2023**, *97*, 115–124. [CrossRef]
4. Puncreobutr, C.; Phillion, A.; Fife, J.; Rockett, P.; Horsfield, A.; Lee, P. In situ quantification of the nucleation and growth of Fe-rich intermetallics during Al alloy solidification. *Acta Mater.* **2014**, *79*, 292–303. [CrossRef]
5. Medvecká, D.; Kuchariková, L.; Uhríčik, M. The failure degradation of recycled aluminum alloys with high content of β-Al5FeSi intermetallic phases. *Trans. Tech. Publ.* **2020**, *403*, 97–102.
6. Alvarez-Antolin, F.; Asensio-Lozano, J.; Cofiño-Villar, A.; Gonzalez Pociño, A. Analysis of different solution treatments in the transformation of β-AlFeSi particles into α-(FeMn)Si and their Influence on different ageing treatments in Al-Mg-Si alloys. *Metall. J.* **2020**, *10*, 620. [CrossRef]
7. Wang, X.; Ma, P.-K.; Zhang, S.-Y.; Liu, X.; Wang, C.; Wang, D.-W.; Wang, H.-Y. High-ductility AA6061 alloys produced by combination of sub-rapid solidification and Cr-alloying. *J. Mater. Res. Technol.* **2022**, *18*, 1566–1577. [CrossRef]
8. Liu, X.; Wang, C.; Zhang, S.-Y.; Song, J.-W.; Zhou, X.-L.; Zha, M.; Wang, H.-Y. Fe-bearing phase formation, microstructure evolution, and mechanical properties of Al-Mg-Si-Fe alloy fabricated by the twin-roll casting process. *J. Alloys Compd.* **2021**, *886*, 161202. [CrossRef]
9. Katsivarda, M.; Vazdirvanidis, A.; Pantazopoulos, G.; Kolioubas, N.; Papadopoulou, S.; Rikos, A.; Spiropoulou, E.; Papaefthymiou, S. Investigation of the effect of homogenization practice of 6063 alloy billets on beta to alpha transformation and of the effect of cooling rate on precipitation kinetics. *Mater. Sci. Forum* **2018**, *941*, 884–889. [CrossRef]
10. Uttarasak, K.; Chongchitnan, W.; Matsuda, K.; Chairuangsri, T.; Kajornchaiyakul, J.; Banjongprasert, C. Evolution of Fe-containing intermetallic phases and abnormal grain growth in 6063 aluminum alloy during homogenization. *Results Phys.* **2019**, *15*, 102535. [CrossRef]
11. Bayat, N.; Carlberg, T.; Cieslar, M. In-situ study of phase transformations during homogenization of 6005 and 6082 Al alloys. *J. Alloys Compd.* **2017**, *725*, 504–509. [CrossRef]
12. Liu, C.L.; Azizi-Alizamini, H.; Parson, N.C.; Poole, W.J.; Du, Q. Microstructure evolution during homogenization of Al-Mg-Si-Mn-Fe alloys: Modelling and experimental results. *Trans. Nonferrous Met. Soc. China* **2017**, *27*, 747–753. [CrossRef]

13. Kemsies, R.H.; Milkereit, B.; Wenner, S.; Holmestad, R.; Kessler, O. In situ DSC investigation into the kinetics and microstructure of dispersoid formation in Al-Mn-Fe-Si(-Mg) alloys. *Mater. Des.* **2018**, *146*, 96–107. [CrossRef]
14. Warmuzek, M. Solidification path of the AlFeMnSi alloys in a stage of primary intermetallic phases precipitation. *Trans. Foundry Res. Inst.* **2015**, *55*, 51–60.
15. Gao, T.; Hu, K.; Wang, L.; Zhang, B.; Liu, X. Morphological evolution and strengthening behavior of α-Al(Fe,Mn)Si in Al–6Si–2Fe–xMn alloys. *Results Phys.* **2017**, *7*, 1051–1054. [CrossRef]
16. Österreicher, J.A.; Kumar, M.; Schiffl, A.; Schwarz, S.; Bourret, G.R. Secondary precipitation during homogenization of Al-Mg-Si alloys: Influence on high temperature flow stress. *Mater. Sci. Eng. A* **2017**, *687*, 175–180. [CrossRef]
17. Zhao, Q.; Qian, Z.; Cui, X.; Wu, Y.; Liu, X. Influences of Fe, Si and homogenization on electrical conductivity and mechanical properties of dilute Al-Mg-Si alloy. *J. Alloys Compd.* **2016**, *666*, 50–57. [CrossRef]
18. Liu, X.; Jia, H.L.; Wang, C.; Wu, X.; Zha, M.; Wang, H.Y. Enhancing mechanical properties of twin-roll cast Al–Mg–Si–Fe alloys by regulating Fe-bearing phases and macro-segregation. *Mater. Sci. Eng. A* **2022**, *831*, 142256. [CrossRef]
19. Baldan, R.; Malavazi, J.; Couto, A. Microstructure and mechanical behavior of Al9Si0.8Fe alloy with different Mn contents. *Mater. Sci. Technol.* **2017**, *33*, 1192–1199. [CrossRef]
20. Rosefort, M.; Matthies, C.; Buck, H.; Koch, H. Using SEM and EDX for a simple differentiation of α- and β-AlFeSi-Phases in Wrought Aluminum Billets. In *Light Metals 2011: The Minerals*; Lindsay, S.J., Ed.; Metals & Materials Society: Pittsburgh, PA, USA, 2011; pp. 711–716.
21. Liu, F.; Zhu, X.; Qin, J.; Zhou, W.; Ling, J.; Dong, Q.; Yu, J.; Nagaumi, H.; Zhang, B. Effect of Mn/Cr ratio on precipitation behaviors of α-Al(FeMnCr)Si dispersoids and mechanical properties of Al–Mg–Si–Cu alloys. *Mater. Sci. Eng. A* **2022**, *860*, 144269. [CrossRef]
22. Liu, X.; Ma, Y.-L.; Wang, X.; Zhang, S.-Y.; Zhang, M.-X.; Wang, H.-Y. Enhanced long-term thermal stability and mechanical properties of twin-roll cast Al–Mg–Si alloys with Mn and Cu additions. *Mater. Sci. Eng. A* **2023**, *872*, 144945. [CrossRef]
23. Alexander, D.T.L.; Greer, A.L. Nucleation of the Al6(Fe,Mn)- to-α-Al–(Fe,Mn)–Si transformation in 3XXX aluminium alloys. I. Roll-bonded diffusion couples. *Philos. Mag.* **2004**, *84*, 3051–3070. [CrossRef]
24. Trømborg, E.; Dons, A.L.; In, L.A.; Lohne, O.; Nes, E.; Ryum, N. (Eds.) *Aluminium Alloys: Their Physical and Mechanical Properties: ICAA3*; SINTEF Metallurgy: Trondheimin, Norway, 1992; pp. 270–275.
25. Alexander, D.; Greer, A. Solid-state intermetallic phase tranformations in 3XXX aluminium alloys. *Acta Mater.* **2002**, *50*, 2571–2583. [CrossRef]
26. Anyalebechi, P.N. Effects of silicon, iron, and chromium on the response of aluminum alloy 3004 ingots to homogenization. In Proceedings of the TMS Annual Meeting, New Orleans, LA, USA, 9–13 March 2008; pp. 201–218.
27. Qian, X.; Parson, N.; Chen, X.G. Effect of homogenization treatment and microalloying with Mn on the microstructure and hot workability of AA6060 aluminum alloys. *J. Mater. Eng. Perform.* **2019**, *28*, 4531–4542. [CrossRef]
28. Trink, B.; Weißensteiner, I.; Uggowitzer, P.J.; Strobel, K.; Pogatscher, S. High Fe content in Al-Mg-Si wrought alloys facilitates excellent mechanical properties. *Scr. Mater.* **2022**, *215*, 114701. [CrossRef]
29. Song, D.F.; Zhao, Y.L.; Wang, Z.; Jia, Y.-W.; Li, D.-X.; Fu, Y.-N.; Zhang, D.-T.; Zhang, W.-W. 3D Fe-Rich Phases Evolution and Its Effects on the Fracture Behavior of Al–7.0Si–1.2Fe Alloys by Mn Neutralization. *Acta Metall. Sin.* **2022**, *35*, 163–175. [CrossRef]
30. Song, D.F.; Jia, Y.; Li, Q.; Zhao, Y.; Zhang, W. Effect of Initial Fe Content on Microstructure and Mechanical Properties of Recycled Al-7.0Si-Fe-Mn Alloys with Constant Mn/Fe Ratio. *Materials* **2022**, *15*, 1618. [CrossRef]
31. Aguilar, I.G.; Torres, J.T.; Valdés, A.F.; Saldivar, A.A.F. The effect of interrumped homogenization on β-Al5FeMnSi→α-Alx(Fe and Mn)Si transformation in the A6063 aluminum alloy. *Metals* **2022**, *12*, 2117. [CrossRef]

metals

MDPI

Article

Effect of Gating System Design on the Quality of Aluminum Alloy Castings

Marek Brůna *, Marek Galcik, Richard Pastircak and Elena Kantorikova

Department of Technological Engineering, Faculty of Mechanical Engineering, University of Žilina, Univerzitná 8215/1, 010 26 Zilina, Slovakia; marek.galcik@fstroj.uniza.sk (M.G.); richard.pastircak@fstroj.uniza.sk (R.P.); elena.kantorikova@fstroj.uniza.sk (E.K.)

* Correspondence: marek.bruna@fstroj.uniza.sk

Abstract: In this paper, a naturally pressurized gating system has been designed to reduce the turbulence of the melt during casting. The influence of gate dimensions, foam filters, a trident gate and a vortex element were evaluated. Their effect on melt velocity, flow characteristics, number of oxides, casting properties and mechanical properties were observed. ProCAST Simulation software v.2023 and a water flow test were also evaluated to assist in the experimental evaluation of the castings. Melts showed a relationship between melt velocity and porosity of castings. Quantitative evaluation of the surface porosity showed a trend of decreasing porosity with decreasing melt velocity. The greatest reduction in the melt velocity was achieved by a M4 design, which was associated with the highest reduction in the oxides. The pores analyzed proved the presence of oxide layers on their inner surface and a possible theory of pore formation when the initiator of porosity is entrained double oxide layers. The best metal yield was achieved with M1, but the difference between M2 and M4 was negligible (2–5% yield difference), so it can be stated that the beneficial effect of the M4 design in providing the best quality castings is not negated by the increase in metal yield.

Keywords: gating system; melt velocity; foam filter; melt flow; aluminum alloy casting quality

Citation: Brůna, M.; Galcik, M.; Pastircak, R.; Kantorikova, E. Effect of Gating System Design on the Quality of Aluminum Alloy Castings. *Metals* **2024**, *14*, 312. https://doi.org/10.3390/met14030312

Academic Editor: Carlos Alexandre Dos Santos

Received: 7 February 2024
Revised: 1 March 2024
Accepted: 5 March 2024
Published: 7 March 2024

1. Introduction

The surface of the molten aluminum alloy reacts with the surrounding atmosphere to form a surface oxide layer due to the high affinity of aluminum for oxygen. The surface oxide layer is not harmful. The problem arises when the surface oxide layer is subjected to entrainment forces. When entrained into the bulk of the metal volume, a new oxide layer will form at the surface in a matter of milliseconds. Turbulence causes the entrained oxide layer to fold, trapping air between the layers. The oxide layer can deteriorate the final quality of the casting if it is entrained into the melt by the turbulence of the melt surface and solidifies in the casting [1–5]. Several studies in recent years indicate that entrained double oxide layers in the casting act as porosity initiators and that hydrogen contributes to pore formation by diffusing between the layers, thereby enlarging the pores [6–8].

In order to prevent such defects, it is important to understand the mechanism of their occurrence. The effort is to prevent their occurrence by choosing an appropriate design for the gating system. The design of a naturally pressurized gating system appears to be a suitable solution from the point of view of suppressing reoxidation during the filling of the gating system [9–14]. Its disadvantage is the high velocity of the melt entering the mold cavity through the gate. Supercritical velocity is a consequence of the fact that a naturally pressurized system has approximately the same ratio of cross-section area for all individual parts of the gating system, which means that there is no velocity reduction as, for example, with a negative pressure gating system, where the speed is reduced due to the law of conservation of energy (parts of the gating system closer to casting have higher cross-section areas).

The supercritical velocity causes a large amount of spattering of the melt in the cavity, which in turn causes a large number of entrained oxides in the casting, thus negating the advantage of a naturally pressurized gating system.

The high melt velocity exiting the gates in the naturally pressurized gating system needs to be reduced by a suitable solution. Research has shown that by introducing a filter into the gating system, it is possible to adjust the flow of the melt and reduce its velocity, but it is also important to pay attention to the location of the filter in the gating system, as inappropriate placement of the filter will have a negative effect [15–21]. In recent years, researchers have attempted to solve the problem of a naturally pressurized gating system by unconventional modification of its individual elements. Such possibilities include the introduction of a vortex element at the end of the runner [10,17], the use of the so-called vortex gate with an inserted filter in the upper part of the gate [2,10], or the "trident gate" with two filters and a bubble trap [10,21].

This paper deals with the use of filters and unconventional elements in the gating system and their influence on the melt flow in the gate area, melt velocity, number of oxides, foundry properties, porosity and mechanical properties of castings. Reoxidation cannot be completely eliminated from the casting process. However, the aim is to use the mentioned elements to reduce it as far as possible and to improve the quality of the castings from the aluminum alloy.

2. Materials and Methods

A specific casting was designed for the purpose of the experiment (Figure 1). Based on the dimensions and weight of the casting, a naturally pressurized gating system was calculated. At this stage of the experiment, the effect of dimensioned gates, foam filters, trident gates and a combination of trident gates with a vortex element at the end of the runner were observed. The foam filters' dimensions were 50 × 50 × 20 mm. To compare the influence of each element, a gating system without a filter with a gate of the same cross-section as the runner was used as a reference sample.

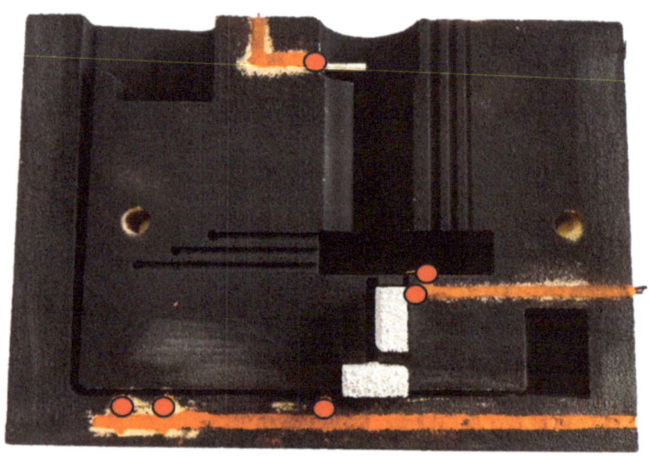

Figure 1. Mold with foam filter and highlighted measurement points.

An $AlCu_4Ti$ aluminum alloy with a chemical composition shown in Table 1 (measured on a Q2 ION spectrometer) was used for casts. This $AlCu_4Ti$ aluminum alloy is widely used in the automotive industry for components of spark-ignition and diesel engines of vehicles and was used based on the knowledge from other works [22–24].

Table 1. Casting chemical composition.

Element	Si	Fe	Cu	Mn	Zn	Ti	Al
w.%	0.05	0.11	5.03	0.42	0.01	0.19	bal.

The pouring height was 100 mm above the mold level. The maintained height of the melt level in the pouring basin was 30 mm above the level of the sprue. The casting temperature was 745 ± 5 °C, and the temperature of the molds and the environment were 17 ± 2 °C. Melting was conducted in an electric-resistant furnace without a degassing process. For the most relevant evaluation of the individual modifications to the gating systems, it was necessary to ensure that the quality of the melt prepared was approximately the same for each casting. Therefore, the density index (DI) was measured immediately before and after casting. The average DI values were 2.36–3.63%. The basic boundary conditions for the numerical simulation were set as follows: pouring height of 100 mm, casting temperature of 745 °C, initial mold temperature of 17 °C, resin-bonded sand used for the molds (with the appropriate setting of the heat transfer coefficient) and cast alloy $AlCu_4Ti$. A tensile tests and impact tests were carried out in accordance with EN standards. The test equipment used for the tensile test was an Inspektdesk 50 kN and the impact test was performed on a PSW 30 Charpy impact tester.

The ProCAST simulation software v.2023 was used to evaluate the melt velocity in the gating system and in the mold cavity, the melt flow behind the gate, the number of oxides and the fluidity. The two-phase filling module was used during the simulations. The flow in the cavity was observed using a water experiment and compared with the simulations. During casting, the actual melt velocity in the runner and at the cavity entrance was evaluated using measurement points. After casting, mechanical properties were evaluated, and a fluidity test, porosity of castings, microscopic analysis and metal yield evaluation were also performed.

Design of Casting and Gating System

The proposed shape of the casting shown in Figure 2 allows for a comprehensive evaluation as it includes areas for fluidity testing, hot tear testing, mechanical properties testing and an area for evaluating melt flow in the mold cavity. The optimal design of the gating system can reduce the turbulent flow of the molten metal, minimizing the amount of gas and trapped impurities. Based on the dimensions and weight of the casting, the design of a naturally pressurized gating system was adapted, including the calculation of the cross-sectional dimensions of its individual parts. Figure 3 shows every modification of the gating system used in experimental works. A reference design without a filter (modification M1), where the cross-sectional area of the gate is the same as the cross-sectional area of the runner, is shown in Figure 3a. The next modification is a gating system with an extended gate (modification M2), shown in Figure 3b. The area of this gate has been dimensioned so that the melt velocity at this point is a maximum of 0.5 $m \cdot s^{-1}$, neglecting hydraulic losses, friction losses, etc. Figure 3c shows a modification of the gating system in which a 10 ppi foam filter (modification M3) and a 30 ppi foam filter (modification M4) were placed horizontally above the runner in the gate area. In the next modification (Figure 3d), the effect of the trident gate (modification M5) was evaluated. This type of gate contains the so-called bubble trap and chambers for one filter placed horizontally above the runner and the other filter placed vertically. Foam filters with a density of 20 ppi were inserted into the trident gate. The gate area remains unchanged compared to the M1 gate area. The final modification (Figure 3e) used a combination of a trident gate with two 20 ppi foam filters and a cylindrical vortex element at the end of the runner (modification M6).

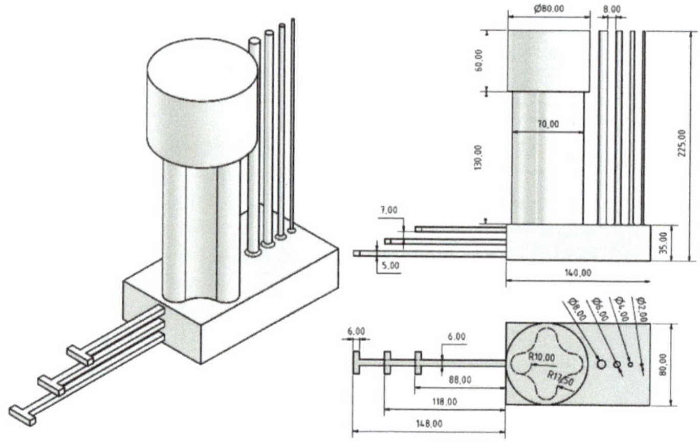

Figure 2. Casting construction.

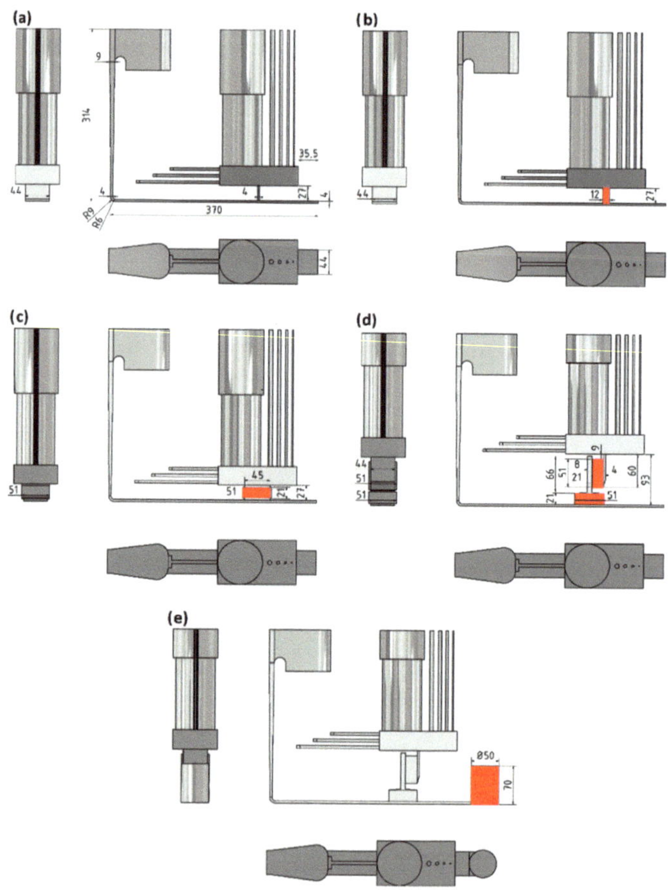

Figure 3. Gating system modifications: (**a**) M1, (**b**) M2, (**c**) M3 and M4, (**d**) M5 and (**e**) M6.

3. Results and Discussion

3.1. Melt Velocity Determination

Melt velocity during pouring was evaluated by measurement points (1.4 mm diameter copper wires) placed in the mold as shown in Figure 4. For all gating system modifications, points RE, E1 and E2 were placed in the runner and point E4 was placed above the gate. In M1 and M2, point E3 was located approximately at the center of the gate height, and in M3 and M4, approximately at the center of the filter. In M5 and M6, E3 was located at the top of the trident gate. The melt velocity was only measured at points E2 and E4; points RE, E1 and E3 were only used as an aiding spots. Point E2 was located in front of the gate and E4 approximately 5 mm above the gate. Measuring the melt velocity at these points was chosen to assess the effectiveness of the gate system used in reducing the velocity of the melt entering the mold cavity. The measurement points were connected to the NATIONAL INSTRUMENTS NI USB-6008 module. The distance between the points and the time taken for the melt to travel that distance was known. The average melt velocity was calculated from the difference in distance between sensors and the voltage peaks reached over time at the measuring points. Time and voltage data for the measurement points were recorded using LabVIEW software—version 2018.

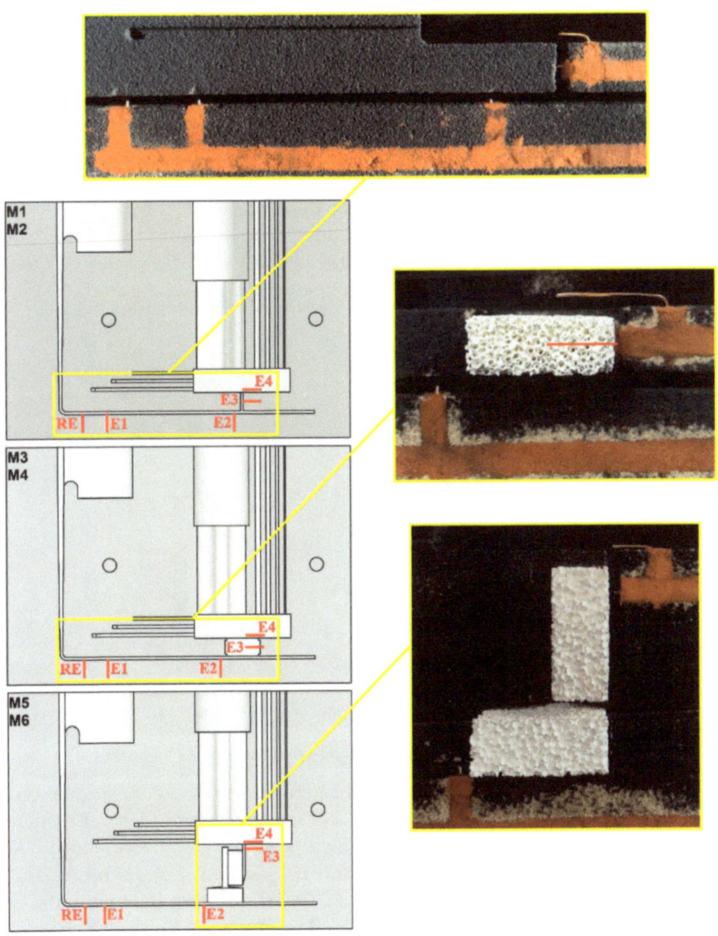

Figure 4. Measurement points placement in the molds.

During pouring, it is not possible to maintain identical filling conditions every time, but the velocity values at point E2 show that the filling conditions were approximately the same for all modifications. Point E4 was the second point where the melt velocity was measured. The results show a decrease in melt velocity at this point in M2 to M5 compared to the reference gating system M1. At M6, the melt velocity could not be evaluated during pouring, because an unknown measurement error occurred. M4 and M3 achieved the best results when the melt velocity at point E4 decreased by approximately 85% compared to the reference gating system M1. In M2, the velocity decreased by 16% and in M5 by 47% compared to M1.

Using simulation software, the melt velocity in the region above the gate was analyzed using a color scale (Figure 5). From the results, it can be concluded that the melt velocities in the individual modifications are largely consistent with the values from the measurement points. For a better quantitative evaluation and comparison of the melt velocity, the velocity measurement in the simulations was carried out at points E2 and E4. The location of these points corresponded approximately to the location of the measurement points in the molds. The instantaneous maximum velocity of the melt was measured at these points. The results also show the highest reduction in velocity for the gate in M4 and M3, where there was a reduction of approximately 89% compared to the reference modification. Compared to M1, the reduction in velocity above the gate was 35% in M2, 63% in M5 and 74% in M6. The graph comparing the melt velocity values obtained from the simulations and those measured during casting are shown in Figure 6. From the graph it can be seen that there is a visible correlation between the values measured at the mold and those obtained from the simulations. The widening of the gate in M2 is not very effective in reducing the melt speed. The use of a foam filter in M3 and M4 are beneficial as it increases the gate area and radically reduces the melt velocity. The trident gate in M5 and M6 did not reduce the melt velocity as expected due to an undersized gate area. As can be seen from the simulations, even when the melt is slowed down using two filters, the velocity increases again in the region of the small cross-section gate. The vortex element at the end of the runner in M6 reduced the melt velocity after the gate compared to M5.

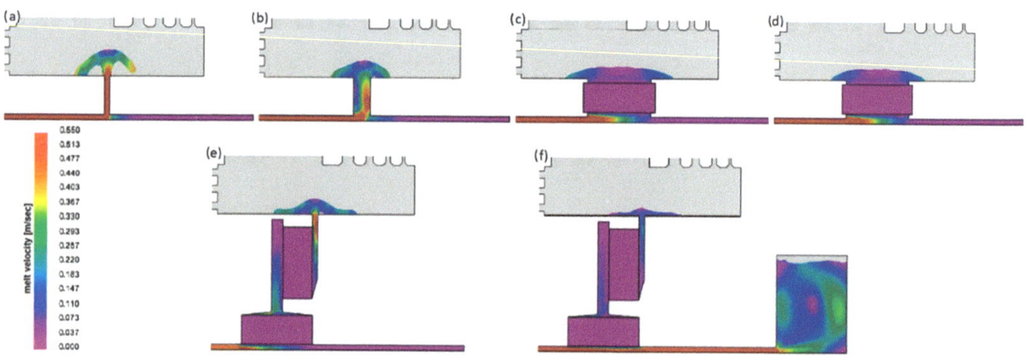

Figure 5. Simulation of melt velocity: (**a**) M1, (**b**) M2, (**c**) M3, (**d**) M4, (**e**) M5 and (**f**) M6.

3.2. *Melt Flow Determination*

The flow behind the gate was analyzed by simulation and also by physical experiment with the inspection fluid. The inspection fluid experiment was conducted based on previous findings [17,18,25]. In these studies, the authors used water to observe the flow in a gate system. Most of the works use water at a temperature of 20 °C for experiments, which has a kinematic viscosity of 0.01 $m^2 \cdot s^{-1}$, and the kinematic viscosity of the liquid aluminum alloy is 0.00959 $m^2 \cdot s^{-1}$. The study [26] showed that even the Reynolds number is an important parameter in water experiments to obtain a reliable analogy when observing the nature of flow between metal and water, and that the Froude number has almost no effect. The study [27] found only small differences in the observed flow of water and steel, despite a difference in kinematic viscosity of about 18%. The authors of [25] found in their experimental conditions that water flow at a temperature of 53.1 °C results in a Reynolds number similar to that of an aluminum alloy melt.

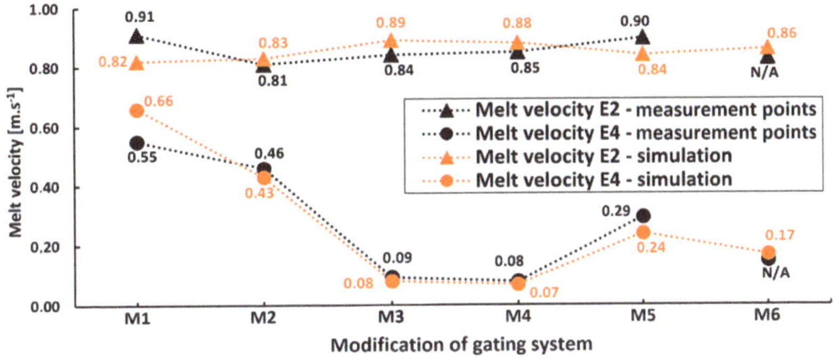

Figure 6. Graph of melt velocity evaluation.

Molds (Figure 7a) were made from sand blocks (Figure 7b) covered by PMMA sheets for the purpose of observing the flow using the inspection liquid experiment (Figure 7c). A gating system was milled into the sand block and the surface of the block was impregnated to prevent water seepage. As both the simulations and the experimental melts were cast in sand molds, the use of a sand block for the water experiment was also advantageous from the point of view of maintaining identical flow conditions, such as friction, etc. Based on the research work, water with a temperature of 53.1 ± 0.3 °C was chosen for the purpose of the water experiment. In order to maintain the same conditions as during the pouring of the liquid metal, the water was poured from a height of 100 mm and the water level in the pouring basin was maintained at the same level as during the pouring of the metal. A blue dye was added to the water to provide contrast between the flowing medium and the mold wall, and an orange dye was added to highlight any turbulence in the inspection liquid. The flow of the liquid in the mold was recorded on camera in ultra-slow motion.

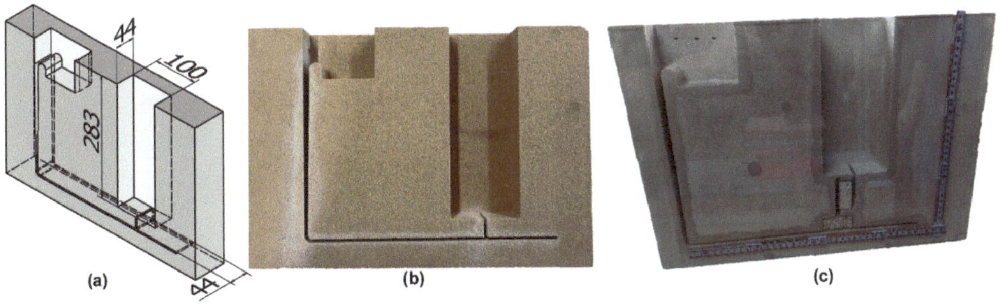

Figure 7. Mold for inspection liquid test, (**a**) mold dimensions, (**b**) mold sand blocks, (**c**) mold covered with PMMA.

Both the simulations and the water experiment showed a large spattering in M1 and the formation of the so-called fountain effect (Figure 8). The filling of the mold cavity was turbulent in the following stages. In contrast, in M2, there was less spattering at the start of filling, but the flow was also turbulent and uncontrolled during filling (Figure 9). The foam filters in M3 and M4 ensured a much smoother filling of the mold cavity without spattering, which was associated with a significant reduction in velocity (Figures 10 and 11). The higher density of the foam filter gave better results, which could be observed by simulations, and also by image analysis of the water test. In M5 and M6, a reduction in spatter was observed compared to the reference gating system, but the filling of the mold cavity was still turbulent (Figures 12 and 13). This was a surprising finding, because according to the study [11], the most optimal system is the combined system with trident gates and a bubble trap. In the design with a vortex element at the end of the runner, a calmer cavity filling process was observed compared to the M5. As mentioned above, the trident gate was expected to have a greater effect on the velocity reduction behind the gate. Therefore, the advantage of the trident gate can be suppressed if the gate size is inappropriately designed. Based on the comparison of the flow analysis behind the gate using simulations and the water experiment, it is possible to see a major correlation between the results obtained.

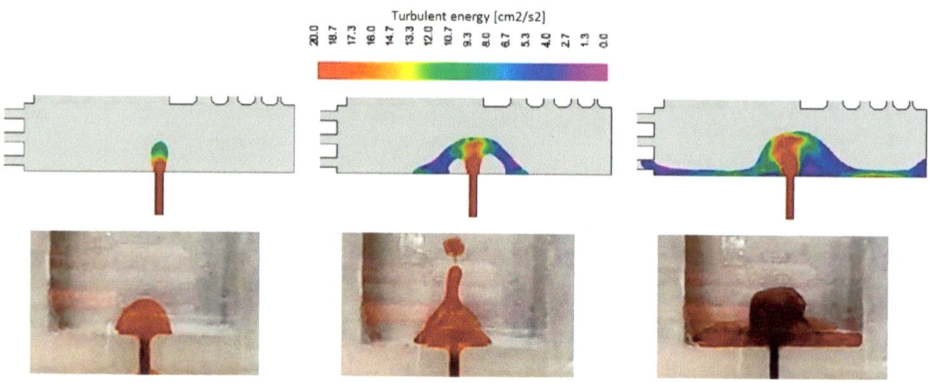

Figure 8. Fluid flow in M1, the color spectrum represents the amount of turbulent energy.

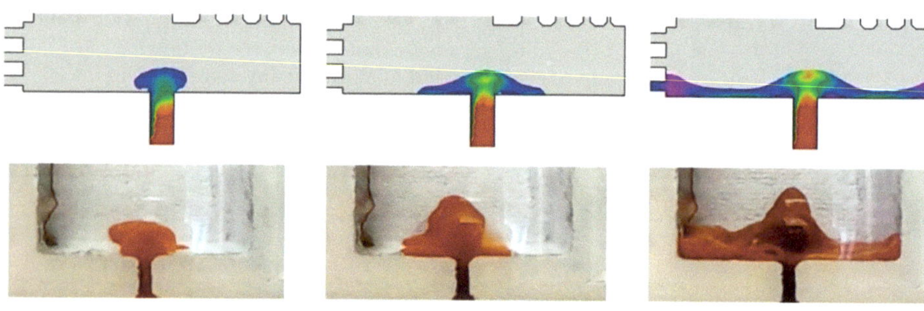

Figure 9. Fluid flow in M2, the color spectrum represents the amount of turbulent energy.

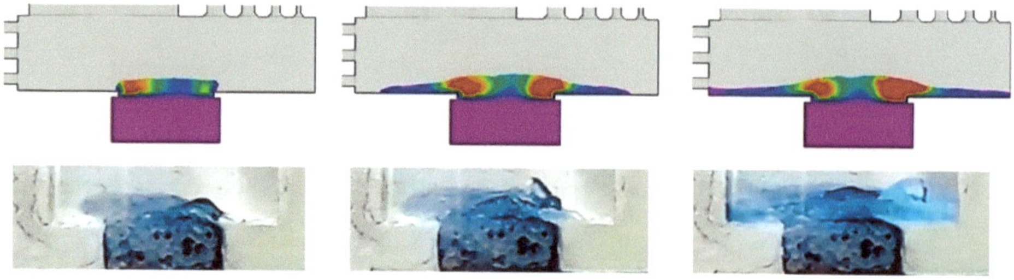

Figure 10. Fluid flow in M3, the color spectrum represents the amount of turbulent energy.

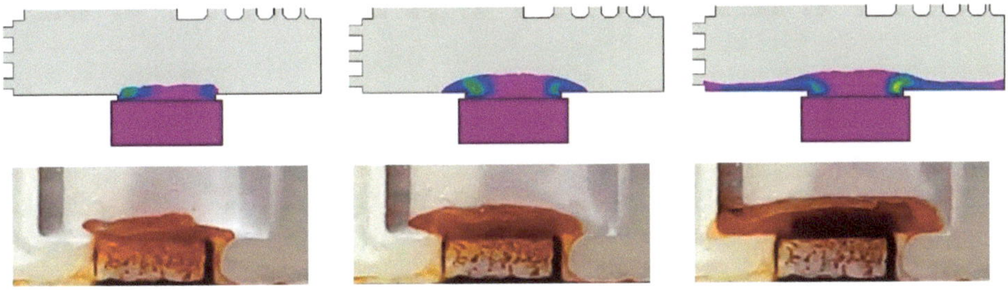

Figure 11. Fluid flow in M4, the color spectrum represents the amount of turbulent energy.

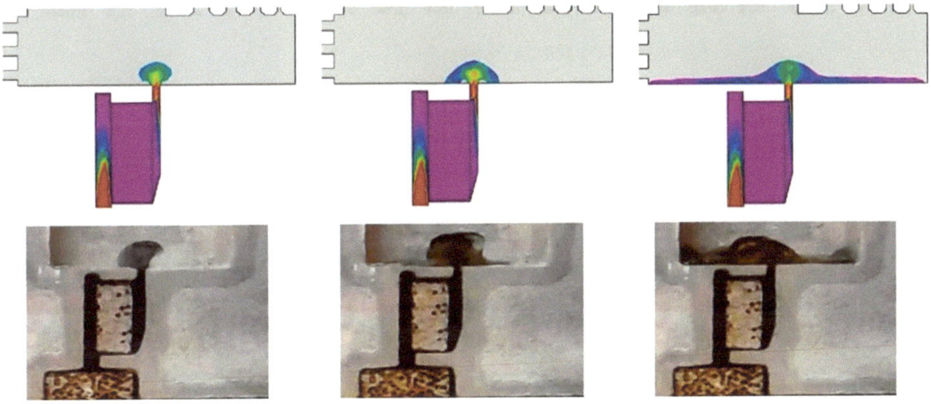

Figure 12. Fluid flow in M5, the color spectrum represents the amount of turbulent energy.

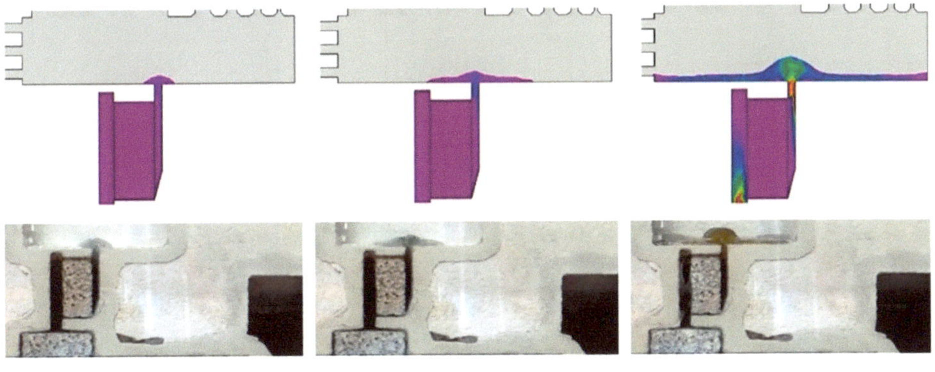

Figure 13. Fluid flow in M6, the color spectrum represents the amount of turbulent energy.

3.3. Amount of Oxides Determination

The analysis of the trapped oxides by means of simulations (Figure 14) showed that their amount in the casting was highest in M1. An improvement can be seen in the M2 design compared to the reference design. M5 and M6 showed a significant reduction in their amount, but the best results were obtained in M3 and M4. When comparing the measured values of the melt velocity, it was possible to observe a tendency for the number of oxides in the castings to decrease as the melt velocity at the entrance to the mold cavity decreases, which is associated with a calmer filling process.

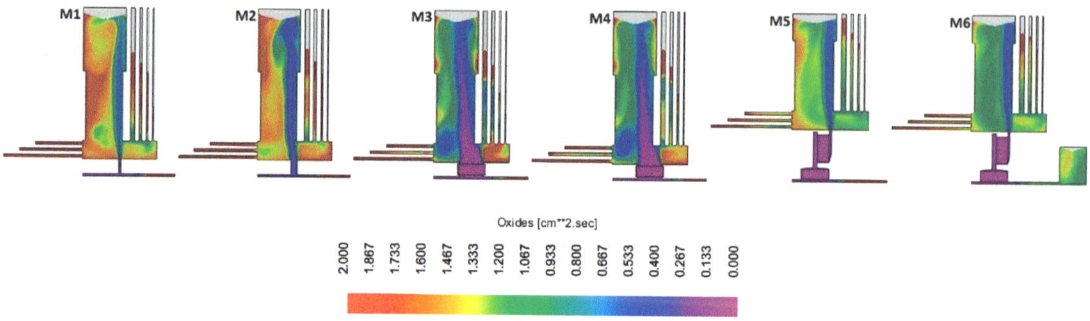

Figure 14. Simulation of number of oxides.

3.4. Casting Characteristics Evaluation

The casting contains parts for assessing fluidity (area A in Figure 15) and cracking susceptibility (area B in Figure 15). The part of the casting used for the fluidity evaluation consists of four test rods of circular cross-section with graduated diameters. The diameters of the test rods were 8 mm, 6 mm, 4 mm and 2 mm. The aim of this evaluation was to analyze the suitability of using individual elements in the gating system for casting thin-walled castings. The results obtained were also compared with the simulation. The rectangular section test rods used to evaluate the susceptibility to cracking had graduated arm lengths (142 mm, 112 mm and 82 mm). The end of the test bars was in the shape of a "T", the purpose of which was to prevent the metal from shrinking and thereby initiating cracking.

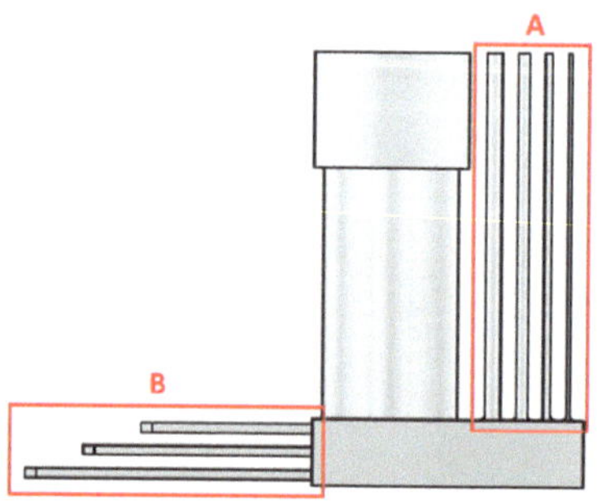

Figure 15. Location of fluidity test and hot tearing evaluation.

Figure 16 shows the effect of each modification on fluidity. The best fluidity test results with test rod diameters of 8 and 6 mm were obtained with M2, and the worst results were obtained with M5 and M6, which was also consistent with the results of the simulations. With test rod diameters of 4 and 2 mm, the best results were obtained with M3 and M4. On the contrary, the lowest fluidity was observed in M5 and M6. The results of the fluidity of the 4 and 2 mm diameter test bars did not agree with the assessment made by the simulation software, where the best results were observed in M1 and M2 and, conversely, the worst in M3 and M4. The best combination of the ability to fill parts of the mold cavity with smaller and larger parts can be attributed to the M3 and M4 designs. In these, despite a significant reduction in melt speed after the gate, there was no significant reduction in melt flow in the 8 and 6 mm diameter rods compared to M1 and M2, and the best results were obtained in the 4 and 2 mm diameter rods. The good results obtained in M3 and M4 in all the test rods could be due to the reduced number of impurities in the melt after the gate. These are created during the turbulent filling of the mold cavity and oxides are entrained in the melt, while any impurity in the melt degrades its fluidity. The lowest fluidity of all test bars in the M5 and M6 designs could be due to the lower hydrostatic pressure compared to other designs; also, the trident gate and vortex element at the end of the runner prolong the time the melt takes to enter the mold cavity. As a result, the melt flowed into the cavity at a lower temperature compared to other modifications, which significantly reduced the fluidity. The influence of the gate design on the susceptibility to cracking was not observed. In the test bars used to assess cracking susceptibility, there was no violation in the integrity of the arms in any of the bars, nor was there any initiation of cracks.

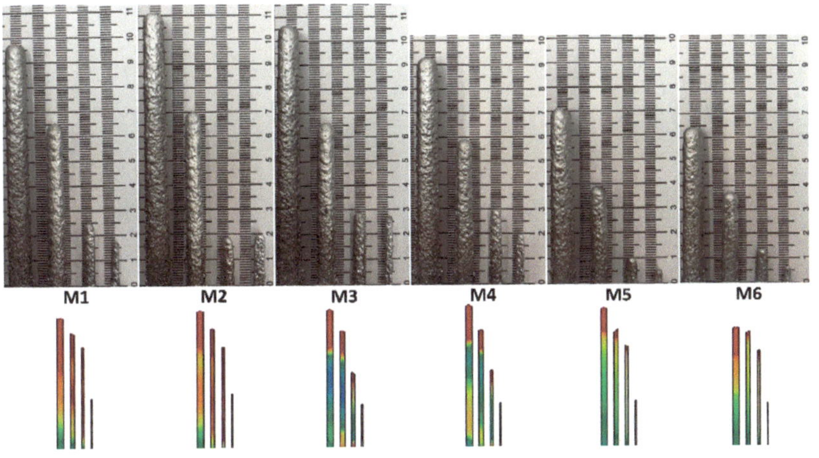

Figure 16. Fluidity test, the color spectrum represents susceptibility to misrun (red = maximum).

3.5. Porosity Evaluation

In order to assess the quality of the castings, their porosity was evaluated on block-shaped samples with the dimensions of 125 × 40 × 5 mm, as shown in Figure 17. As part of the evaluation, the samples were subjected to surface porosity analysis. The result was a percentage expression of the area of the pores in relation to the total area of the section of the casting. The surface porosity of the castings was measured using the QuickPHOTO INDUSTRIAL 3.1 software based on the color contrast between the pore and the surface of the sample, which allowed the porosity to be quantified.

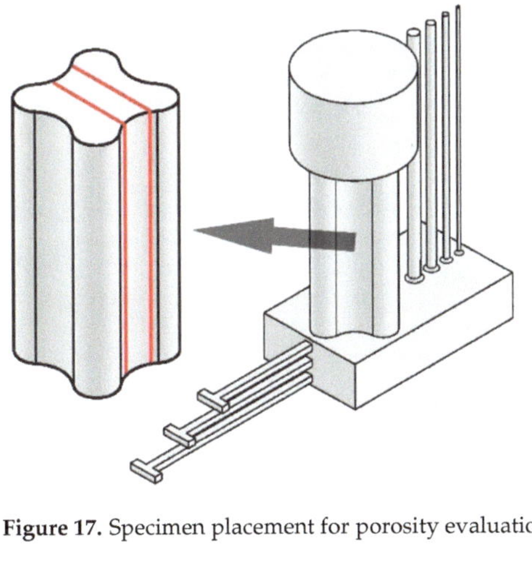

Figure 17. Specimen placement for porosity evaluation.

The highest rate of surface porosity was measured in construction M1. A significant reduction in porosity occurred in M3, where the surface porosity was reduced by 64% compared to M1. Similar results were seen in M4 where the results were improved by 61%. Compared to the reference gating system, the pore content was reduced by 26% in M2, 33% in M5 and 49% in M6. By analyzing the melt velocity and flow behind the gate, it can be seen that the cavity filling process becomes smoother as the velocity decreases. This is related to a smaller number of air bubbles trapped in the melt. From this, it is possible to observe a trend where the porosity of the castings decreases with a decrease in the speed of the melt entering the cavity. Figure 18 shows samples from the surface porosity evaluation and Figure 19 shows the surface porosity evaluation graph.

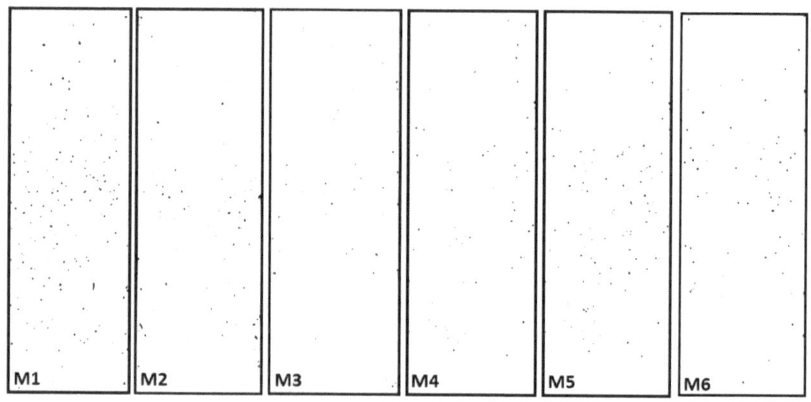

Figure 18. Porosity evaluation.

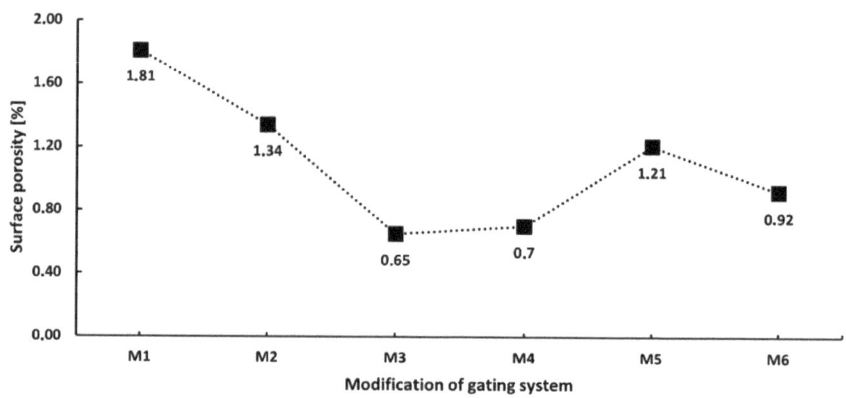

Figure 19. Graph of porosity evaluation.

From the measured values, it is possible to observe a parabolic dependence between the surface porosity (SP) and the melt velocity in the gates v (Figure 20). This dependence can be described by the equation:

$$SP = 2.4\sqrt{v} \quad (1)$$

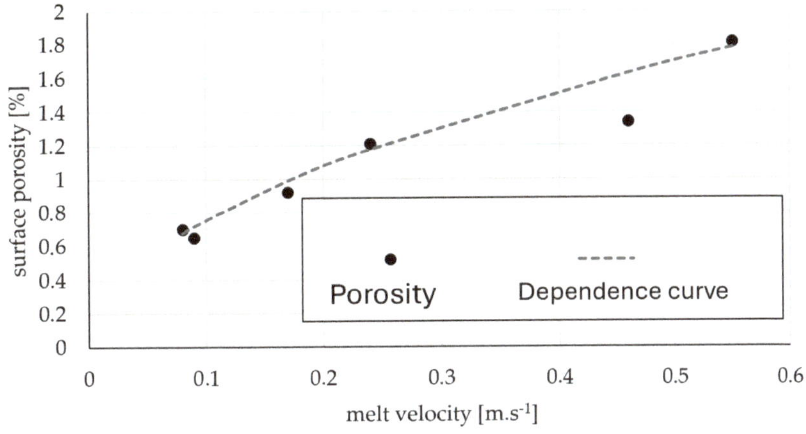

Figure 20. Parabolic dependence between the surface porosity and the melt velocity in the gates3.6 Mechanical properties evaluation.

The direct dependence between the porosity and the speed of the melt in the gates documents its more significant influence than the conditions and the course of solidification of the casting.

Samples were taken from the center sections of the castings for tensile and impact toughness tests. The arrangement of the specimens for each test is shown schematically in Figure 21. A total of four tensile test specimens and four impact toughness test specimens were taken from the casting. After the tensile tests, the elongation was evaluated and the yield strength agreed.

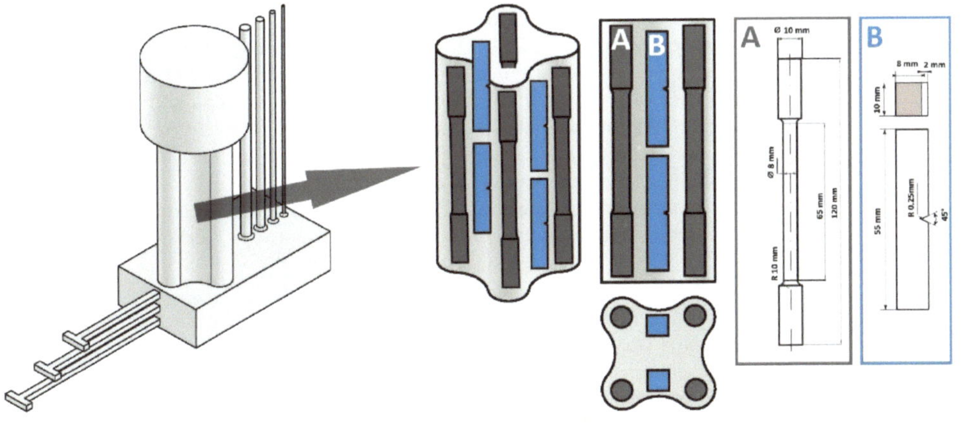

Figure 21. Specimen placement for mechanical properties evaluation.

Figure 22 shows a graph of the mean values of the mechanical properties as a function of the modification of the gating system used. It can be seen from the graph that the modification of the gating system used did not have a significant effect on the impact toughness, except for M4 where it increased by 15% compared to M1. From the tensile and yield strength results, it can be seen that the greatest increase was achieved with the foam filter modifications M3 and M4, where better results were obtained with a higher filter density. Compared to the reference gating system M1, the tensile strength of M4 was improved by approximately 18% and the yield strength by 70%. The tensile strength limit and yield strength of M3 increased by 12% and 58%, respectively, and M6 increased by 9% and 43%, respectively. The results are relatively consistent with research in study [28], in which the application of 10 PPI filters had remarkable positive standardized effects on the UTS and % elongation. Filtration of the melt using foam filters has a beneficial effect on the mechanical properties. There is a breakdown of oxide films into smaller formations that do not have such a harmful effect, leading to an increase in the strength of the material. By using a permanent mold, an increase in ductility, mainly by changing the morphology of the eutectic, during solidification can occur. The increase in elongation in the mentioned study was surprising because of the use of sand molds. In the case of materials cast in sand molds, the increase in strength properties is accompanied by a decrease in plastic properties.

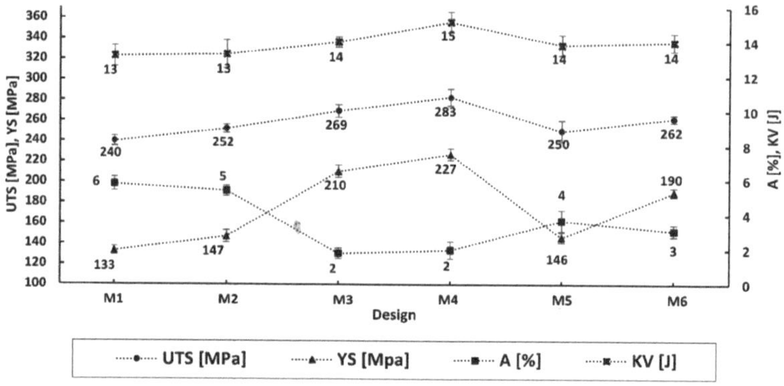

Figure 22. Graph of mechanical properties evaluation.

The M2 and M5 gating system designs achieved approximately the same values, with an increase in tensile strength of approximately 5% and an increase in yield strength of approximately 11%. Elongation decreased as tensile strength increased. From the evaluation, a certain relationship can be observed between the melt velocity, the flow behind the gate, the number of oxides in the castings, the porosity and the mechanical properties. As the melt velocity behind the gate decreases, the mold cavity fills more smoothly. As the simulations showed, this may also be related to the reduced number of solidified oxides in the castings. Porosity analysis also shows a trend towards a reduction in porosity. Modifications M3 and M4 obtained the lowest porosity values; therefore, the functional section of the samples is the least affected among them, which can also be attributed to obtaining the best values in mechanical properties.

3.6. Microscopic Analysis

A high-quality primary alloy was chosen for the experiments so that the porosity in the casting was not affected by the nucleation of pores on internal impurities. The structure of the casting is formed by typical structural components for this type of alloy. It consists of dendrites of the alpha phase, eutectic (mechanical mixture of alpha phase and Al2Cu phase) and complex phases based on AlMnCu and AlFeCu (Figure 23 left). The matrix is formed by dendrites of the alpha phase with an SDAS index value for individual modifications in the range of 37 to 52 μm. These values are characteristic for solidification in sand molds. Pores were not observed in the close vicinity of complex phases (Figure 23 right); it can be assessed that the influence of intermetallic phases as initiators of porosity in casts was not confirmed. There was also no influence of the melt flow rate when filling the mold on the structure of the casting.

Results of the structure evaluation did not confirm the effect of structure formation on porosity; therefore, an evaluation using SEM-EDX analysis was performed. Since the initiator of the formation of porosity in our case was not individual structural components or impurities, the evaluation of the pore surface focused on the formation of oxides. The relationship between pores and double oxide layers was observed using SEM-EDX analysis because several studies have suggested that the mechanism of pore formation in castings is initiated by entrained double oxide layers. This mechanism suggests the presence of oxide layers on the inside of the pore surface. This theory and the presence of oxides in the pore region are also confirmed by other works [5,7,11]. Figure 24 shows SEM images of the inner surface of pores taken from a random area of a cast sample (M1 on the left and M4 design on the right) to assess areal porosity. The entire pore region was also subjected to EDX element map analysis (Figure 25). EDX analysis of the element map and point EDX analysis showed that there was a certain concentration of oxygen in the surface region of the pores, indicating the presence of an oxide layer. The other pores could also have been formed by this mechanism, as the presence of oxygen was also detected in the other pores analyzed by EDX analysis. The readings for the presence of oxygen may not be accurate, as EDX analyzers can have problems detecting the presence of elements with atomic numbers below 10, but they are able to detect its presence in the sample. In addition to the oxygen content, folds were also observed on the inner surface of the pores (Figure 26), which is a typical morphology of the oxide layer according to works [5] and, therefore, another sign of its presence.

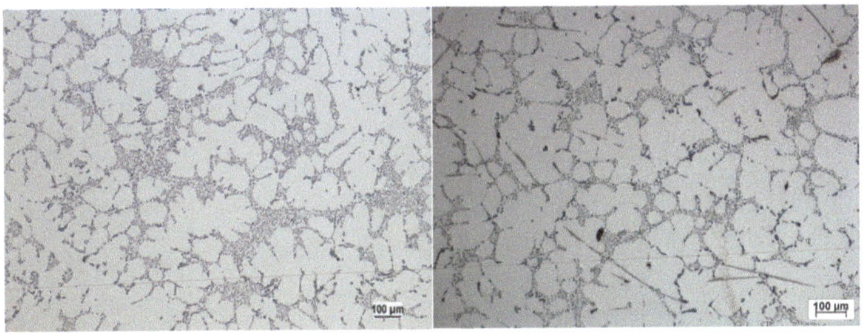

Figure 23. Structural components of investigated alloy.

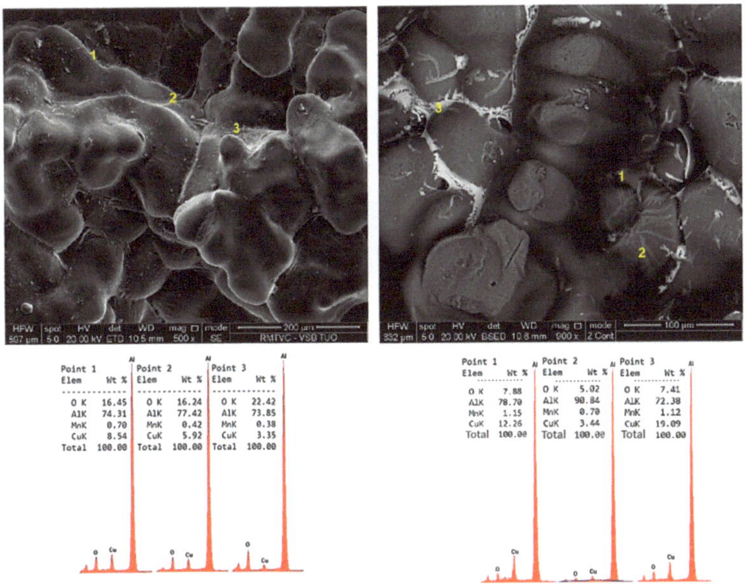

Figure 24. SEM analysis of inner pore surface with point EDX analysis (**Left** for M1, **Right** for M4).

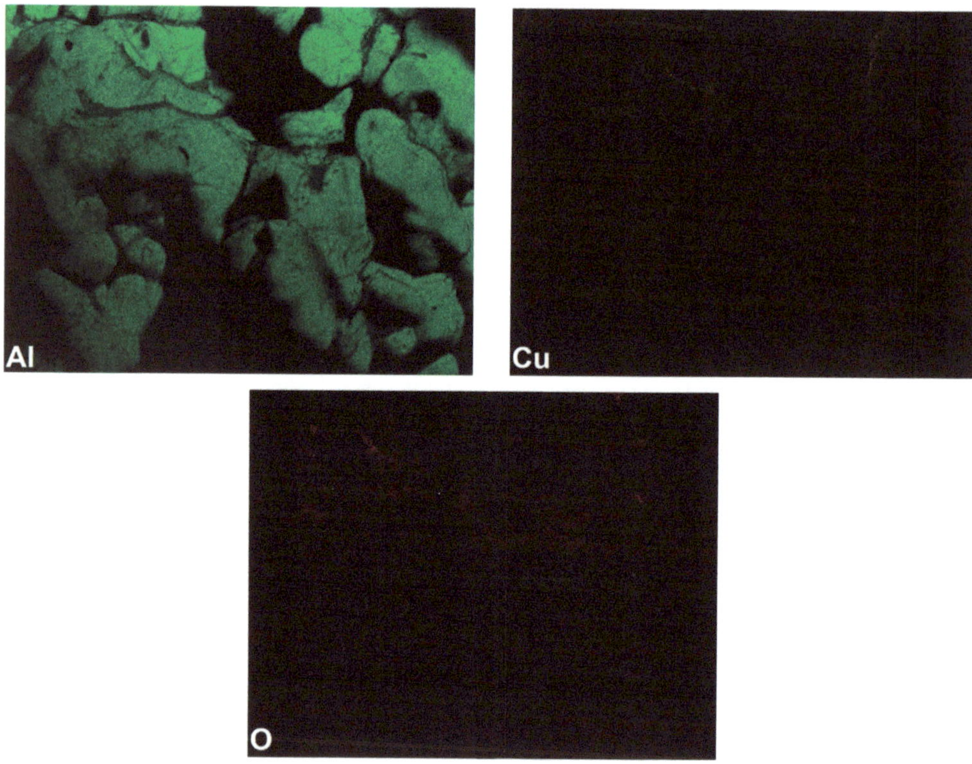

Figure 25. EDX map analysis of inner pore surface (M1 design).

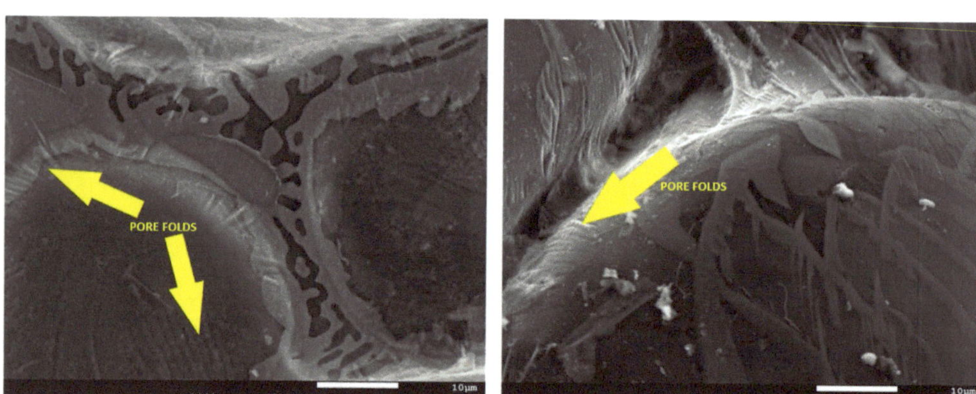

Figure 26. Oxide film folds in the inner pore surface.

3.7. Metal Yield Determination

The metal yield of each gating system modification was determined by weighing after removal from the mold. The best metal yield was achieved with M1, but the difference between M2 and M4 was negligible (2–5% yield difference). The metal yield of M5 was 11% lower than M1. The worst results were achieved by design M6, where the yield was 55% worse than M1. Even if the quality of the castings were improved with M6, such an increase in metal yield would not be negligible and would significantly increase the cost of the castings. The beneficial effect of the M4 design in providing the best quality castings is not negated by the increase in metal yield.

4. Conclusions

The following conclusions can be drawn from the results obtained:

- The melt velocity in the naturally pressurized gating system, without the use of any element to reduce it, reached supercritical values. The modeling and experimental casts confirm the effectiveness of the gating systems presented here as a means of controlling the velocity of metal entering the mold cavity, although they also allow for a much less turbulent flow of metal through the gating system, thus reducing the possibility of double oxide formation caused by the reoxidation process.
- According to the simulations, the melt velocity behind the gate and the associated nature of the filling of the mold cavity affected the number of oxides in the castings. As the melt velocity increased, there was more turbulent filling of the cavity and the formation of a greater number of oxides in the castings.
- Experimental melts showed a relationship between melt velocity and porosity of the castings. Quantitative evaluation of the surface porosity of the castings showed a trend of decreasing porosity with decreasing melt velocity in the mold cavity. The porosity could be the result of the entrapment of air bubbles between the oxide layers during the impact of the melt from a great height and the turbulent nature of filling the mold cavity.
- The direct dependence between the porosity and the speed of the melt in the gates (Figure 20) documents its more significant influence than the conditions and the course of solidification of the casting.
- The use of a foam filter in the gate area provided the most significant reduction in melt velocity behind the gate compared to other gating system modifications. The higher density of the foam filter provided a greater reduction. The greatest reduction in melt velocity in the mold cavity was achieved with M4, which was associated with: the highest reduction in the oxides; the melt entered the mold cavity without spattering and a smooth filling of the mold cavity was ensured without turbulence, obtaining the most advantageous combination of fluidity, and achieving the lowest level of surface porosity in the castings. Tensile strength, yield strength and toughness were significantly increased, and the best casting quality was achieved without a significant increase in metal yield.
- The pores analyzed proved the presence of oxide layers on their inner surface and a possible theory of pore formation when the initiator of porosity is entrained double oxide layers.

Author Contributions: Conceptualization, M.B. and M.G.; methodology, M.B.; software, E.K.; validation, R.P. and E.K.; formal analysis, R.P.; investigation, M.B.; resources, M.G.; data curation, E.K.; writing—original draft preparation, M.G.; writing—review and editing, M.B.; visualization, E.K.; supervision, R.P.; project administration, M.G.; funding acquisition, M.B. All authors have read and agreed to the published version of the manuscript.

Funding: This paper was supported by VEGA 1/0241/23 and KEGA 029ŽU-4/2023 projects.

Data Availability Statement: The raw data supporting the conclusions of this article will be made available by the authors on request.

Conflicts of Interest: The authors declare that they have no known competing financial interests or personal relationships that could have appeared to influence the work reported in this paper.

References

1. Campbell, J. *Complete Casting Handbook*, 2nd ed.; Elsevier: Oxford, UK, 2015; p. 1054.
2. Liu, J.; Wang, Q.; Qi, Y. Atomistic simulation of the formation and fracture of oxide bifilms in cast aluminum. *Acta Mater.* **2019**, *164*, 673–682. [CrossRef]

3. Jalilvand, M.M.; Saghafian, H.; Divandari, M.; Akbarifar, M. Surface oxidation study of molten Mg–Al alloys by oxide/metal/oxide sandwich method. *J. Magnes. Alloy* **2021**, *10*, 1704–1717. [CrossRef]
4. Chen, Q.; Griffiths, W. Modification of Double Oxide Film Defects with the Addition of Mo to An AlSi-Mg Alloy. *Metall. Mater. Trans. B* **2021**, *52*, 502–516. [CrossRef]
5. Neuser, M.; Grydin, O.; Andreiev, A.; Schaper, M. Effect of Solidification Rates at Sand Casting on the Mechanical Joinability of a Cast Aluminium Alloy. *Metals* **2021**, *11*, 1304. [CrossRef]
6. Nunes, H.; Emadinia, O.; Soares, R.; Vieira, M.F.; Reis, A. Adding Value to Secondary Aluminum Casting Alloys: A Review on Trends and Achievements. *Materials* **2023**, *16*, 895. [CrossRef]
7. El-Sayed, M.A.; Essa, K.; Hassanin, H. Effect of runner thickness and hydrogen content on the mechanical properties of A356 alloy castings. *Int. J. Met.* **2022**, *16*, 2175–2186. [CrossRef]
8. Gyarmati, G.; Fegyverneki, G.; Tokár, M.; Mende, T. Investigation on Double Oxide Film Initiated Pore Formation in Aluminum Casting Alloys. *Int. J. Eng. Manag. Sci.* **2020**, *5*, 141–153.
9. Metzloff, K.; Mageza, K.; Sekotlong, D. Velocity measurement and verification with modelling of naturally pressurized gating system. *Int. J. Met.* **2020**, *14*, 610–621.
10. Remišová, A.; Brůna, M. Analysis of Reoxidation Processes with Aid of Computer Simulation. *Arch. Foundry Eng.* **2019**, *19*, 55–60. [CrossRef]
11. Dojka, R.; Jezierski, J.; Campbell, J. Optimized Gating System for Steel Castings. *J. Mater. Eng. Perform.* **2018**, *27*, 5152–5163. [CrossRef]
12. Marquardt, W. Continuing the Conversation Naturally Pressurized Fill System. In Proceedings of the SFSA Technical and Operating Conference, Chicago, IL, USA, 11 December 2018; p. 30.
13. Majidi, S.H.; Beckermann, C. Effect of pouring conditions and gating system design on air entrainment during mold filling. *Int. J. Met.* **2018**, *13*, 255–272. [CrossRef]
14. Poweleit, D. Naturally Pressurized Gating Systems in Investment Castings. In Proceedings of the 73rd SFSA Technical and Operating Conference, Chicago, IL, USA, 11–14 December 2019; p. 17.
15. Yavuz, H.; Kara, A.; Çubuklusu, H.E.; Aybarç, U. The effect of filter type on filling during Low Pressure Die Casting: A numerical study. *J. Achiev. Mater. Manuf. Eng.* **2016**, *75*, 71–77. [CrossRef]
16. Hwang, H.Y.; Nam, C.H.; Choi, Y.S.; Hong, J.-H.; Sun, X. A study on effects of ceramic foam filter on flow aspect through water modeling experiment. *China Foundry* **2017**, *14*, 216–225. [CrossRef]
17. Kheirabi, A.; Baghani, A.; Bahmani, A.; Tamizifar, M.; Davami, P.; Shabani, M.O.; Mazahery, A. Understanding the occurrence of the surface turbulence in a nonpressurized bottom gating system: Numerical simulation of the melt flow pattern. *J. Mater. Des. Appl.* **2015**, *232*, 230–241. [CrossRef]
18. Lyu, C.; Papanikolaou, M.; Jolly, M. Numerical Process Modelling and Simulation of Campbell Running Systems Designs. In *Shape Casting*; Springer: Berlin/Heidelberg, Germany, 2019; pp. 53–64.
19. Papanikolaou, M.; Pagone, E.; Jolly, M.; Salonitis, K. Numerical Simulation and Evaluation of Campbell Running and Gating Systems. *Metals* **2020**, *10*, 68. [CrossRef]
20. Jezierski, J.; Dojka, R.; Janerka, K. Optimizing the Gating System for Steel Castings. *Metals* **2018**, *8*, 266. [CrossRef]
21. Brůna, M.; Galčík, M.; Sládek, A.; Martinec, D. Possibilities of bifilm amount reduction in Al castings by gating system design optimization. *Arch. Metall. Mater.* **2021**, *66*, 549–559. [CrossRef]
22. Bate, C.; King, P.; Sim, J.; Manogharan, G. A Novel Approach to Visualize Liquid Aluminum Flow to Advance Casting Science. *Materials* **2023**, *16*, 757. [CrossRef]
23. Rajkumar, I.; Rajini, N.; Alavudeen, A.; Prabhu, T.R.; Ismail, S.O.; Mohammad, F.; Al-Lohedan, H.A. Experimental and simulation analysis on multi -gate variants in sand casting process. *J. Manuf. Process.* **2021**, *62*, 119–131. [CrossRef]
24. Yousefian, P. Pore Formation in Aluminum Castings: Theoretical Calculations and the Extrinsic Effect of Entrained Surface Oxide Films. Master's Thesis, University of North Florida, Jacksonville, FL, USA, 2017.
25. Xinyi, Z.; Ning, Z.L.; Cao, F.Y.; Liu, S.-G.; Huang, Y.-J.; Liu, J.-S.; Sun, J.-F. Effect of Double Oxide Film Defects on Mechanical Properties of As-Cast C95800 Alloy. *Acta Metall. Sin.* **2017**, *30*, 541–549.
26. Farhoodi, B.; Raiszadeh, R.; Ghanaatian, M.H. Role of double oxide film defects in the formation of gas porosity in commercial purity and Sr-containing Al alloys. *J. Mater. Sci. Technol.* **2014**, *30*, 154–162. [CrossRef]
27. Tiryakioğlu, M.; Yousefian, P.; Eason, P.D. Quantification of entrainment damage in A356 aluminum alloy castings. *Metall. Mater. Trans. A* **2018**, *49*, 5815–5822. [CrossRef]
28. El-Sayed, M.A.; Essa, K.; Hassanin, H. Influence of Bifilm Defects Generated during Mould Filling on the Tensile Properties of Al-Si-Mg Cast Alloys. *Metals* **2022**, *12*, 160. [CrossRef]

metals

MDPI

Article

Electrical Equivalent Circuit Model Prediction of High-Entropy Alloy Behavior in Aggressive Media

Jose Cabrera-Peña [1], Santiago Jose Brito-Garcia [2], Julia Claudia Mirza-Rosca [2,3,*] and Gustavo M. Callico [1]

[1] Institute for Applied Microelectronics (IUMA), University of Las Palmas de Gran Canaria, 35017 Las Palmas de Gran Canaria, Spain; gustavo.callico@ulpgc.es (G.M.C.)
[2] Mechanical Engineering Department, University of Las Palmas de Gran Canaria, 35001 Las Palmas de Gran Canaria, Spain
[3] Materials Engineering and Welding Department, Transilvania University of Brasov, 500036 Brasov, Romania
* Correspondence: julia.mirza@ulpgc.es

Abstract: Due to the optimistic outcomes of the research on high-entropy alloys, new designs of these alloys are being encouraged. We studied the high-entropy CoCrFeMoNi alloy and the CoCrFeMoNi alloy doped with Zr. In order to choose the best electrical equivalent circuit for the prediction of the behavior of these high-entropy alloys at various potentials in artificial seawater, electrochemical impedance spectroscopy (EIS) measurements were conducted on samples with and without Zr-doped CoCrFeMoNi. At various potential levels, the impedance spectra were measured between −1.0 and +0.8 V vs. SCE. The study consists of a preliminary section with microstructure by metallography, open-circuit potential, and linear polarization curves by direct-current tests followed by visual analysis of the impedance spectra, and, finally, the selection of an equivalent electrical circuit model to fit the experimental data. By leveraging the advantages of EIS analysis, the information is essential for materials development, corrosion-mitigation strategies, and the successful implementation of these alloys in practical applications. It is important to note that selecting an equivalent circuit is often an iterative and subjective process, as it involves a balance between model complexity and the ability to accurately represent the system's behavior.

Keywords: EIS; equivalent circuit; Zr-doped; high-entropy alloys

Citation: Cabrera-Peña, J.; Brito-Garcia, S.J.; Mirza-Rosca, J.C.; Callico, G.M. Electrical Equivalent Circuit Model Prediction of High-Entropy Alloy Behavior in Aggressive Media. *Metals* **2023**, *13*, 1204. https://doi.org/10.3390/met13071204

Academic Editor: Carlos Alexandre dos Santos

Received: 31 May 2023
Revised: 21 June 2023
Accepted: 26 June 2023
Published: 29 June 2023

1. Introduction

Electrochemical impedance spectroscopy (EIS) is a technique used to analyze the electrical properties of electrochemical systems. It provides valuable information about the interfacial processes occurring at the electrode–electrolyte interface [1]. In EIS, a small sinusoidal perturbation signal is applied to the electrochemical system over a range of frequencies and the response of the system is then measured, typically in terms of impedance, which is the ratio of the applied voltage to the resulting current. The impedance measured in EIS consists of two components: real (resistance) and imaginary (reactance) parts. The real part represents the system's resistance to the flow of charge, while the imaginary part reflects the system's capacitive or inductive behavior.

By varying the frequency of the applied signal, EIS generates an impedance spectrum that provides insights into various electrochemical processes. This includes information about charge transfer reactions, adsorption/desorption processes, diffusion of species, and other phenomena occurring at the electrode–electrolyte interface. EIS is widely used in various fields, including biomaterials [2–4], fuel cells [5–7], the chemical industry [8–11], etc., to characterize and understand the behavior of electrochemical systems. The obtained impedance data can be further analyzed and modeled to extract valuable parameters and gain insights into the underlying electrochemical processes. The use of advanced impedance measurement setups and dedicated software has accelerated experiments in the field of EIS. As a result, data acquisition is no longer the time-consuming aspect of the process; rather, the analysis of experimental data has become the primary bottleneck. Therefore, impedance modeling plays a crucial role in the EIS method [12–15].

The kinetic approach to EIS modeling, which is well-established and documented in various textbooks [16,17], is commonly used. This approach involves linearizing nonlinear equations of electrochemical kinetics by utilizing small perturbation signals. The linearized equations are then subjected to Laplace transformation, along with differential equations and partial derivatives equations derived from adsorption and mass transport processes, respectively [18].

The analysis of corrosion behavior of a new high-entropy alloy with electrochemical impedance spectroscopy (EIS) offers several advantages:

Non-destructive characterization: EIS is a non-destructive approach, allowing for the analysis of corrosion behavior without causing harm to or changing the material being investigated. When examining innovative or expensive high-entropy alloys, this is especially helpful because it allows for many measurements to be taken on the same sample, producing useful information while preserving the material for additional study or applications.

Comprehensive electrochemical information: In-depth electrochemical knowledge on the corrosion behavior of high-entropy alloys is provided by EIS. It provides information about a number of electrochemical processes that take place at the alloy's surface, including mass transport, double-layer capacitance, and adsorption/desorption phenomena. This enables a thorough comprehension of the mechanisms causing corrosion and the variables affecting an alloy's corrosion resistance.

Quantitative assessment of corrosion resistance: The quantitative evaluation of corrosion resistance characteristics is possible with EIS. Important corrosion parameters, such as polarization resistance, corrosion rate, capacitance, and impedance components, can be retrieved by fitting the experimental impedance data to the relevant equivalent circuit models. These variables offer measurable indicators of the alloy's resistance to corrosion, facilitating performance assessment and material comparison.

Sensitivity to localized corrosion: For high-entropy alloys, localized corrosion processes, such as pitting or crevice corrosion, are of significant importance and EIS can shed light on these phenomena. EIS can identify and describe localized corrosion processes by examining the impedance spectrum and keeping track of changes in the low-frequency range, which helps determine the alloy's susceptibility to these types of corrosion.

Evaluation of protective coatings or treatments: EIS is frequently used to assess the efficacy of surface treatments or protective coatings on high-entropy metals. It is feasible to evaluate the barrier qualities, adherence, and long-term performance of the protective layers by contrasting the impedance responses of coated and untreated samples. This knowledge is essential for boosting high-entropy alloy durability and corrosion protection strategies for practical applications.

Fast and efficient data acquisition: EIS experiments can be performed relatively quickly, allowing for efficient data acquisition. With modern high-performance impedance-measurement setups and dedicated software, a significant number of experiments can be conducted within a few laboratory days. This rapid data acquisition facilitates the screening and comparison of different high-entropy alloys or the evaluation of corrosion behavior under various conditions.

Taking into account all these, the purpose of our study is to select the proper electrical equivalent circuit for the prediction of the behavior of two high-entropy alloys at different potentials in artificial seawater. The study consists of three sections: (1) a preliminary section with microstructure by metallography, open-circuit potential, and linear polarization curves by direct current tests; (2) visual analysis of the impedance spectra; and (3) selection of an equivalent electrical circuit model to fit the experimental data. By leveraging the advantages of EIS analysis, the information is essential for materials development, corrosion-mitigation strategies and the successful implementation of these alloys in practical applications.

2. Materials and Methods

2.1. Samples Preparation

The CoCrFeMoNi alloy (Sample 1) and the Zr-doped CoCrFeMoNi alloy (Sample 2) were investigated in this study. The samples were produced by the ERAMET Laboratory at the Politehnica University of Bucharest using a vacuum arc remelting (VAR) system, model MRF ABJ 900. The laboratory carefully considered potential losses due to vaporization and the expected incorporation of chemical elements during the melting process to design the metallic charge.

To ensure high purity, raw materials with a minimum purity of 99.7% for Co, Cr, Fe, Mo, Ni, and Zr were utilized. To achieve uniformity, the resulting alloys underwent eight cycles of flipping and remelting in the VAR equipment, with four repetitions for each part.

The elemental composition of both Sample 1 and Sample 2 was determined through energy-dispersive X-ray analysis (EDS). This analysis was conducted using a Fei XL30 ESEM scanning electron microscope (MTM, Leuven, Belgium) equipped with an EDAX Sapphire detector (EDAX Inc, Pleasanton, CA, USA), allowing for accurate characterization of the alloy's elemental information, presented in Table 1. The steps of the sample´s preparation are presented in Figure 1 and explained in our work where also the mechanical properties and corrosion behavior of the samples at corrosion potential were previously studied [19].

Table 1. The results in wt% of EDS analysis of the alloy Sample 1 and Sample 2.

wt%	Sample 1	Sample 2
Co	19.81	19.53
Cr	19.98	19.02
Fe	18.04	17.84
Mo	25.18	26.36
Ni	16.99	17.12
Zr	-	0.13

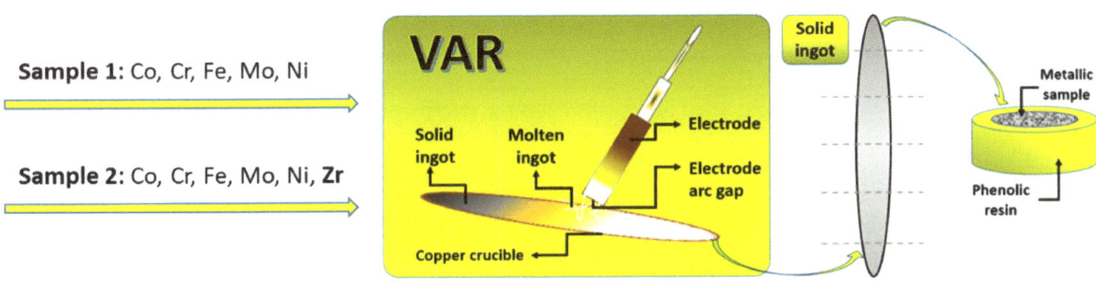

Figure 1. Sample obtention and preparation for the tests.

2.2. Electrochemical Measurements

Three electrodes were used in a typical electrochemical cell setup for all the tests that were performed. A working electrode, a reference electrode, and a counter electrode are included in this arrangement. The working electrode (Samples 1 and 2) is constructed of the alloy under investigation and is where the electrochemical reaction of interest takes place. It provides a platform for studying the behavior of the material. The reference electrode serves as a reference point for potential measurements and maintains a stable potential. In this study, a saturated calomel electrode (SCE) was employed as the reference electrode due to its well-known potential (+0.242 V). The counter electrode is an inert electrode that completes the electrical circuit and facilitates the balanced flow of electrons. Platinum, an inert material that does not participate in the electrochemical reaction, was chosen as the counter electrode in this case. The working electrode and the counter electrode conduct current via the electrolyte solution, which, in this case, is a 3.5% NaCl solution replicating a marine environment, throughout the electrochemical reaction. To ascertain the system's electrochemical properties, the potential difference between the working electrode and the reference electrode is measured. The reference electrode ensures correct potential measurements, while the three-electrode arrangement ensures exact evaluations of the electrochemical behavior. By utilizing this configuration and employing the appropriate electrolyte solution, the study aimed to investigate and characterize the electrochemical response and corrosion behavior of the alloy samples in a simulated seawater environment. The tests were conducted using a Biologic SP-150 potentiostat, manufactured in Seyssinet-Pariset, France. To control the experimental conditions and establish the necessary process parameters, the EC-Lab® v-9.55 program (BioLogic Science Instruments SAS, Seyssinet-Pariset, France) was employed. This software provided the required functionalities for executing the corrosion techniques and ensuring accurate measurements and data acquisition during the tests.

2.3. Preliminary Tests

The microstructure of high-entropy alloys was investigated using optical microscopy. To analyze the surface of the samples, an electrochemical etching was carried out. The samples were immersed in a solution containing 10% oxalic acid and a 5 V current was applied to them for a duration of 5 to 25 s. To observe the etched surface, the OLYMPUS PME 3 microscope manufactured by Olympus Corporation in Tokyo, Japan was utilized. This optical microscope allowed for detailed and accurate observations of the microstructure of the alloy samples after the electrochemical etching process.

The open-circuit potential (OCP) of the samples was measured by immersing them in the salt solution for a duration of 24 h. This extended immersion time allowed the samples to reach a stable potential. To ensure the reliability of the measurements, these tests were repeated three times, aiming to achieve reasonably reproducible results. This repetition helped assess the consistency and quality of the obtained OCP values for the samples.

Potentiodynamic polarization is a corrosion-testing technique used to investigate the behavior of a material in different potential ranges. This method involves varying the electrode potential of the sample over a specified range, following the guidelines outlined in the ASTM standard [19]. The potential values used in this study ranged from −1.5 V to +0.8 V relative to the saturated calomel electrode (SCE). During potentiodynamic polarization, the current passing through the electrode is continuously monitored as the potential is swept. A polarization curve can be constructed by graphing the resultant current as a function of the electrode potential. This curve provides valuable information about the material's corrosion characteristics and can help identify key parameters such as corrosion potential, corrosion current density, and passivation behavior. In this study, potentiodynamic polarization was carried out with a scan rate of approximately 0.166 mV/s [20]. This scan rate determines the speed at which the potential is changed during the measurement and influences the resolution and sensitivity of the obtained polarization curve.

2.4. Electrochemical Impedance Spectroscopy

The potentiostatic electrochemical impedance spectroscopy (PEIS) experiment involved conducting impedance tests in a potentiostatic configuration. A sinusoidal signal was applied at a constant potential E, which was adjusted to specific values ranging from −1.0 V to +0.8 V in steps of 200 mV. The potential was maintained at each fixed value for a duration of 30 min, allowing the cell current to stabilize [21]. During this stabilization period, no impedance measurements were conducted. At each potential value, a frequency scan ranging from 100 kHz to 100 mHz was performed. The scan included six points per decade, providing a comprehensive coverage of the frequency range. The amplitude of the sinusoidal signal was set to 10 mV peak-to-peak. By performing these PEIS measurements, the impedance response of the system at different potentials and frequencies could be obtained, offering valuable insights into the electrochemical behavior and characteristics of the studied material or system.

The impedance spectra obtained from the experiments were analyzed using ZSimpWin 3.6 software developed by Informer Technologies, located in Los Angeles, CA, USA. The software allowed for the fitting of the experimental data with simulated data, enabling a comparison between the two. To assess the quality of the fit, the chi-square value was calculated. The chi-square value is a measure of the discrepancy between the experimental and simulated data. A lower chi-square value indicates a closer match between the experimental and simulated impedance spectra. In this study, a chi-square value around 10–5 was considered indicative of a highly accurate fit with a minimal number of components [18]. This suggests that the simulated data closely resemble the experimental data, providing reliable and precise information about the system's impedance behavior.

3. Results and Discussion

3.1. Preliminary Studies

Figure 2 displays optical images showcasing the microstructure of both analyzed alloys. The images reveal a compact and crack-free microstructure in both cases. Interestingly, both alloys exhibit a dendritic microstructure, characterized by branching patterns. However, it is notable that the Sample 2 demonstrated a distinct tendency towards grain refinement compared to the other alloy. This implies that the Sample 2 possesses a more refined and smaller grain size, which can have significant implications for its mechanical and functional properties [22].

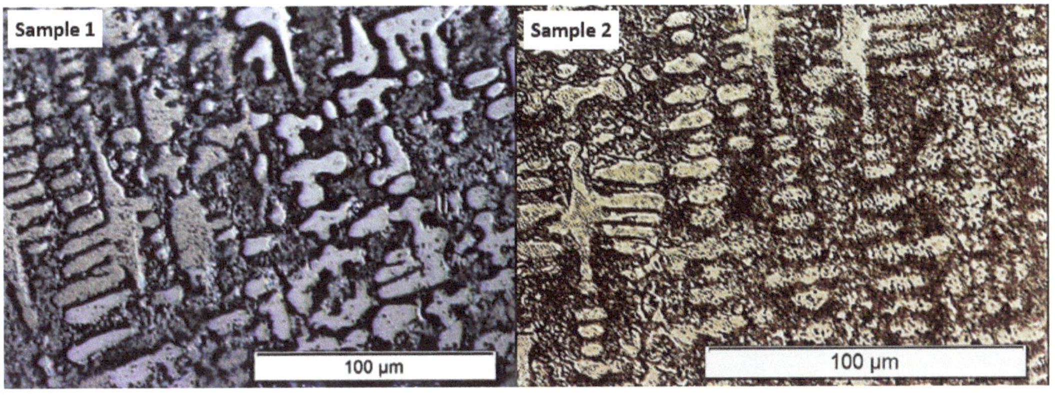

Figure 2. Images obtained by optical microscopy showing the microstructure of the samples.

Figure 3 displays SEM images of the two HEAs. Both alloys had a dendritic microstructure, with Sample 2 showing a trend towards grain refinement. To determine the chemical constituents of the investigated materials on the microareas designated Zones A and B, a semi-quantitative examination was carried out. The quantifiable values and estimated uncertainties for the components of the alloys are shown in Figure 3. It can be shown from the EDS analysis that the dendritic zone, designated Zone A in both samples, was abundant in Ni, Fe, and Co. The alloy in Zone B, which corresponds to the interdendritic region, was high in Cr and Mo, with significantly lower concentrations of Co, Fe, and Ni. In Sample 2, low levels of Zr were found, with slightly greater levels in the interdendritic regions. When Zr was added to the CoCrFeMoNi alloy, the volume fraction of the interdendritic areas reduced, according to a comparison of the two alloys' microstructures. The σ phase frequently coexisted with the FCC structure in this kind of alloy. In our alloys, the 0.48 at% Zr addition reduced the Mo concentration from 20 at% to 17 at% in the interdendritic areas while maintaining a nearly constant Cr content of 25 at%. This reduced the tendency for the sigma phase to form.

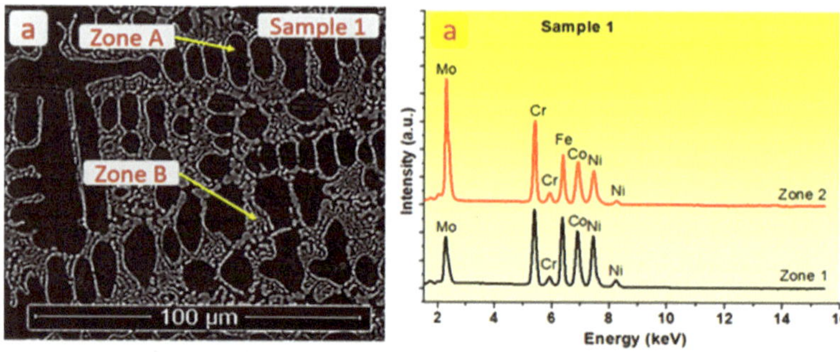

	Sample 1			
	Zone A		Zone B	
El.	Wt. %	At. %	Wt. %	At. %
Mo	13.86	8.85	30.55	20.61
Cr	19.23	22.07	20.42	25.12
Fe	22.02	23.53	16.29	18.66
Co	22.49	22.78	18.34	19.91
Ni	22.41	22.78	14.40	15.69
Zr	-	-	-	-

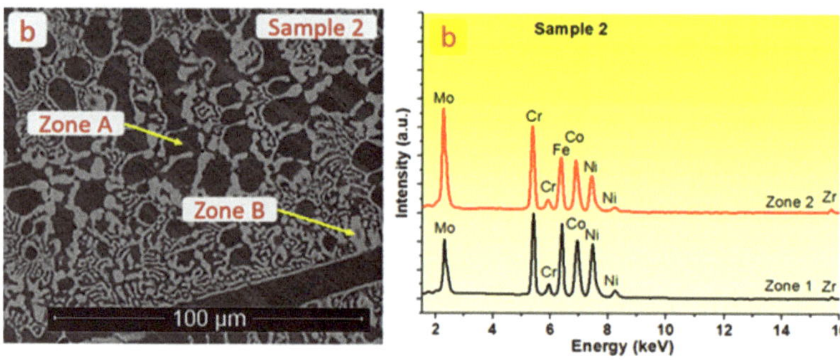

	Sample 2			
	Zone A		Zone B	
El.	Wt. %	At. %	Wt. %	At. %
Mo	13.64	8.51	26.07	17.13
Cr	19.28	22.2	20.58	24.96
Fe	22.21	23.81	17.97	20.29
Co	21.74	22.09	18.92	20.25
Ni	22.54	22.99	15.55	16.82
Zr	0.59	0.39	0.8	0.55

Figure 3. SEM micrographs, EDS spectra, and composition for Sample 1 (**a**) and Sample 2 (**b**).

The difference between the potential of a metallic electrode and the potential of the reference electrode when there is no current coursing between them is known as the open-circuit potential (OCP). This potential can change as the metal goes through corrosion processes and is affected by the chemical conditions that it is in. An OCP diagram can be created by tracking the OCP over time to show how the potential changes with regard to time (see Figure 4). The amount of oxygen that is dissolved, the pH value, and the existence of corrosive ions are some of the variables that can affect a metal's OCP in a corrosive environment. Figure 4 presents the OCP diagrams for both Sample 1 and Sample 2. It can be observed that the potential of both samples exhibits fluctuations and a gradual decrease over time. This behavior suggests that the samples are susceptible to corrosion, resulting in the formation of a corrosion product layer on their surfaces. However, after approximately 12 h, the OCP diagrams indicate a stable potential that remains constant over time. This stability suggests that the alloys are no longer undergoing active corrosion and demonstrate a certain level of resistance to corrosion under the given conditions.

The potentiodynamic polarization curves exhibited a noteworthy trend in which the addition of zirconium to Sample 2 resulted in an elevation of anodic current densities (see Figure 5).

This increase in current densities signifies a reduction in corrosion resistance when subjected to the simulated seawater conditions employed in the experimental tests. Within the anodic range of both curves, there were noticeable minor increments in the current. These observations suggest the occurrence of localized corrosion and subsequent repassivation processes. This phenomenon indicates the presence of localized areas of corrosion activity within the overall corrosion behavior of the samples (see Figure 6). The electrochemical corrosion resistance (polarization resistance) obtained from Tafel analysis was 150 ± 16 kΩ·cm^2 for Sample 1 and 113 ± 12 kΩ·cm^2 for Sample 2, which denotes that the alloy without zirconium is more resistant to corrosion. Therefore, doping with Zr, even with produced grain refinement, decreases the corrosion resistance of the CoCrFeNiMo alloy in 3.5% NaCl environment.

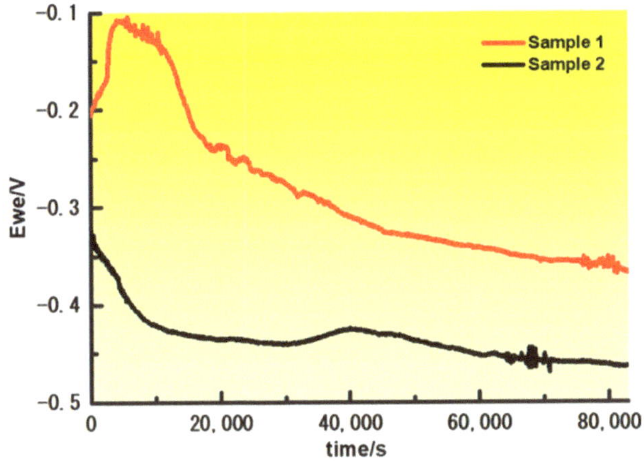

Figure 4. Open-circuit corrosion potential diagrams of Samples 1 and 2.

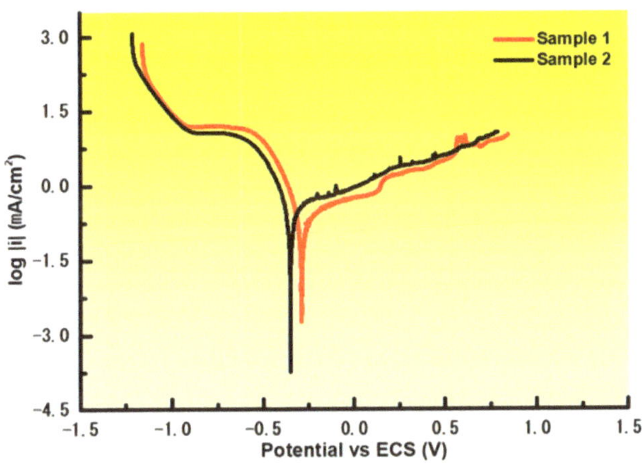

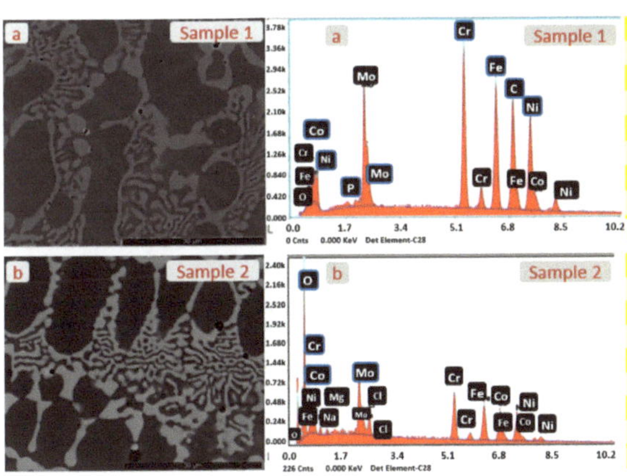

Element	Weight %	Atomic %
Cl	0.46	0.62
Co	21.24	21.46
Cr	19.61	22.46
Fe	20.57	21.93
Mo	17.15	10.64
Na	1.01	1.23
Ni	20.62	20.91
O	0.59	2.2

Element	Weight %	Atomic %
Cl	1.92	1.24
Co	7.98	3.09
Cr	8.57	3.76
Fe	8.28	3.39
Mo	10.52	2.5
Na	2.02	2
Ni	7.26	2.82
O	34.79	49.63

Figure 5. Linear polarization curves of the analyzed samples.

Figure 6. After corrosion, SEM micrographs, EDS spectra, and composition for Sample 1 (**a**) and Sample 2 (**b**).

3.2. Visual Analysis of the Impedance Spectra

Interpreting experimental data in electrochemical impedance spectroscopy (EIS) involves analyzing the impedance spectrum and extracting valuable information about the electrochemical system under investigation. It starts by visually inspecting the impedance spectra, looking for distinct features such as peaks, slopes, and deviations from ideal behavior. These features can provide initial insights into the underlying electrochemical processes. The Nyquist plots of Sample 1 and Sample 2 are presented in Figure 7.

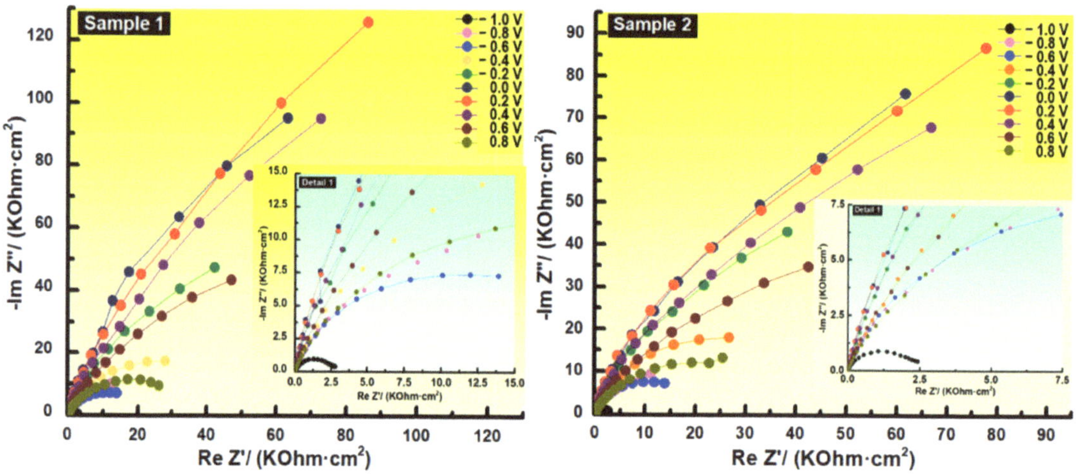

Figure 7. Nyquist plots of Sample 1 and Sample 2.

It becomes clear from examining the Nyquist plots that the samples exhibit three distinct zones. Low impedances seen at high frequencies correspond to the first region. The ratio in the second zone, which represents medium impedances, is larger than in the high-frequency sector. Finally, a line of high impedances at low frequencies comprises the third zone. This distinct pattern strongly suggests the involvement of, at least, two frequency-dependent processes with varying time constants contributing to the overall feedback. The presence of these different processes is evident from the impedance data, indicating that multiple underlying mechanisms are at play, each exhibiting its own characteristic response at different frequencies. This information underscores the complex nature of the electrochemical system under investigation, highlighting the need to consider multiple contributing factors and their respective time scales for a comprehensive understanding of the observed impedance behavior.

The phase displacement versus frequency for various potentials relative to the reference electrode is displayed on the Bode phase plots for Samples 1 and 2 (see Figure 8). The existence of a single peak in phase shift within the potential range of −1.0 V to −0.2 V shows that any time constants connected to the corrosion process are rather frequent. The maximum phase shift constantly moves to lower frequencies as the potential rises. This movement indicates an increase in polarization resistance without any significant alteration in interfacial capacitance (as observed in Figure 8a,c) [16]. Notably, at potentials of 0.4 V, 0.6 V, and 0.8 V for Sample 2 (Figure 8d) and at 0.8 V for Sample 1 (Figure 8b), a sharp variation in phase occurs in the low-frequency region. This signifies the occurrence of pitting corrosion, which is further supported by the observations in Figure 5 when the potential exceeds approximately 0.25 V.

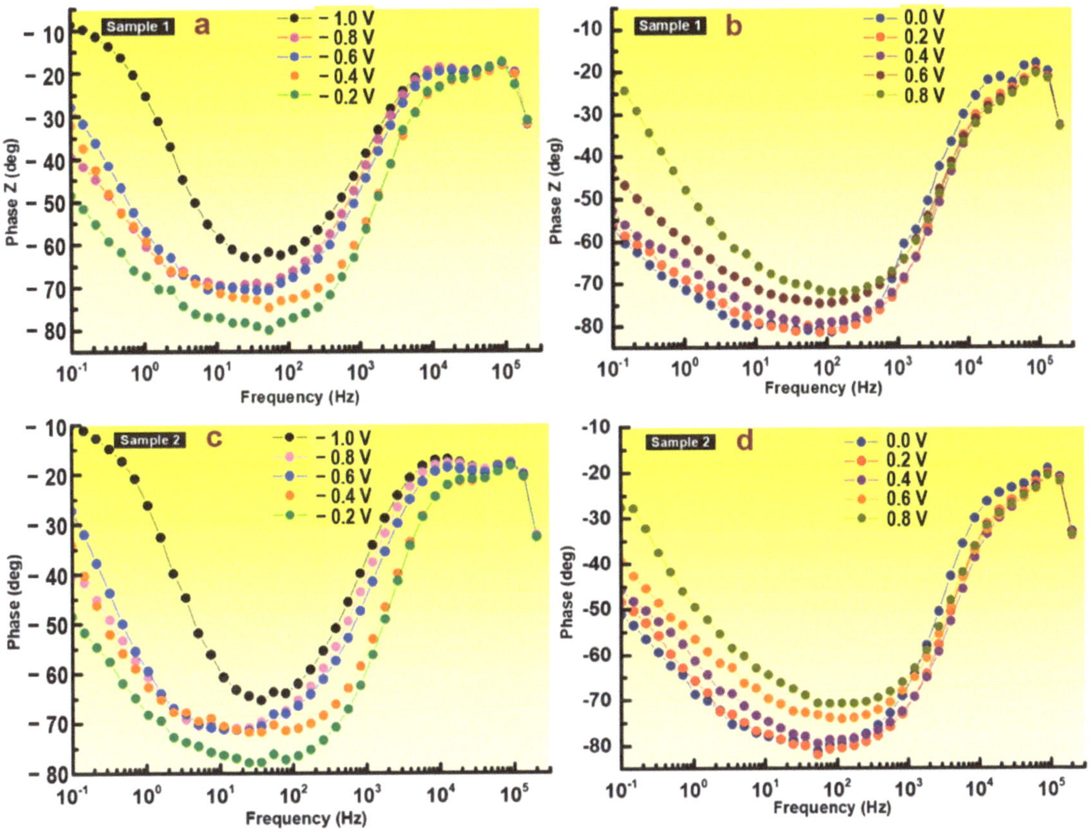

Figure 8. Bode-phase plots of Sample 1 (**a**,**b**) and Sample 2 (**c**,**d**).

The selective solubility of iron ions into the electrolyte for Sample 2 is the cause of pitting corrosion. When zirconium is doped into the high-entropy alloy, the Fe segregation ratio increases, causing this phenomenon to take place. Previous studies on stainless steels have indicated that Fe is more susceptible to selective dissolution compared to other elements, such as Cr, Mo, Ni, and Co, owing to the lower stability of iron species [23,24].

For Bode plots, it is very important to identify the frequencies at which significant changes occur in the impedance spectra. These frequencies correspond to specific electrochemical processes or phenomena. For example, high-frequency behavior generally is associated with charge transfer processes, while low-frequency behavior may be related to mass transport or diffusion.

The impedance spectra of both samples show overlapping curves at high and mid frequencies in the Bode-IZI plots (see Figure 9). Between −0.2 V and 0.0 V, a discernible drop is seen that lasts the entire time the experiment is running. This loss translates into a considerable, more than 10-fold, drop in film resistance across all samples. The phase data show the appearance of a second peak at low frequencies within the theoretical range of 0.0 V to 0.8 V. This suggests that the impedance spectrum can be separated into two time constants that are virtually similar. This observation shows that the electrochemical reaction taking place at the interface between the electrolyte and the alloy involves a two-step mechanism.

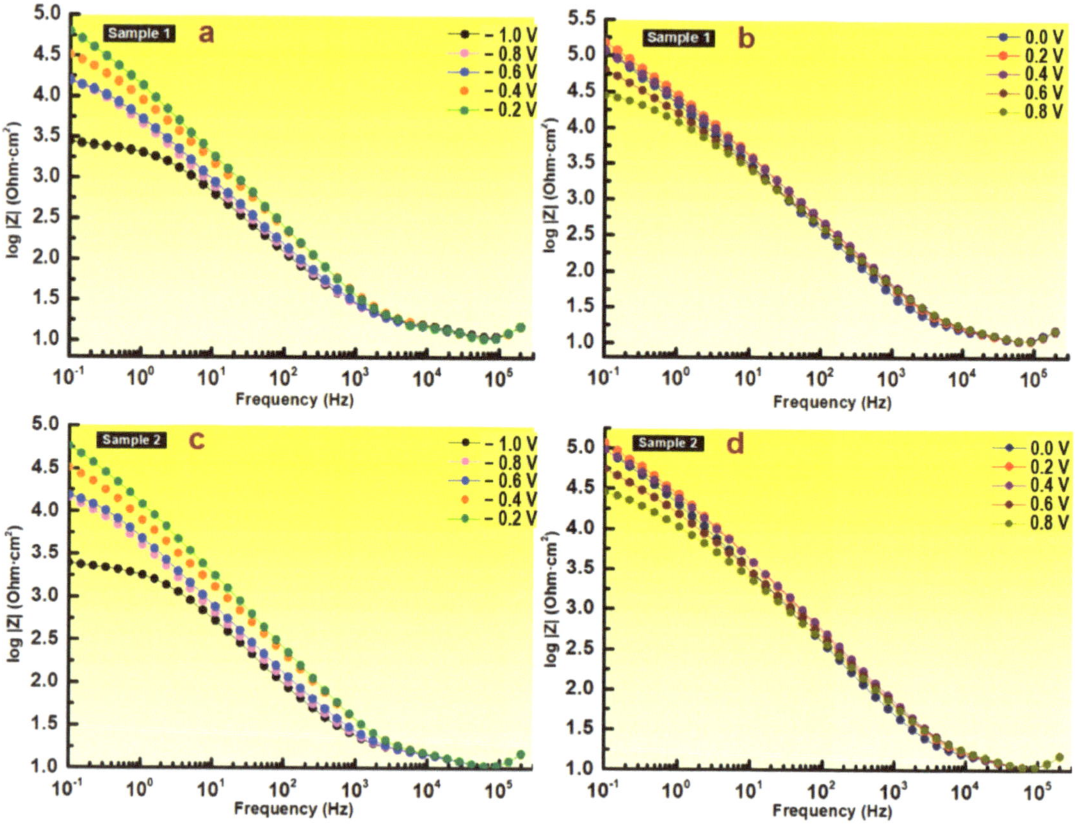

Figure 9. Bode |Z| plots of Sample 1 and Sample 2 at negative potentials (**a**,**c**) and positive potentials (**b**,**d**).

The observed overlapping curves and the distinct decline in film resistance indicate changes in the electrochemical behavior and the formation of a more conductive interface. The presence of two nearly identical time constants suggests a complex reaction mechanism involving multiple steps in the electrochemical process. These findings contribute to a deeper understanding of the electrochemical behavior of the samples and the processes occurring at the electrolyte/alloy interface.

The Bode-IZI graphs, particularly in Figure 9, provide insights into the impedance characteristics of the smooth surface at low frequencies. The combined resistance of the passive film, charge transfer resistance, and ohmic resistance of the electrolyte is reflected in this low-frequency area. Any changes in the low-frequency area can be attributed to modifications in the film itself, possibly as a result of the creation of conductive pathways within the film, since the charge transfer and electrolyte resistances for a particular sample remain rather steady. The rise in low-frequency impedance with voltage suggests that the layer that has formed on the surface of the HEA has a stronger corrosion resistance. Notably, among the investigated potentials, Sample 1 at 0.2 V has the strongest corrosion resistance. Figure 9b,d show that a diffusion-controlled process becomes a considerable portion of the total impedance between 0.2 V and 0.8 V.

These results suggest that corrosion is primarily governed by a diffusion process controlled by ionic conductivity. In this case, the transfer resistance becomes the dominant factor affecting film resistance. The inconsistent behavior observed may be attributed to the pore-filling process occurring in the outer layer of the film. Overall, the analysis of the Bode-IZI graphs provides valuable insights into the impedance characteristics of the film, shedding light on the corrosion behavior and mechanisms occurring at the surface of the HEA samples.

3.3. Selection of an Equivalent Electrical Circuit Model

Choosing the right equivalent circuit is of utmost importance when analyzing EIS data. The selected circuit should have a clear physical interpretation and relevance to the electrochemical system under investigation. Simplicity is key, as the circuit should capture the essential characteristics of the system without unnecessary complexity. Including too many circuit elements can lead to overfitting, which hinders accurate model predictions and complicates the interpretation of fitted parameters.

An effective equivalent circuit should provide a good fit to the experimental data. The quality of the fit can be assessed by calculating the root mean square error, which measures the deviation between the experimental data and the model predictions. By minimizing the root mean square error, the circuit can be optimized to accurately represent the electrochemical system and provide meaningful insights into its behavior.

Choosing an equivalent electrical circuit to fit experimental data in EIS involves a systematic approach. The first step is to gain a comprehensive understanding of the electrochemical system under investigation, including the electrode materials, electrolyte composition, and expected electrochemical processes.

This knowledge will help guide the selection of an appropriate equivalent circuit. For this reason, we will begin by considering basic circuit elements that represent the fundamental electrochemical processes observed in the system. For example, if there is expected charge transfer resistance, a resistor (R) can be used. If there is a double-layer capacitance, a capacitor (C) can be included, or better, a constant-phase element (Q) can be considered. We have tested different electrical equivalent circuits presented in Figure 10.

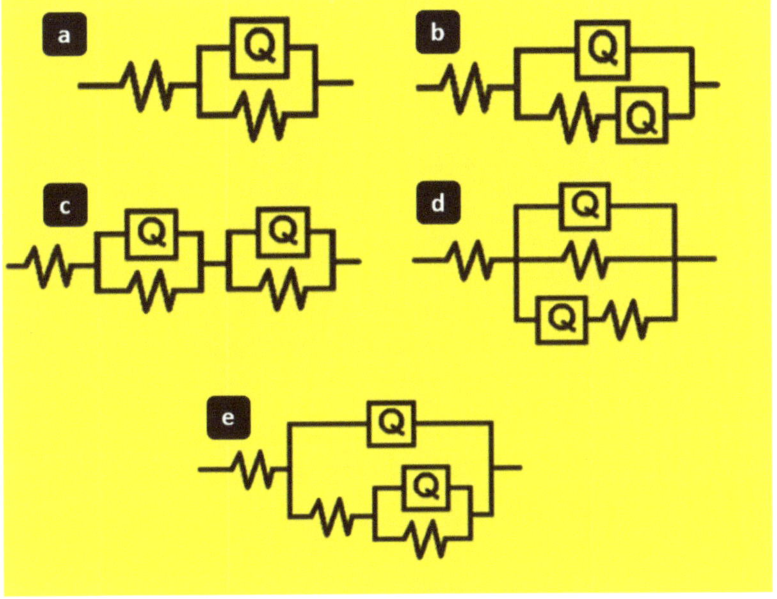

Figure 10. Different electrical equivalent circuits used for fitting the experimental data: (**a**) R(QR), (**b**) R(Q(RQ)), (**c**) R(QR)(RQ), (**d**) R(QR(QR)), and (**e**) R(Q(R(QR))).

The first selected circuit is the simplest circuit, the Randles circuit (see Figure 11). It provides a simplified representation of the electrical behavior of an electrochemical cell or system. The Randles circuit consists of three basic elements: two resistors (R), and a capacitor (C). Each of these elements represents a specific electrical component in the electrochemical system.

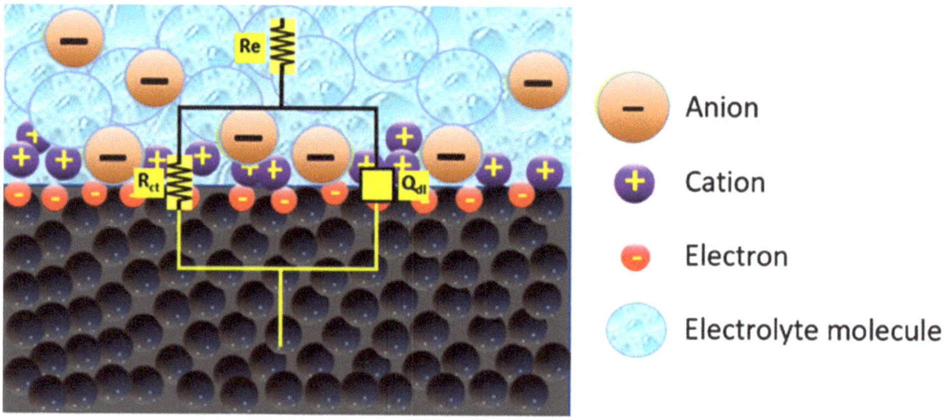

Figure 11. Randles circuit used to fit the experimental data.

In Figure 11, R_e means "electrolyte resistance", R_{ct} is "charge transfer resistance", and C_{dl} is "electrochemical double-layer capacitance". The working electrode and counter electrode are the two endpoints of this circuit that receive current flowing from the counter electrode (on the upper side) to the working electrode (on the bottom side). Visually, it is obvious that the solution resistance R_e must be crossed before the current can proceed. However, there are two conceivable directions for the current to travel after passing through R_e. Because the current always chooses the path of least resistance, or lowest impedance, it can pass via the capacitor related with the electrochemical double-layer C_{dl} (the high-frequency path) or it can move through the resistor associated with charge transfer R_{ct} (the low-frequency path). In this instance, the frequency affects the impedance of C_{dl} and R_{ct}.

According to Kirchhoff's circuit principles, the combined impedance of two elements in series (Re noted Z_1 and the parallel combination of R_{ct} and C_{dl} noted with Z_2) is equal to the sum of the two impedances:

$$Z_{series} = Z_1 + Z_2 \tag{1}$$

On the other hand, according to Equation (2), the inverse of the combined impedance of two parallel circuit elements is equal to the sum of the inverses of each impedance.

$$\frac{1}{Z_{parallel}} = \frac{1}{Z_1} + \frac{1}{Z_2} \tag{2}$$

Equation (2) can be arranged to form the following relation:

$$Z_{parallel} = \frac{Z_1\, Z_2}{Z_1 + Z_2} \tag{3}$$

As a result, by substituting each circuit component and combining them in accordance with Equation (1) through Equation (3), the total impedance of our Randles circuit (Figure 8) may be determined.

$$Z_{circuit\ 1} = R_e + \frac{Z_{C_{dl}}\, Z_{R_{ct}}}{Z_{C_{dl}} + Z_{R_{ct}}} \tag{4}$$

Taking into account that Z_R = R and Z_C = 1/(jwC), where w is the frequency and j indicate the imaginary number $(-1)^{1/2}$, Equation (5) is obtained by substituting the corresponding impedance equations for each circuit component.

$$Z_{\text{circuit 1}} = R_e + \frac{\frac{R_{ct}}{jwC_{dl}}}{R_{ct} + \frac{1}{jwC_{dl}}} \quad (5)$$

The total impedance of our electrochemical circuit is, thus, given in Equation (6) after reorganization and simplification.

$$Z_{\text{circuit 1}} = R_e + \frac{R_{ct}}{1 + R_{ct}C_{dl}(jw)} \quad (6)$$

The denominator of Equation (6) enlarges and the proportion becomes closer to zero when w is large or w $\rightarrow\infty$. This results in Z = R_e. At the highest frequencies (1 MHz–100 kHz), it can sometimes be assumed that the impedance is identical to the solution resistance R_e, even though there is no potentiostat that can apply a frequency so high. The solution resistance must be crossed by the current at high frequencies, but it will do so via C_{dl} rather than R_{ct}. This is due to the fact that, theoretically, the capacitor's impedance lowers as w grows (recall the equation for a capacitor's impedance, Equation (7)). As a result, the capacitor becomes the path of least resistance at high frequencies.

$$Z_c = \frac{1}{jwC} = \frac{-j}{wC} \quad (7)$$

Analyzing the impedance data and looking for additional features that are not adequately captured by the basic circuit elements, a constant-phase element (CPE) has been chosen instead of the capacitor C_{dl} in order to be able to take into consideration the passivated surface's heterogeneities. A CPE's impedance is provided by [18]:

$$Z_{CPE} = \frac{1}{Y^0(jw)^n} \quad (8)$$

The coefficient of ideality, or "n", is one of the characteristics acquired when modeling the system and the response of the real system is closer to the ideal as the value of n is closer to the unit and the surface is more homogeneous. Consequently, the CPE element becomes a simple resistance for n = 0 and a capacitor with capacitance Y^0 for n = 1.

We began fitting the experimental data with the selected circuit elements using the specialized software and comparing the simulated impedance with the experimental data, which can assess the quality of the fit. We adjusted the circuit elements as needed and repeated the fitting process until a satisfactory match between the simulated and experimental data was achieved, and the results are presented in Tables 2 and 3.

Table 2. Circuit parameters calculated from the fitting of the EIS spectra till 0 V vs. SCE with the circuit from Figure 8 for Sample 1.

V (Volt)	R_e ($\Omega\cdot cm^2$)	Y/CPE ($S\cdot s^n/cm^2$)	n/CPE -	R_{ct} ($\Omega\cdot cm^2$)
−1.0	8.36	8.900×10^{-5}	0.7624	1.87×10^5
−0.8	7.36	8.996×10^{-5}	0.7415	1.73×10^4
−0.6	7.93	5.747×10^{-5}	0.7981	1.33×10^4
−0.4	7.39	3.321×10^{-5}	0.8135	2.89×10^4
−0.2	7.77	2.201×10^{-5}	0.8657	6.87×10^4
0.0	7.55	1.394×10^{-5}	0.8787	1.64×10^5

Table 3. Circuit parameters calculated from the fitting of the EIS spectra till 0 V vs. SCE with the circuit from Figure 8 for Sample 2.

V (Volt)	R_e ($\Omega\cdot cm^2$)	Y/CPE ($S\cdot s^n/cm^2$)	n/CPE -	R_{ct} ($\Omega\cdot cm^2$)
−1.0	9.90	8.469×10^{-5}	0.7838	1.93×10^3
−0.8	9.76	6.961×10^{-5}	0.7933	1.78×10^4
−0.6	9.73	5.460×10^{-5}	0.5046	1.61×10^4
−0.4	9.21	3.318×10^{-5}	0.8051	3.69×10^4
−0.2	9.45	2.072×10^{-5}	0.8528	7.59×10^4
0.0	9.29	1.246×10^{-5}	0.8699	1.24×10^5

A circuit with two time constants that takes into account the structure of the passive layer already formed on the surface of the materials has been employed because the single circuit has unacceptable setting errors for potentials higher than the corrosion potential. It should be kept in mind that the primary goal of this study is to determine whether the EIS technique is appropriate for this kind of system. As a result, the choice of equivalent circuit is not as crucial as the fact that it reveals whether variations in the material's electrochemical behavior exist depending on the applied passivation potential or the material's composition. In these instances, we have tested all of the circuits shown in Figure 10 and the provided circuit in Figures 10e and 12 is the one that most closely matches the experimental results.

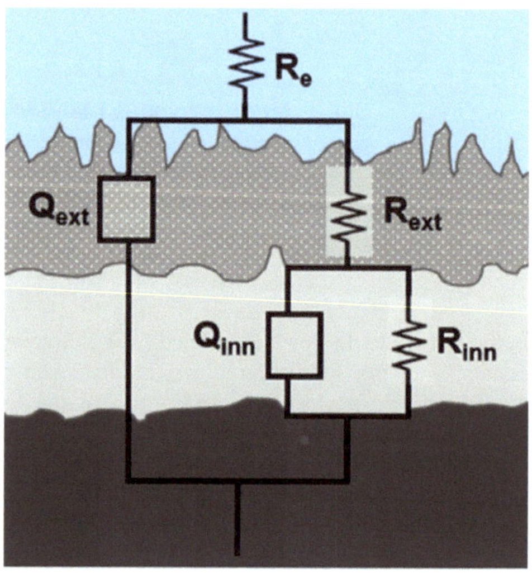

Figure 12. The second electrical equivalent circuit used for EIS data simulation.

The component elements are:

- R_e—electrolyte resistance;
- Q_{ext}—the CPE of the porous external passive layer;
- R_{ext}—resistance of the external porous layer;
- Q_{inn}—CPE of the inner passive layer;
- R_{inn}—resistance of the inner passive layer.

Applying the Kirchhoff's circuit laws for this new circuit enable us to obtain:

$$Z_{circuit\ 2} = R_e + \frac{1}{jwC_{ext} + \frac{1}{R_{ext} + \frac{1}{\frac{1}{R_{int}} + jwC_{int}}}} \quad (9)$$

The parameter R_e, which measures electrolyte resistance and ranges in value from 7.28 to 8.36 $\Omega \cdot cm^2$ for Sample 1 and from 9.12 to 9.94 $\Omega \cdot cm^2$ for Sample 2, displays no notable variation during the studies, indicating that no ions were discharged into the solution within the passive potential range.

The outer porous film exhibits lower resistance than the inner barrier layer, as shown by the fact that R_{ext} values are lower than R_{inn} values (see Figure 13a,b). The presence of high R_{inn} values at all potentials, as well as a decrease with Zr-addition, even support the establishment of a passive layer with excellent corrosion protection.

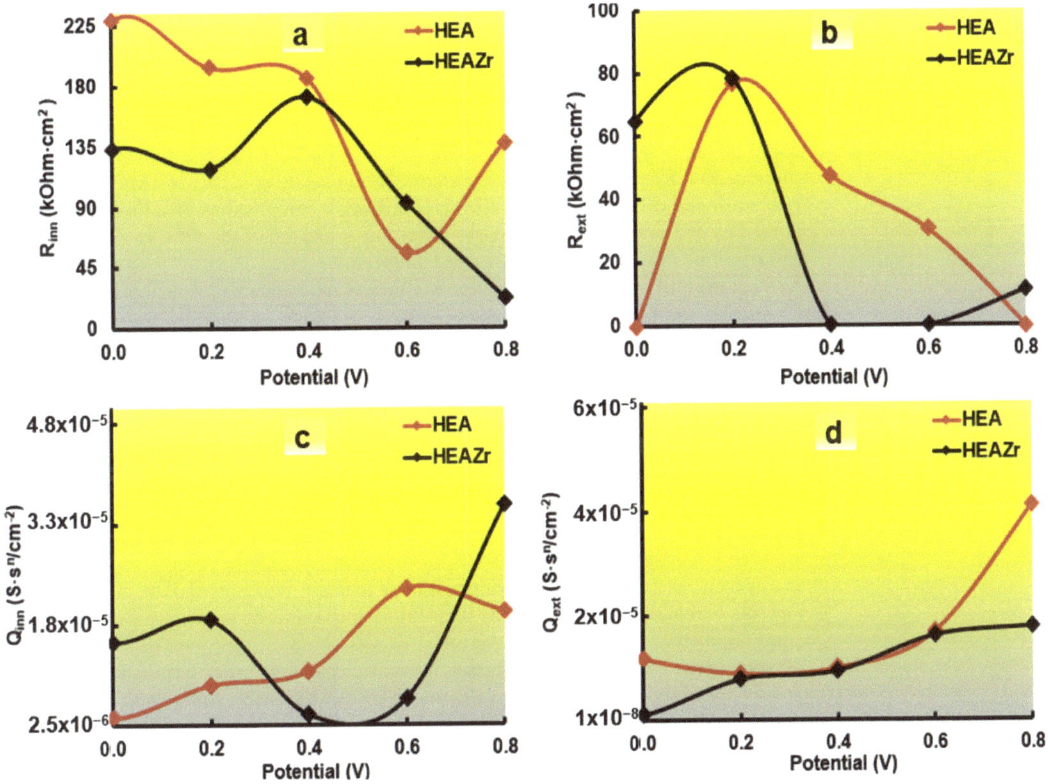

Figure 13. Electrical-circuit parameters: resistance of the inner layer (**a**) and external layer (**b**), and the CPE magnitude for the inner layer (**c**) and external layer (**d**).

The relations between the constant-phase element (CPE, in terms of Q), the capacitance, and the thickness of the oxide film [25] are:

$$C = gQ(\rho\varepsilon\varepsilon_0)^{1-n} \tag{10}$$

$$d = \frac{\varepsilon\varepsilon_0 S}{C} \tag{11}$$

where g is a parameter equal to $1 + 2.88(1 - n)^{2.375}$, ρ is the electrical resistivity, ε and ε_0 represent the dielectric constant of the oxide film and the permittivity of free space, respectively; and S is the effective area of the oxide film, which is about three times the geometric area [26].

It can be observed that the protective oxide film's thickness decreased within potential range because it can be seen that the fitted values of CPE magnitude is increasing (see Figure 13c,d).

At potentials higher than +0.6V, to obtain a better quality of the fitting process, a Warburg element was introduced into the circuit [27]. The Warburg element represents the movement/diffusion of electrons/ions in solid or liquid phases and its impedance is:

$$Z_W = \frac{W}{(jw)^{0.5}} \quad (12)$$

where W is the Warburg coefficient and has the unit $\Omega/s^{1/2}$.

The Warburg element at 0.8 V vs. SCE is almost 10 times higher for Sample 2 ($1.09 \cdot 10^{-4}$ S-sec$^{0.5}$/cm^2) in comparison with this element of Sample 1 ($1.17 \cdot 10^{-5}$ S-sec$^{0.5}$/cm^2). A higher value of the Warburg element suggests that the diffusion process is slower, which could be due to a longer diffusion path. If the diffusion path length is increased, for example, due to the presence of a thicker electrode or a more complex electrode structure, it can lead to slower diffusion and an elevated Warburg element value. Another factor to consider is the lower concentration of species: if the concentration of the diffusing species is lower, it can result in slower diffusion rates and a higher Warburg element value. The EDS [22] and XPS results [28] indicate that the passive film mainly consists of Cr_2O_3 and MoO_3, and both Cr and Mo are found in lower concentrations in Sample 2 than in Sample 1 and, thus, can explain the difference between the Warburg element values for the two samples analyzed.

4. Conclusions

In the present work, high-entropy alloys CoCrFeMoNi and Zr-doped CoCrFeMoNi were prepared using vacuum arc remelting. Electrochemical impedance spectroscopy (EIS) measurements were performed in order to select the proper electrical equivalent circuit for the prediction of the behavior of two high-entropy alloys at different potentials in artificial seawater because choosing the right equivalent circuit is of outmost importance when analyzing EIS data.

By fitting the circuit frequency response to the experimental data, circuit models with various levels of complexity and for various potential ranges were proposed and their compatibility with the experimental data was assessed. Simple electrical components were used to mimic fundamental processes, such as charge transport in bulk solutions, diffusion at low frequencies, and the development of double-layer charges at high frequencies. Additionally, it was demonstrated that various components might be used to express electrochemical events. Depending on how the parameters defining its impedance are chosen, a constant-phase element (CPE) might be used to depict both the diffusion process and the double-layer effect.

Till corrosion potential in the experimental data can be well simulated with a simple Randles circuit displaying a highly accurate fit. After this potential, the electrochemical reaction taking place at the interface between the electrolyte and the alloy involves a two-step mechanism and a proper equivalent electrical circuit was selected.

The electrolyte resistance during the experiments, for both alloys, shows no notable variation, indicating that no ions were discharged into the solution within the passive potential range. The Zr-doped high-entropy alloy sample experiences pitting corrosion over 0.20 V vs. SCE. The preferential solubility of iron ions into the electrolyte caused by the rise in the Fe segregation ratio when the high-entropy alloy is doped with Zr can be used to explain the pitting corrosion in the HEAZr alloy. At potentials higher than +0.6V, to obtain a better quality of the fitting process, a Warburg element was introduced into the circuit suggesting that the diffusion process became important.

It is important to note that selecting an equivalent circuit is often an iterative and subjective process, as it involves a balance between model complexity and the ability to accurately represent the system's behavior. Additionally, the choice of circuit elements may vary depending on the specific electrochemical system and the phenomena of interest.

Author Contributions: J.C.-P., writing—original draft preparation, investigation, and management; S.J.B.-G., conceptualization and validation; J.C.M.-R., investigation, writing—review and editing; G.M.C., methodology, investigation, and data curation. All authors have read and agreed to the published version of the manuscript.

Funding: This work was supported by the Romanian National Authority for Scientific Research, CNDI–UEFISCDI, through project number PN-III-P2-2.1-PED-2019-3953, contract 514PED/2020: "New ceramic layer composite material processed by laser techniques for corrosion and high-temperature applications—LASCERHEA", within PNCDI III and by Cabildo de Gran Canaria, project number CABINFR2019-07.

Data Availability Statement: All data provided in the present manuscript are available to whom it may concern.

Acknowledgments: The authors thank Ionelia Voiculescu y Victor Geanta of ERAMET Laboratory of the Politehnica University of Bucharest, Romania, for providing the high-entropy alloys and for their assistance and comments regarding this work.

Conflicts of Interest: The authors declare no conflict of interest.

References

1. Barsoukov, E.; Macdonald, J.R. (Eds.) *Impedance Spectroscopy: Theory, Experiment, and Applications*; John Wiley &Sons, Inc.: Hoboken, NJ, USA, 2005. [CrossRef]
2. Socorro-Perdomo, P.P.; Florido-Suárez, N.R.; Mirza-Rosca, J.C.; Saceleanu, M.V. EIS Characterization of Ti Alloys in Relation to Alloying Additions of Ta. *Materials* **2022**, *15*, 476. [CrossRef] [PubMed]
3. López Ríos, M.; Socorro Perdomo, P.P.; Voiculescu, I.; Geanta, V.; Crăciun, V.; Boerasu, I.; Mirza Rosca, J.C. Effects of nickel content on the microstructure, microhardness and corrosion behavior of high-entropy AlCoCrFeNix alloys. *Sci. Rep.* **2020**, *10*, 21119. [CrossRef]
4. Jimenez-Marcos, C.; Mirza-Rosca, J.C.; Baltatu, M.S.; Vizureanu, P. Experimental Research on New Developed Titanium Alloys for Biomedical Applications. *Bioengineering* **2022**, *9*, 686. [CrossRef]
5. Avram, D.N.; Davidescu, C.M.; Dan, M.L.; Mirza-Rosca, J.C.; Hulka, I.; Pascu, A.; Stanciu, E.M. Electrochemical Evaluation of Protective Coatings with Ti Additions on Mild Steel Substrate with Potential Application for PEM Fuel Cells. *Materials* **2022**, *15*, 5364. [CrossRef] [PubMed]
6. Avram, D.N.; Davidescu, C.M.; Dan, M.L.; Mirza-Rosca, J.C.; Hulka, I.; Stanciu, E.M.; Pascu, A. Corrosion resistance of NiCr(Ti) coatings for metallic bipolar plates. *Mater. Today Proc.* **2023**, *72*, 538–543. [CrossRef]
7. Avram, D.N.; Davidescu, C.M.; Dan, M.L.; Stanciu, E.M.; Pascu, A.; Mirza-Rosca, J.C.; Iosif, H. Influence of titanium additions on the electrochemical behaviour of nicr/ti laser cladded coatings. *Ann. Dunarea Jos Univ. Galati Fascicle XII Weld. Equip. Technol.* **2022**, *33*, 107–111. [CrossRef]
8. Perdomo-Socorro, P.P.; Florido-Suárez, N.R.; Verdú-Vázquez, A.; Mirza-Rosca, J.C. Comparative EIS study of titanium-based materials in high corrosive environments. *Int. J. Surf. Sci. Eng.* **2021**, *15*, 152–164. [CrossRef]
9. Arabzadeh, H.; Shahidi, M.; Foroughi, M.M. Electrodeposited polypyrrole coatings on mild steel: Modeling the EIS data with a new equivalent circuit and the influence of scan rate and cycle number on the corrosion protection. *J. Electroanal. Chem.* **2017**, *807*, 162–173. [CrossRef]
10. Lopez-Dominguez, D.; Gomez-Guzman, N.B.; Porcayo-Calderón, J.; Lopez-Sesenes, R.; Larios-Galvez, A.K.; Sarmiento-Bustos, E.; Rodriguez-Clemente, E.; Gonzalez-Rodriguez, J.G. An Electrochemical Study of the Corrosion Behaviour of T91 Steel in Molten Nitrates. *Metals* **2023**, *13*, 502. [CrossRef]
11. Larios-Galvez, A.K.; Vazquez-Velez, E.; Martinez-Valencia, H.; Gonzalez-Rodriguez, J.G. Effect of Plasma Nitriding and Oxidation on the Corrosion Resistance of 304 Stainless Steel in LiBr/H_2O and $CaCl_2$-LiBr-$LiNO_3$-H_2O Mixtures. *Metals* **2023**, *13*, 920. [CrossRef]
12. Revilla, R.I.; Wouters, B.; Andreatta, F.; Lanzutti, A.; Fedrizzi, L.; De Graeve, I. EIS comparative study and critical Equivalent Electrical Circuit (EEC) analysis of the native oxide layer of additive manufactured and wrought 316L stainless steel. *Corros. Sci.* **2020**, *167*, 108480. [CrossRef]
13. Zhao, Z.; Zou, Y.; Liu, P.; Lai, Z.; Wen, L.; Jin, Y. EIS equivalent circuit model prediction using interpretable machine learning and parameter identification using global optimization algorithms. *Electrochim. Acta* **2022**, *418*, 140350. [CrossRef]
14. Martinez, S.; Šoić, I.; Špada, V. Unified equivalent circuit of dielectric permittivity and porous coating formalisms for EIS probing of thick industrial grade coatings. *Prog. Org. Coat.* **2021**, *153*. [CrossRef]
15. Diard, J.P.; Montella, C. Non-intuitive features of equivalent circuits for analysis of EIS data. The example of EE reaction. *J. Electroanal. Chem.* **2014**, *735*, 99–110. [CrossRef]
16. Scully, J.; Silverman, D.; Kendig, M. (Eds.) *Electrochemical Impedance: Analysis and Interpretation*; ASTM International: West Conshohocken, PA, USA, 1993. ISBN 978-0-8031-1861-4.

17. Wang, S.; Zhang, J.; Gharbi, O.; Vivier, V.; Gao, M.; Orazem, M.E. Electrochemical impedance spectroscopy. *Nat. Rev. Methods Prim.* **2021**, *1*, 41. [CrossRef]
18. Boukamp, B.A. A Nonlinear Least Squares Fit procedure for analysis of immittance data of electrochemical systems. *Solid State Ion.* **1986**, *20*, 31–44. [CrossRef]
19. Material, C.; Databases, P. Standard Reference Test Method for Making Potentiostatic and Potentiodynamic Anodic. *Annu. B ASTM Stand.* **2004**, *94*, 1–12.
20. *ASTM Standard G102-89*; Standard Practice for Calculation of Corrosion Rates and Related Information from Electrochemical Measurements. ASTM International: West Conshohocken, PA, USA, 2006.
21. *ISO 16773-1-4:2016*; Electrochemical Impedance Spectroscopy (EIS) on Coated and Uncoated Metallic Specimens. International Organization for Standardization: Geneva, Switzerland, 2016.
22. Brito-Garcia, S.; Mirza-Rosca, J.; Geanta, V.; Voiculescu, I. Mechanical and Corrosion Behavior of Zr-Doped High-Entropy Alloy from CoCrFeMoNi System. *Materials* **2023**, *16*, 1832. [CrossRef]
23. Wang, Z.; Feng, Z.; Zhang, L. Effect of high temperature on the corrosion behavior and passive film composition of 316 L stainless steel in high H_2S-containing environments. *Corros. Sci.* **2020**, *174*, 108844. [CrossRef]
24. Wang, Z.; Zhang, L.; Zhang, Z.; Lu, M. Combined effect of pH and H_2S on the structure of passive film formed on type 316L stainless steel. *Appl. Surf. Sci.* **2018**, *458*, 686–699. [CrossRef]
25. Orazem, M.E.; Frateur, I.; Tribollet, B.; Vivier, V.; Marcelin, S.; Pébère, N.; Bunge, A.L.; White, E.A.; Riemer, D.P.; Musiani, M. Dielectric Properties of Materials Showing Constant-Phase-Element (CPE) Impedance Response. *J. Electrochem. Soc.* **2013**, *160*, C215–C225. [CrossRef]
26. Kocijan, A.; Merl, D.K.; Jenko, M. The corrosion behaviour of austenitic and duplex stainless steels in artificial saliva with the addition of fluoride. *Corros. Sci.* **2011**, *53*, 776–783. [CrossRef]
27. Jáquez-Muñoz, J.M.; Gaona-Tiburcio, C.; Méndez-Ramírez, C.T.; Baltazar-Zamora, M.Á.; Estupinán-López, F.; Bautista-Margulis, R.G.; Cuevas-Rodríguez, J.; Flores-De los Rios, J.P.; Almeraya-Calderón, F. Corrosion of Titanium Alloys Anodized Using Electrochemical Techniques. *Metals* **2023**, *13*, 476. [CrossRef]
28. Wang, W.; Wang, J.; Sun, Z.; Li, J.; Li, L.; Song, X.; Wen, X.; Xie, L.; Yang, X. Effect of Mo and aging temperature on corrosion behavior of (CoCrFeNi)100-xMox high-entropy alloys. *J. Alloys Compd.* **2020**, *812*, 152139. [CrossRef]

metals

Article

Microstructure, Hardness, and Linear Reciprocating Sliding Wear Response of Directionally Solidified Al–(2.5, 3.5, 4.5)Cu–(0.25, 0.50)Cr Alloys

Rafael V. Lantmann, André M. S. Mariante, Tiago V. Pinheiro, Eleani M. da Costa and Carlos A. dos Santos *

School of Technology, Pontifical Catholic University of Rio Grande do Sul (PUCRS), 6681, Porto Alegre 90619-900, RS, Brazil; rafael.lantmann@edu.pucrs.br (R.V.L.); a.mariante@edu.pucrs.br (A.M.S.M.); tiago.pinheiro@edu.pucrs.br (T.V.P.); eleani@pucrs.br (E.M.d.C.)

* Correspondence: carlos.santos@pucrs.br; Tel.: +55-51-98166-5777

Abstract: Aluminum alloys containing transition metal alloying elements have attracted interest from researchers. The effect of Cr additions of 0.25 and 0.50% on the thermal profile, microstructure, hardness, and linear reciprocating sliding wear response of as-cast hypoeutectic Al–Cu alloys with 2.5, 3.5, and 4.5% Cu (wt.%) was investigated. The binary Al–Cu and ternary Al–Cu–Cr alloys were directionally solidified under upward non-steady state heat transfer conditions using a dedicated solidification apparatus. Thermal analysis based on differential thermal analysis (DTA) and cooling curve profiles was performed to determine solidification thermal parameters such as Liquidus temperature (T_L), transformation enthalpy (ΔH), and liquid cooling rate ($\dot{T}_L$). Samples extracted from the solidified ingots were submitted to optical microscopy, hardness measurement, and linear reciprocating sliding wear test using a high-frequency reciprocating rig (HFRR). The results showed a decrease at the beginning of solidification (T_L) and of the transformation enthalpy (ΔH) when both alloy Cu and Cr contents increased, with a higher influence of Cu. The addition of Cu decreased cooling rates, whereas the increase in the alloy Cr concentration showed an opposite behavior, increasing cooling rates. The refinement of the primary dendrite arm spacing (λ_1), as a consequence of the increase in alloying elements and solidification cooling rates, enhanced the hardness of the alloys, with the maximum value of 58 HB achieved in the ternary Al–4.5Cu–0.50Cr alloy. The wear tests indicated a better response to wear associated with microstructure refinement for the alloys with 2.5% Cu, for both Cr contents, an almost constant behavior for the 3.5% Cu alloys, and an opposite performance for the alloys with 4.5% Cu alloys that showed better wear resistance with coarsening of the λ_1 and with the increase in the amount of the eutectic microconstituent.

Keywords: hypoeutectic Al–Cu–Cr alloys; solidification; microstructure; hardness; sliding wear

check for updates

Citation: Lantmann, R.V.; Mariante, A.M.S.; Pinheiro, T.V.; da Costa, E.M.; dos Santos, C.A. Microstructure, Hardness, and Linear Reciprocating Sliding Wear Response of Directionally Solidified Al–(2.5, 3.5, 4.5)Cu–(0.25, 0.50)Cr Alloys. *Metals* **2023**, *13*, 1178. https://doi.org/10.3390/met13071178

Academic Editor: Yanxin Qiao

Received: 26 May 2023
Revised: 19 June 2023
Accepted: 22 June 2023
Published: 24 June 2023

1. Introduction

It is well-known that the microstructure and mechanical, corrosion, and wear properties of aluminum alloys are affected by adding alloying elements. Some elements improve alloy strength by solid solution strengthening, while others allow for heat treatment to enhance mechanical properties by precipitation hardening mechanisms [1]. The most common alloying elements added to aluminum alloys are Cu, Mn, Si, Mg, and Zn. Some transition metal elements, such as Ti, Cr, V, Fe, Co, Ni, and Sc, are also added to increase strength and physical properties or to act as microstructure modifiers and/or grain refiners, as is the case of Ti, B, V, Sc, and Sr [2,3].

In general, transition metal elements have limited solid solubility in the α–Al matrix, with a tendency to form secondary intermetallic phases and improve specific properties. Another benefit of adding Ti, Mn, Cr, and Zr is to minimize the susceptibility to solidification cracking during casting and welding processes, in particular interdendritic cracking, attributed to both presence of intermetallic particles in the interdendritic regions and a decrease in the solidification temperature interval (between Liquidus and Solidus temperatures). Furthermore, the improvement of grain boundary cohesion, especially at elevated temperatures, is of interest when transition metal alloying elements are added to aluminum alloys [4]. Elements such as Cr, Ni, and Zr are beneficial, whereas P is harmful, and Si and Mg are considered neutral elements in this context. Alloying elements such as Mn and Sc induce eutectic transformations with Al, being segregated in the dendritic peripheral regions, while Cr, Ti, and Zr promote peritectic transformations, enriching the central region of the interdendritic zones [5]. The formation of small and distributed dispersoids in the α–Al matrix is another advantage of using transition metal elements such as Sc and Zr. These dispersoids can act as preferred sites for nucleation and stabilization of hardening precipitates during aging after solutioning, inducing strength increment [6].

Some works have investigated the influence of Cr additions on the microstructure formation, mechanical properties, and manufacturing processes of cast Al alloys. In binary Al–Cr alloys, small additions of Cr predominantly form the second phase $CrAl_7$ due to its low equilibrium solid solubility limit in the α–Al matrix. During solidification or heat treatment, this precipitate acts as an obstacle to grain growth [7]. An important concern in this topic, in special in near-net-shape processes, such as additive manufacturing by laser powder bed fusion (LPBF), is that the addition of 3%Cr to Al has shown significant high-temperature strength by solid-solution hardening after aging heat treatment [8], demonstrating the potential of Al–Cr alloys to produce three-dimensional metal components from metal powders by additive manufacturing processes based on laser irradiation. When Cr content is higher, quasicrystals are formed over a wide composition range. These quasicrystals can be decagonal quasicrystals (DQCs—periodic along the 10-fold axis and quasiperiodic on the perpendicular plane) or icosahedral quasicrystals (IQCs—three-dimensional quasi-periodic), depending on the Cr content [9,10]. In ternary Al–Cu–Cr alloys, AlCuCr intermetallic particles can be found depending on the alloy Cu and Cr contents, solidification conditions, and heat treatment processes. With high Cr content, the Al–Cu–Cr system has been considered one of the most complex alloy systems, since this system includes some uncharacterized phases such as icosahedral and decagonal quasicrystals [11–16].

The effect of Cr addition (0.9 and 1.24 wt.%) on microstructure, mechanical properties, wear resistance, and formability of an Al–2.3%Cu alloy was analyzed by El-Nasser et al. [17]. Both as-cast and heat-treated (solutioning and aging) conditions were investigated. From the results, the authors observed a decrease in the strength with the increase of chromium addition. However, when higher Cr content was added, the best formability was achieved. Ravikumar et al. [18] investigated the effect of different Cr additions (0.1, 1, and 2%) on the microstructure, hardness, tensile strength, and dry sliding wear of the aluminum-copper casting alloy A206, containing 4.5%Cu, 0.22%Mn, 0.19%Mn, and 0.19%Ti. The alloys were analyzed in the as-cast condition, and after solutioning and aging heat treatments. In the as-cast condition, the results showed that the effect of Cr addition on hardness was insignificant. However, in the heat-treated condition, Cr acted as a stronger solid-solution strengthener in the alloy content of 0.1%. Ribeiro et al. [19] observed an identical positive behavior of 0.25 and 0.50%Cr additions in an Al–3.8%Cu alloy (wt%). The addition of Cr promoted the formation of Al–Cu–Cr–Fe precipitates, preventing the formation of detrimental Al–Fe phases. When the Cr content was increased, the hardness and tensile strength also increased. The influence of 0.3% Cr addition on microstructure, phase composition, and mechanical properties of the Al–5.1%Cu–1.7%Y–0.3%Zr alloy (wt%) in the as-cast, homogenized, and hot-rolled conditions was investigated by Amer et al. [20]. The results demonstrated that the addition of Cr promotes the precipitation of an Al–Cu–Y–Cr phase, improving mechanical responses, especially after rolling and annealing processes. Monti et al. [21] investigated the effect of Cr and Fe addition on the hot cracking formation in Al–Cu alloys processed by laser powder bed fusion (LPBF). Secondary phase formation and microstructure refining were examined under high cooling rates. The results demonstrated that under high cooling rates, Cr remained the majority in solid solution, and a small amount was observed in some precipitates with other alloying elements.

Despite the aforementioned works, the literature is still scarce in information related to the effect of small Cr additions on solidification, microstructure formation, and mechanical properties of hypoeutectic Al–Cu alloys. Furthermore, features such as macrostructural morphologies, microstructural length scales, and the presence and distribution of secondary phases can be used to correlate the positive effect of the Cu:Cr ratio on mechanical properties and wear behavior. In the present work, the influence of Cr additions (0.25 and 0.50%) to Al–(2.5, 3.5, 4.5%)Cu alloys on the solidification condition, microstructure, hardness, and linear reciprocating sliding wear response was investigated. Upward directional solidification was applied to minimize natural convection and solute segregation due to different densities between elements to be minimized, as well as to obtain a high spectrum of cooling rate during solidification, inducing diverse microstructural formation. Based on the results, interrelationships between microstructure features and mechanical properties are discussed.

2. Materials and Methods

The proposed Al–Cu–Cr alloys investigated in this work are listed in Table 1. The alloys containing 2.5, 3.5, and 4.5%Cu and 0.25 and 0.50%Cr (wt.%) were prepared from 99.9% pure Al ingots, 99.99% electrolytic Cu wires, and 99.0% pure Cr chips. The metals were melted in an electric resistance soak furnace, using silicon carbide crucibles covered with boron nitride coating. Initially, pure Al was melted at 900 °C; subsequently, Cu was added to the melt, and after 5 min., Cr was added. The molten metal was kept for 3 h under mechanical stirring at intervals of 15 min to ensure the complete dissolution of solutes. No inert atmosphere was used. Approximately 4 kg of each alloy was produced. The chemical compositions were confirmed by an arc/spark optical emission spectrometry (OES) analyzer (SpectroMaxx, Spectro Ametek, Kleve, Germany), and the average values of six measurements are presented in Table 1.

Table 1. Chemical composition of the alloys (wt.%).

Alloys	Cu	Cr	Fe	Others	Al
2.5Cu	2.57	-	0.057	0.072	97.3
2.5Cu–0.25Cr	2.48	0.264	0.060	0.095	97.1
2.5Cu–0.50Cr	2.55	0.464	0.070	0.115	96.8
3.5Cu	3.56	-	0.062	0.177	96.2
3.5Cu–0.25Cr	3.49	0.255	0.063	0.191	96.0
3.5Cu–0.50Cr	3.52	0.490	0.084	0.105	95.8
4.5Cu	4.54	-	0.065	0.094	95.3
4.5Cu–0.25Cr	4.48	0.247	0.062	0.110	95.1
4.5Cu–0.50Cr	4.58	0.485	0.074	0.160	94.7

Differential thermal analyzes at heating and cooling rates of 10 °C/min were carried out using a simultaneous thermal analyzer (SDT 650 Discovery, TA Instruments) in samples extracted from the alloys. The analyses were performed using alumina crucibles 90 µg, heating/cooling steps in the ranges of 40–750 °C and 750–40 °C, respectively, under a protective N_2 atmosphere with a 100 mL/min flow rate. TRIOS 5.2 software (TA Instruments, New Castle, DE, USA) was used for processing the results. Liquidus temperatures (T_L) and transformation enthalpies (ΔH) were determined from the cooling curves.

Subsequently, the alloys were directionally solidified in the solidification apparatus shown in Figure 1a, which consists of a resistive tubular furnace having two independent heating zones, a cylindrical AISI 1020 carbon steel mold instrumented with six type-K thermocouples (1.6 mm diameter) positioned along the height of the mold cavity, and a water jet cooling system with a flow rate of 3 L/min located at the bottom of the mold. Our previous works present details about the solidification experimental system [19,22,23].

The alloys were melted at about 720 °C (approximately 10% above the alloy's Liquidus temperatures). The liquid cooling rates ($\dot{T}_L$) were determined from the cooling curves acquired by the thermocouples, as described in [24,25].

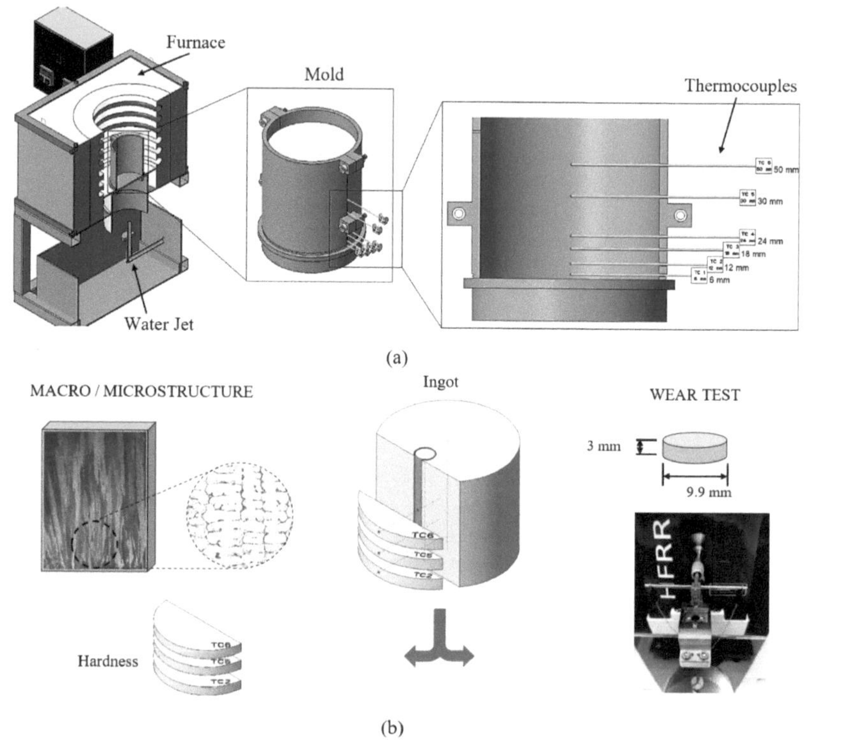

Figure 1. Schematic representation of: (**a**) solidification apparatus and (**b**) samples extracted for microstructural analysis and wear test.

Macrostructure and microstructure analyses were performed following the recommendation of ASTM E3 and E407 standard methods [26,27] For longitudinal macrostructure revelation, the samples were etched with a fluorite aqua regia solution (270 mL DI H_2O + 20 mL HNO_3 + 60 mL HCl + 10 mL HF). The electro polisher/etcher power supply (ElectroMet 4, Buehler) was used with an electrolyte (800 mL ethanol 95% + 140 mL DI H_2O + 60 mL $HClO_4$ 60%), 1.8 A current, and 25–30 s dwell time to reveal the microstructures. The microstructures were analyzed along the length of the ingots at 12, 30, and 50 mm distant from the bottom of the ingots, corresponding to the TC2, TC5, and TC6 thermocouple locations, respectively, as illustrated in Figure 1b, using an optical microscope (PMG, Olympus, Tokyo, Japan). The primary dendrite arm spacings (λ_1) were measured at these positions (at least 15 measurements for each position) to estimate the evolution of the dendritic length scale. For the linear reciprocating sliding wear tests, disks of 9.9 mm diameter and 3.0 mm thickness were transversally extracted from the ingots at the same positions where TC2, TC5, and TC6 were located. All disc specimens were ground on one face using sandpapers (#220, 320, 400, 600, 1200 grit meshes), polished with alumina solutions (1.0 and 0.25 μm), cleaned in an ultrasonic bath for 3 min using ethanol, and dried using a hand-held hot-air device. The surface finish of the discs was measured with the surface roughness micrometer (Mitutoyo, Kawasaki, Japan, SJ-210 model, 0.002 μm precision), resulting in a superficial roughness <0.6 μm. Brinell hardness measurements, as recommended by the ASTM E10 standard test method [28], were carried in the transverse section of the ingots at the same thermocouple positions, using 2500 N test load, 5 mm ball diameter, and 30 s dwell time. The average values and standard deviations of six measurements are presented. Linear reciprocating sliding wear tests were performed according to the ASTM G133 standard test method [29] using 2 N normal applied load, 6 mm diameter counter body sphere (Zr-Y material, 1200 HV surface hardness), 50 Hz frequency, 1 mm stroke, 250 mm total sliding distance, 25 °C room temperature, and 60% humidity using a high-frequency reciprocation rig tribometer (HFRR 4.2, Ducom, Bangalore, India). All wear tests were performed in duplicate for each alloy and position (on the same face of the discs) and the average values of the coefficient of friction (COF) were determined.

3. Results and Discussion

3.1. Solidification Thermal Parameters

The thermal curves obtained from DTA analyses for the binary alloys are presented in Figure 2. During the heating of the alloys (red curves), it was noted one slight first peak at 544.7 °C for the Al–2.5Cu alloy (Figure 2a), corresponding to the temperature for the start of melting (Solidus temperature). This peak was observed for the Al–3.5Cu alloy at 544.1 °C (Figure 2b), while the Al–4.5Cu alloy showed this transformation at 543.1 °C (Figure 2c). Still considering the heating curves, the end of melting (corresponding to the Liquidus temperature) was given at 653.7, 649.6, and 648.0 °C for the Al–2.5Cu, Al–3.5Cu, and Al–4.5Cu alloys, respectively. As these curves are related to the heating step, the lowest endothermic peaks correspond to the melting point of the alloys. In all curves, the onset of these peaks ranging from 600 to 610 °C, which represents the temperature at which the material starts to melt. During the cooling of the alloys (blue curves), T_L was determined at 650.7, 647.5, and 644.6 °C for the Al–2.5Cu, Al–3.5Cu, and Al–4.5Cu alloys, respectively. For the Al–3.5Cu alloy, an isolated unexpected peak was observed at 568.1 °C. Related to the Solidus temperature determined in the cooling curve (indicated by the second peak), the values were 544.4, 544.2, and 537.2 °C for the alloys with 2.5, 3.5, and 4.5%Cu contents, respectively. As mentioned by Jaradeh and Carlberg [30], it is typical to determine the Solidus temperature from the heating curve and the Liquidus temperature from the cooling curve due to the accuracy of the obtained values. However, in the present results insignificant differences between T_L (<4 °C) and T_S (<6 °C) during heating and cooling conditions were observed. The two transition points observed in the curves correspond to the formation of the primary α–Al phase (matrix) at T_L and the eutectic microconstituent (α-Al + θ) at T_S. In summary, for the binary alloys, the increase of alloy Cu content decreases both T_L and T_S.

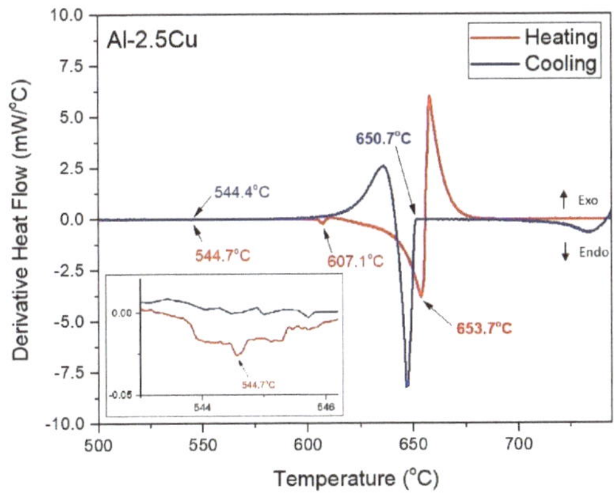

(a)

Figure 2. *Cont.*

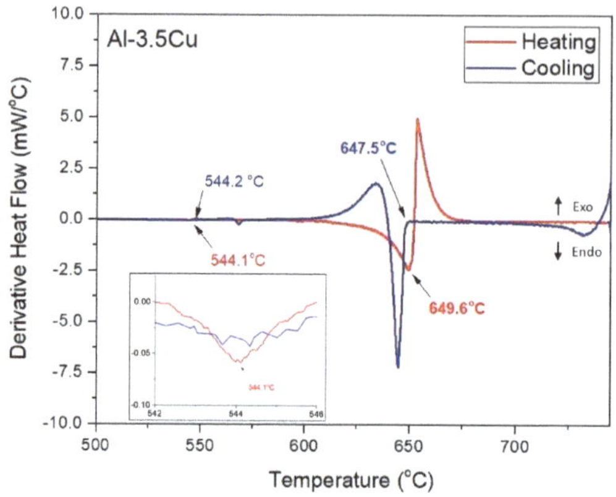

(**b**)

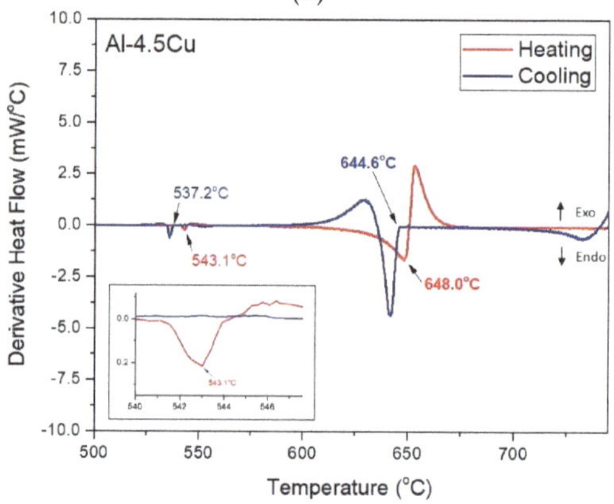

(**c**)

Figure 2. DTA curves during heating and cooling: (**a**) Al–2.5Cu; (**b**) Al–3.5Cu; (**c**) Al–4.5Cu binary alloys.

Figure 3a–f show the results of the ternary alloys with 0.25 and 0.50%Cr contents. The alloys with 2.5%Cu indicated Liquidus temperature values of about 648.9 and 647.4 °C for the alloys with 0.25 and 0.50%Cr contents, respectively (Figure 3a,b). With the increase of the alloy Cu content to 3.5% (Figure 3c,d), the Liquidus temperature decreased from 646.8 to 643.3 °C to those alloys with 0.25 and 0.50%Cr, while the alloys with 4.5%Cu showed values of 644.8 and 643.9 °C with 0.25 and 0.50%Cr, respectively (Figure 3e,f). As observed, the temperature at the beginning of solidification (Liquidus temperature) decreased with the increase of Cu and Cr contents in the alloys. The influence of alloying elements on the Liquidus temperatures was more influenced in the case of Cu addition as compared to the addition of Cr.

As observed in Figure 3, additional peaks were detected in the solidification intervals. For the alloys with 2.5%Cu, during the heating step (red curves), peaks appeared at 607.1 and 604.7 °C when the alloy Cr content was 0.25 and 0.50%, respectively. For the alloys with 3.5%Cu, heating curves showed intermediate peaks at 606.4 and 597.5 °C for the alloys with 0.25 and 0.50%Cr. With 4.5%Cu content, the alloys with 0.25 and 0.50%Cr showed quasi-identical peaks to those alloys with lower Cu contents, with peaks at 603.7 and 601.5 °C in the heating curves. When considering the cooling curves, intermediate peaks were only observed for the Al–3.5Cu–0.25Cr, Al–4.5Cu–0.25Cr, and Al–4.5Cu–0.50Cr alloys, at 573.3, 561.1, and 565.3 °C, respectively. This can be attributed to the diffusion of alloying elements during cooling when the amount of Cu and Cr is higher than a critical value. Aluminum alloys with small Cu and Cr additions can precipitate secondary phases or intermetallic compounds at low temperatures, such as Al–Cr and Al–Cu–Cr, and if the Fe is present, Al–Fe–Cr and Al–Cu–Fe–Cr. However, the formation of these precipitates depends on the solidification conditions and solute diffusion characteristics [17,31].

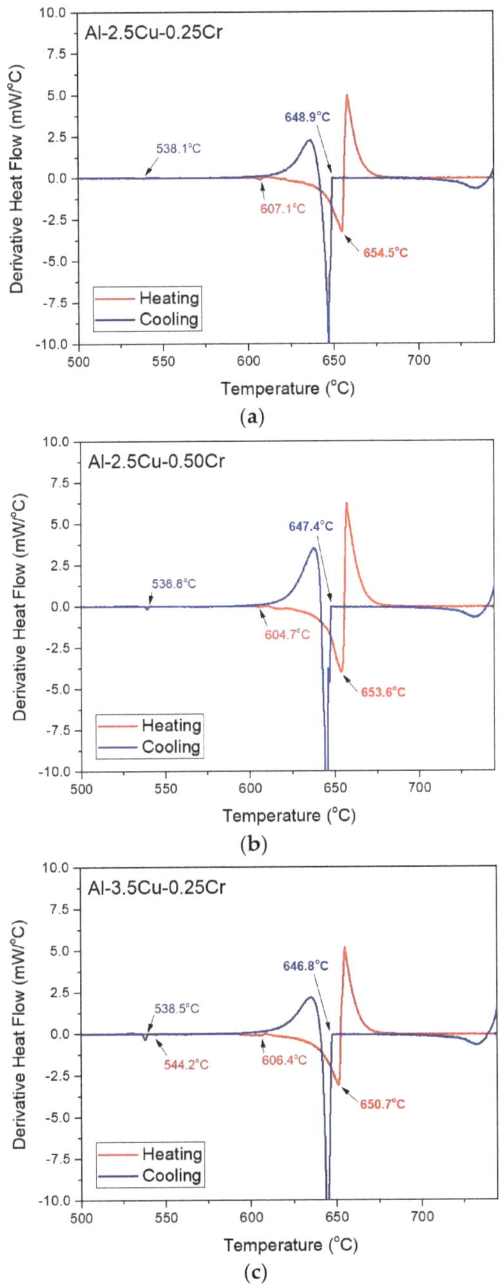

Figure 3. *Cont.*

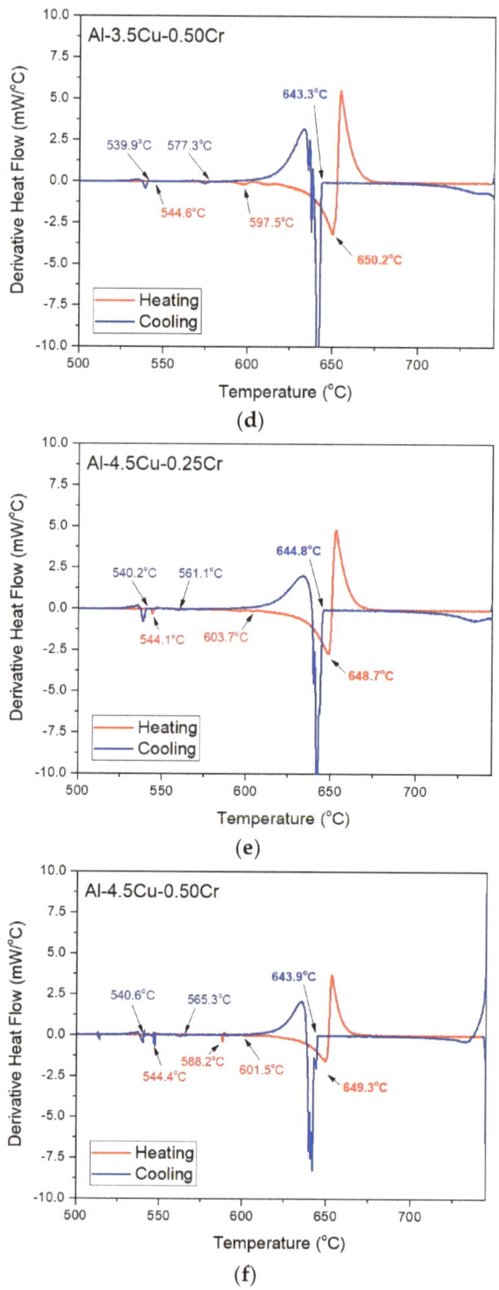

Figure 3. DTA curves during heating and cooling: (**a**) Al–2.5Cu–0.25Cr; (**b**) Al–2.5Cu–0.50Cr; (**c**) Al–3.5Cu–0.25Cr; (**d**) Al–3.5Cu–0.50Cr; (**e**) Al–4.5Cu–0.25Cr; (**f**) Al–4.5Cu–0.50Cr ternary alloys.

The transformation enthalpies (ΔH) determined on the cooling curves are shown in Figure 4a–c. The values were calculated using the TRIOS software and considering the baseline type linear, onset type inflection, applying the command "Execute—Intelligent" in the tool "Analysis". For the Al–2.5Cu–xCr alloys (Figure 4a), the observed values were 313.07, 334.92, and 341.02 J/g for those alloys with 0, 0.25, and 0.50%Cr, respectively, indicating an increase in enthalpy transformation during solidification with the increase of the alloy Cr content. The alloy with 0.50%Cr showed a second peak closed at 533 °C with an enthalpy value of about 2.06 J/g, demonstrating that this transformation is related to a small amount of the transformed material. The alloys with 3.5%Cu presented values of 307.28 J/g, with 0.25 and 0.50%Cr the values were 332.13, and 313.06 J/g, respectively (Figure 4b). Once again, secondary peaks were observed at temperatures of 537 °C indicating transformations with a small amount of the phases. For the Al–3.5Cu–xCr alloys, the enthalpies did not demonstrate a tendency as a function of the alloy Cr content. As the alloy Cu content increased to 4.5% (Figure 4c), the enthalpy value decreased to 188.84 J/g, and 326.41 and 283.36 J/g with 0.25 and 0.50%Cr, respectively. In general, the enthalpy values decreased more significantly with the increase of the alloy Cu content, and less with the alloy Cr content variation.

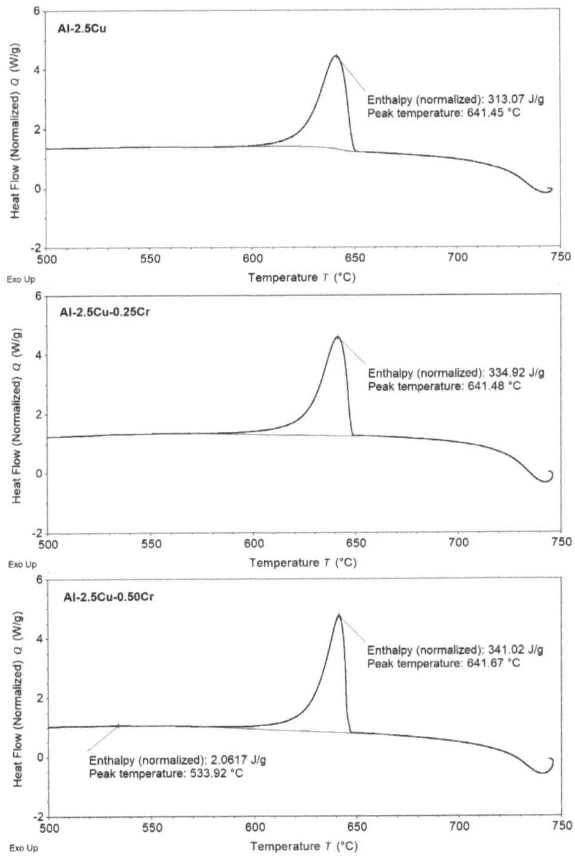

(a)

Figure 4. *Cont.*

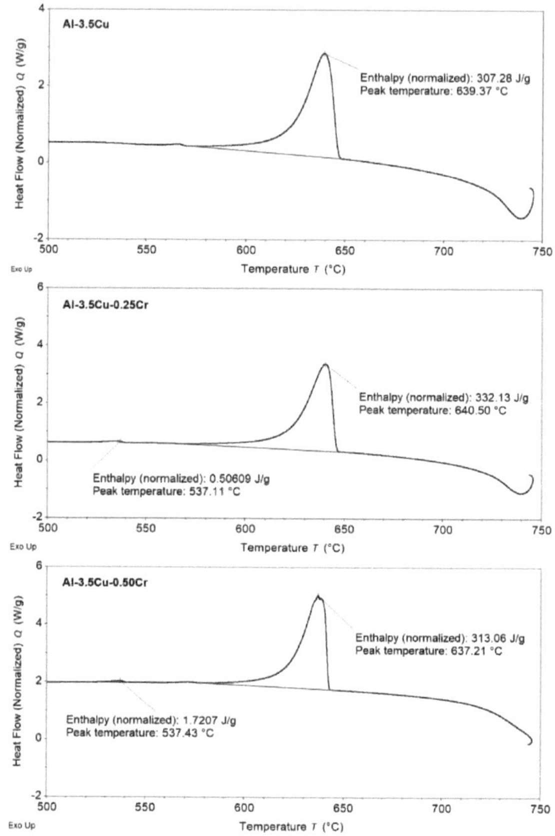

(**b**)

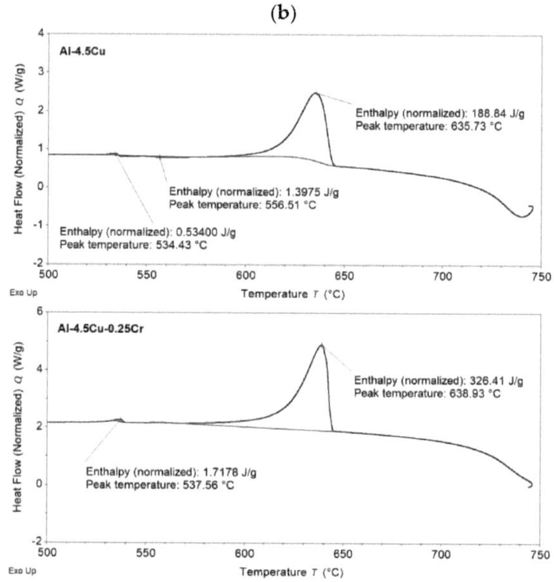

Figure 4. *Cont.*

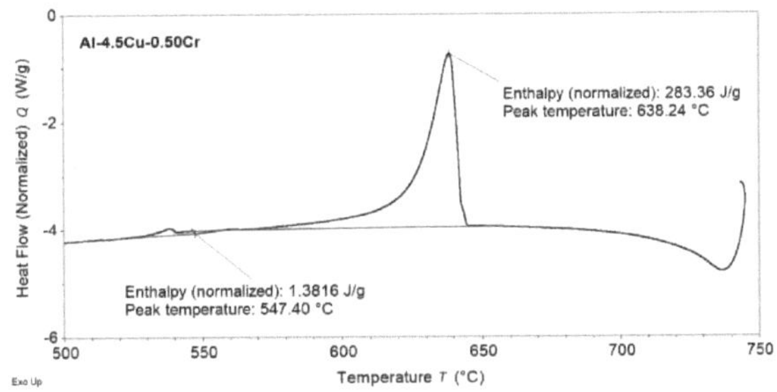

(c)

Figure 4. Transformation enthalpies obtained from DTA curves during cooling: (**a**) Al–2.5Cu–xCr; (**b**) Al–3.5Cu–xCr; (**c**) Al–4.5Cu–xCr alloys. Blue lines: baseline.

3.2. Directional Solidification

Figure 5a,b show examples of the thermal profiles obtained during directional solidification of the Al–3.5Cu binary alloy and Al–3.5Cu–0.50Cr ternary alloy, respectively. Liquidus temperatures determined from DTA data during the cooling step are indicated by horizontal dashed lines on the graphs. The cooling rates $\left(\dot{T} = dT_L/t_L\right)$ for each thermocouple were determined in the time corresponding to the Liquidus temperature.

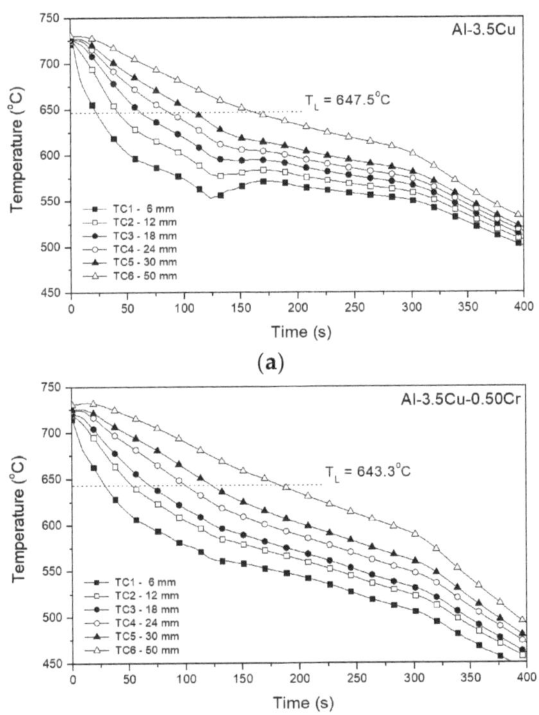

(b)

Figure 5. Solidification thermal profiles: (**a**) Al–3.5Cu; (**b**) Al–3.5Cu–0.50Cr alloys.

The cooling rates determined during the solidification of ternary alloys with 0.25 and 0.50%Cr are shown in Figure 6a,b, respectively. The results showed that the variation of the alloy Cu content influences the cooling rate, decreasing as the Cu concentration increases.

However, the cooling rate is slightly affected by the alloy Cr content, showing a tendency to increase as the Cr concentration increases from 0.25 to 0.50%. These observations agree with those reported in the literature for Al–Cu–Cr, Al–Fe–Cr, and Al–Si–Cr alloys with different Cr contents. With Al–3.8%Cu–(0.25 and 0.50)%Cr alloys, Ribeiro et al. [19] observed an increase in the liquid cooling rate with increasing Cr content. For Al–Fe alloys with 1, 2, and 3% Cr additions, Liu et al. [31] found that small additions of Cr increased both the cooling rate and the compositional undercooling during solidification. In the work reported by Rosso et al. [23] with hypoeutectic Al–Si alloys with small Cr additions (0.25 and 0.40%), it was observed that the cooling rates increased with the increase in the addition of Cr for alloys with lower Si content. However, it is important to note that the cooling rate during solidification depends not only on the chemical composition of the metal, but also on thermal parameters such as the liquid growth rate and liquid thermal gradient, the thermophysical properties of the metal and mold, the conditions of the metal/mold interface, and others [32–39].

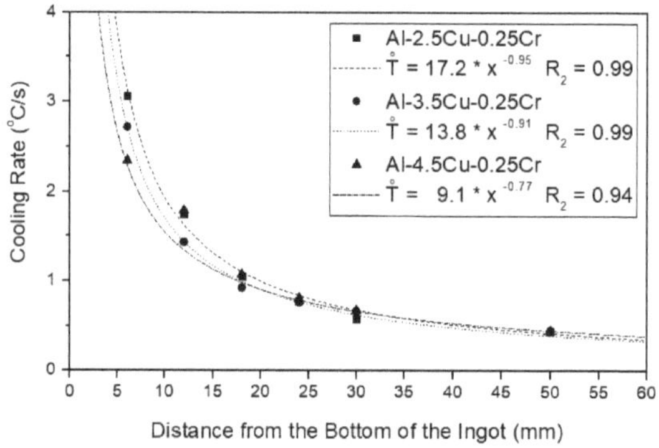

(a)

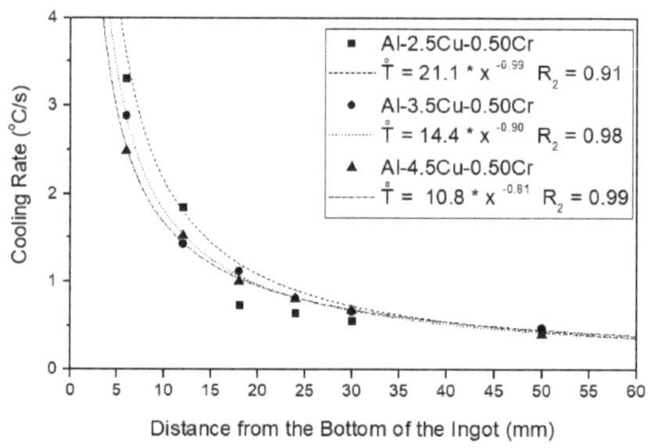

(b)

Figure 6. Liquid cooling rates: (**a**) Al–xCu–0.25Cr; (**b**) Al–xCu–0.50Cr alloys.

3.3. Microstructure and Hardness

Longitudinal section macrographs of the ingots showed a complete columnar macrostructure in all alloys, with fine grains at the bottom of the ingots (under high cooling rates) and coarse grains close to the top of the ingots (under low cooling rates). An example is shown in Figure 7a for the Al–4.5Cu–0.50Cr alloy, which is the most critical alloy composition for exhibiting a columnar-to-equiaxed transition in the macrostructure. In Figure 7b, the microstructures of the Al–3.5Cu–0.25Cr alloy at the positions corresponding to the thermocouples TC2, TC5, and TC6 (located at 12, 30, and 50 mm from the bottom of the ingot, respectively) are shown.

The solidified microstructure is composed of a directional cellular/dendritic matrix (primary α–Al phase) and a eutectic microconstituent (α–Al phase + θ) in the interdendritic regions. At high cooling rates (near the bottom of the ingot) the microstructure is predominately cellular. As the distance from the bottom increases (with low cooling rates), the microstructure changes to a dendritic morphology. Isolated Al–Cr particles were not observed in all samples, evidencing that the Cr concentration of the investigated alloys was insufficient to precipitate this type of compound. This behavior is similar to that reported by Liu et al. [31] with Al–Fe–Cr alloys.

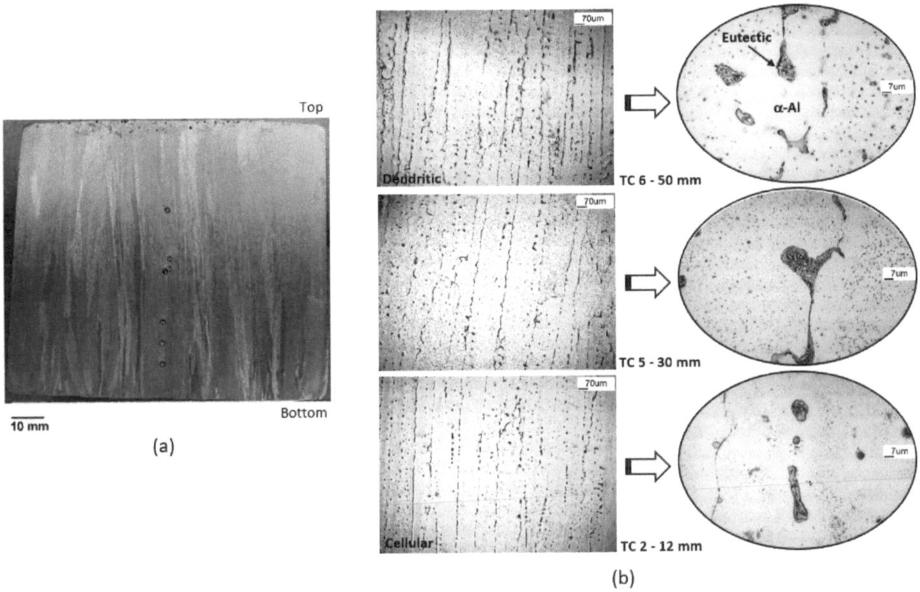

Figure 7. (**a**) Macrostructure of the Al–4.5Cu–0.50Cr alloy, and (**b**) microstructures of the Al–3.5Cu–0.25Cr alloy at 12 mm, 30 mm, and 50 mm from the bottom of the ingot: left side (100×), right side (500×).

Figure 8 shows typical microstructures of the other alloys, at different magnifications to emphasize microstructure details. As observed, the solidified microstructure becomes more dendritic as alloying elements and cooling rates increase, as well as the distance from the bottom of the ingots. For the Al–2.5Cu–xCr alloys, a predominantly cellular microstructure was observed up to 30 mm (TC2 and TC5 positions), with a cellular/dendritic transition at 50 mm from the bottom of the ingot. When the alloy Cu contents were increased to 3.5% and 4.5%, the cellular/dendritic transition occurred at 30 mm, with complete cellular microstructure up to 12 mm and predominant dendritic microstructure after 30 mm. In all alloys, no evidence of macrosegregation was detected in the Cu and Cr profiles, since the chemical composition measurements carried out in the positions corresponding to the thermocouples TC2, TC5, and TC6 showed similar values.

In order to correlate the as-cast microstructures with the cooling rates, the primary dendrite arm spacings (λ_1) as a function of the cooling rates were plotted in Figure 9a,b for the ternary alloys with 0.25 and 0.50%Cr contents, respectively. A trend of increasing λ_1 with decreasing cooling rate and increasing alloying elements was noted, especially for alloys with 0.25%Cr. The highest value of λ_1 was observed at the lowest cooling rate (~0.4 °C/s) for all alloys. Furthermore, alloys with 0.50%Cr showed a smaller variation of λ_1 when compared to alloys with 0.25%Cr.

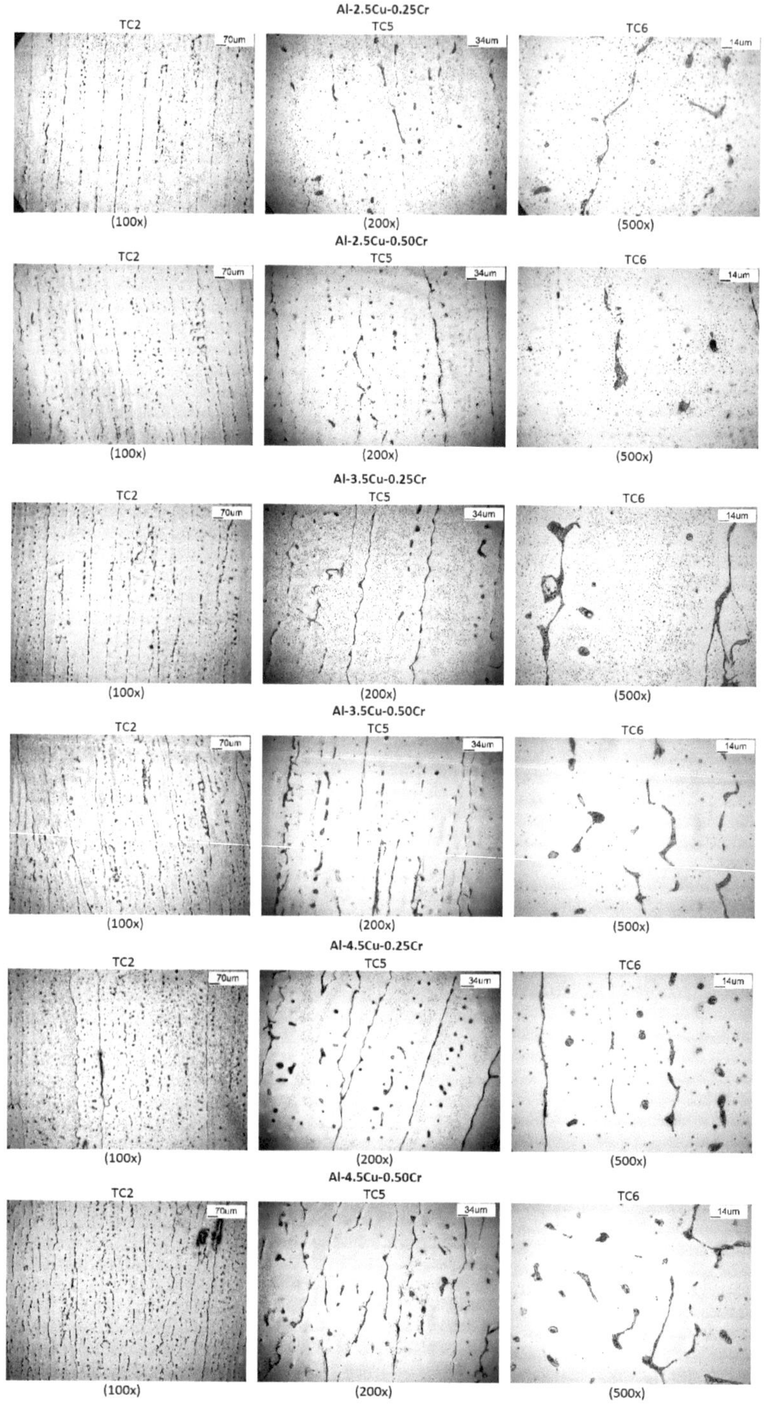

Figure 8. Optical micrographs of the alloys at different magnifications.

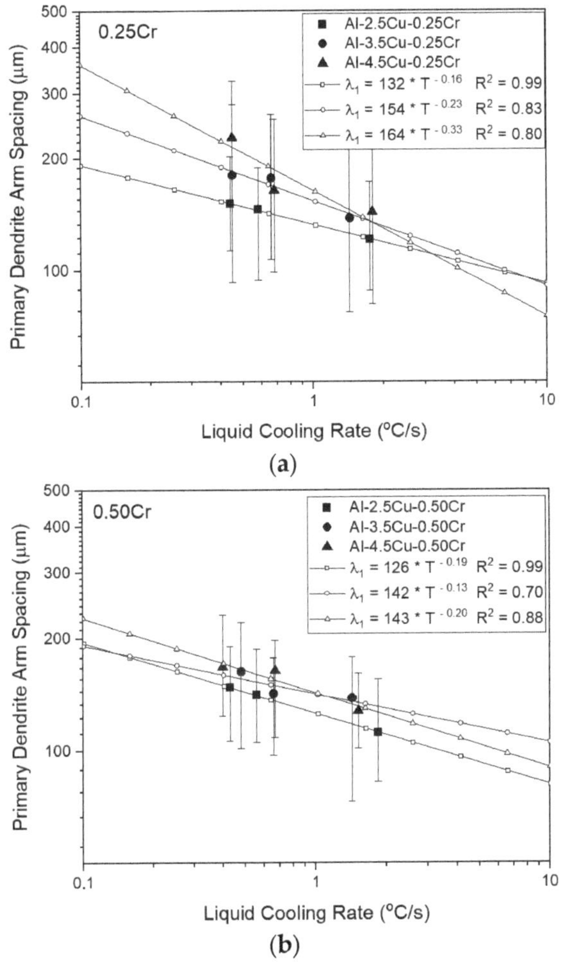

Figure 9. Primary dendrite arm spacing as a function of the liquid cooling rate of the alloy ingots: (**a**) Al–xCu–0.25Cr; (**b**) Al–xCu–0.50Cr.

Brinell hardness measurements obtained along the length of the ingots are shown in Figure 10a–c. All alloys showed similar behavior, with an increase in hardness with an increase in Cu and Cr alloying elements and a decrease with an increase in the distance from the bottom of the ingots. This behavior may be associated with the refinement of the dendritic arm spacing, which minimizes dislocation movement during plastic deformation and improves mechanical strength. The Brinell hardness of the alloys with 2.5%Cu reached the maximum value of 47 HB for the alloy with 0.50%Cr in the position close to the bottom of the ingot, while the minimum value was observed for the binary Al–2.5%Cu alloy at the 50 mm position, with a value of 37.5 HB. The addition of Cu and Cr alloying elements increased the hardness of the alloys, with the maximum value observed for the alloy with 4.5%Cu and 0.50%Cr.

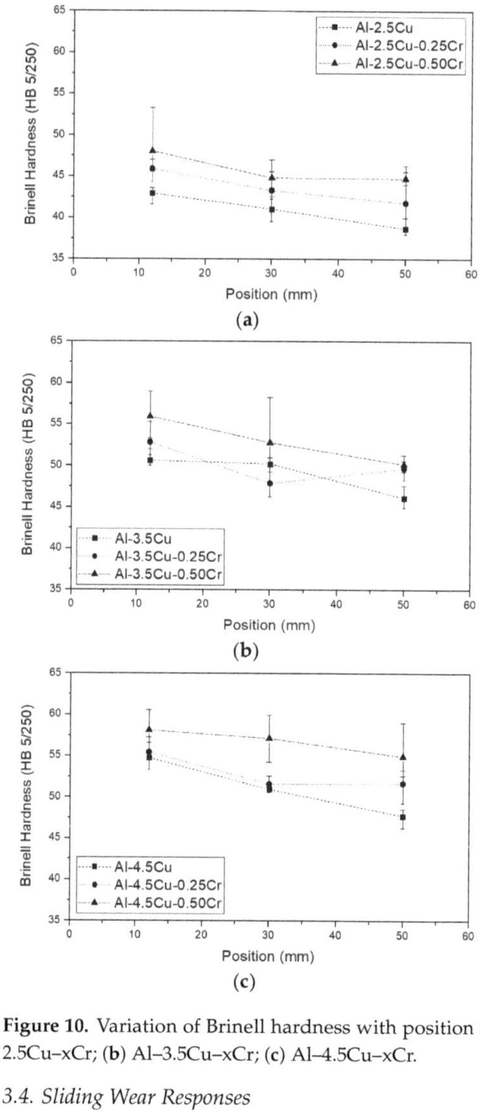

Figure 10. Variation of Brinell hardness with position along the length of the alloy ingots: (**a**) Al–2.5Cu–xCr; (**b**) Al–3.5Cu–xCr; (**c**) Al–4.5Cu–xCr.

3.4. Sliding Wear Responses

The wear tests were performed in duplicate. As examples, Figure 11a shows the duplicate results obtained for the binary Al–2.5Cu alloy, in the position corresponding to the thermocouple TC2 (12 mm from the bottom of the ingot), while Figure 11b presents the behavior of the ternary Al–4.5Cu–0.25Cr alloy at 50 mm from the bottom, corresponding to the position of TC6. The results exhibited similar behavior in both runs with some fluctuation during testing. An initial transient regime was observed at the first sliding distance, followed by a quasi-steady-state behavior until the end of the test. In general, the additions of Cu and Cr to the alloys resulted in a decrease in COF, regardless of the positions. As noted, the results showed that repeatability was achieved, with result variations of less than 5% for different alloys and test conditions.

Figure 12 shows a summary of the COF results obtained for the binary alloys, while the results for the ternary alloys are depicted in Figure 13. As observed in Figure 12, the Al–2.5Cu alloy showed an increased COF as the distance from the bottom of the ingot increased, as a consequence of the coarsening of the microstructure. Furthermore, a higher variation during the tests was noted for those positions near the top of the ingot. For the Al–3.5Cu alloy, the initial transient regime was observed with lower COF values in relation to the Al–2.5Cu alloy, increasing slightly during the tests and with the distance. As the alloy Cu content increased to 4.5%, the COF decreased for all thermocouple positions, with similar values and reduced variation during the tests.

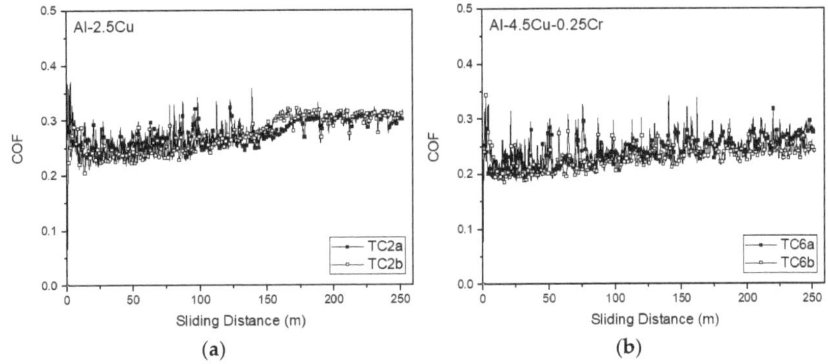

Figure 11. Coefficient of friction (COF) as a function of sliding distance and thermocouple position: (**a**) Al–2.5Cu alloy—TC2 (thermocouple at 12 mm); (**b**) Al–4.5Cu–0.25Cr alloy—TC6 (thermocouple at 50 mm).

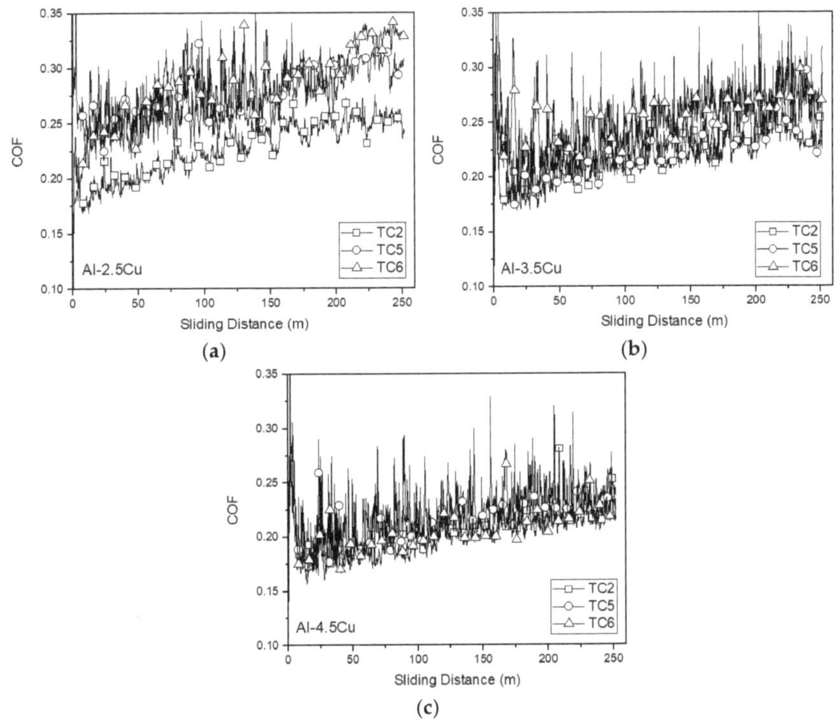

Figure 12. Coefficient of friction (COF) as a function of sliding distance for binary alloys: (**a**) Al–2.5Cu; (**b**) Al–3.5Cu; (**c**) Al–4.5Cu.

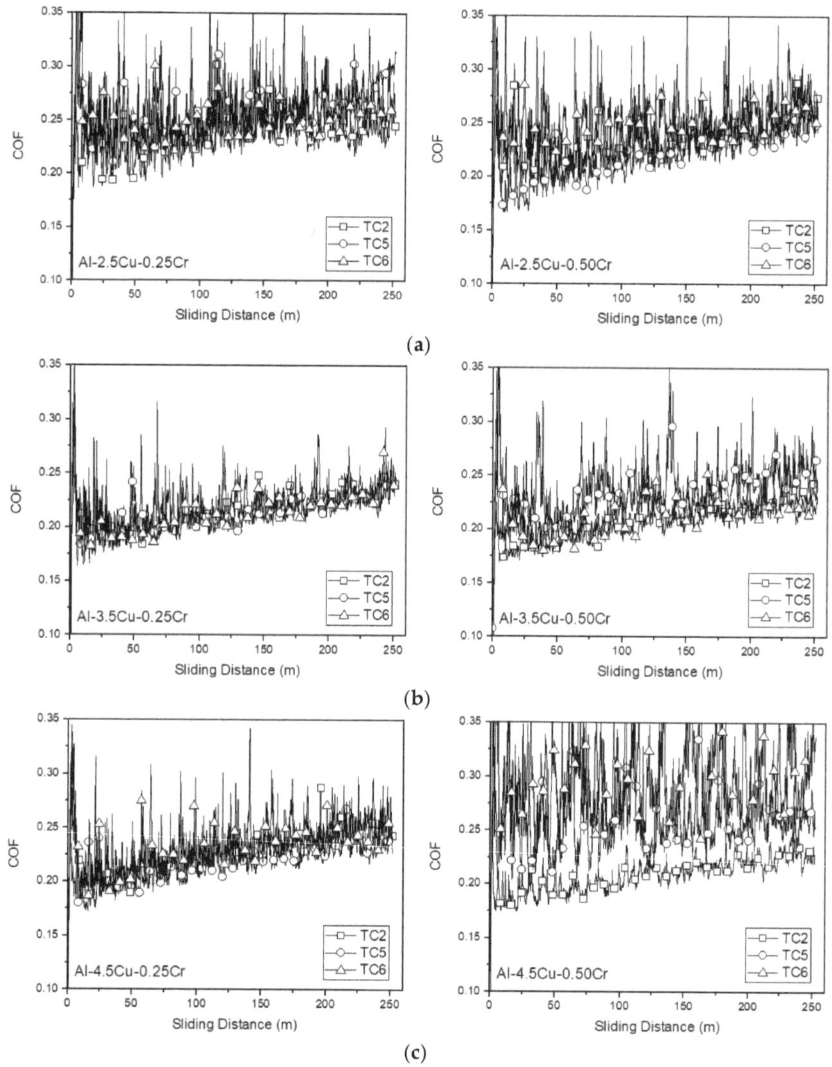

Figure 13. Coefficient of friction (COF) as a function of sliding distance for ternary alloys: (**a**) Al–2.5Cu–0.25Cr and Al–2.5Cu–0.50Cr; (**b**) Al–3.5Cu–0.25Cr and Al–3.5Cu–0.50Cr; (**c**) Al–4.5Cu–0.25Cr and Al–4.5Cu–0.50Cr.

With Cr additions (Figure 13), COF decreased with the increase in the Cr content for the alloys with higher Cu content and remained almost constant for the lower Cu-containing alloys. As also observed during the tests with the binary alloy, there were some fluctuations due to the heterogeneous as-cast microstructure, with the presence of interdendritic microporosity and precipitates. The highest COF fluctuation was observed in the alloy with 4.5%Cu and 0.50%Cr in positions with coarser microstructures.

The calculated wear volume loss at different thermocouple positions is shown in Figure 14. An image analyzer was used to determine the average cross-sectional area of the ellipsoid wear track and to calculate the wear volume considering the length of the stroke, as suggested in ASTM G 133 standard test method [29]. These values are the average of duplicate tests. Analyzing the results in Figure 14b–d, it can be noted that the alloys with 2.5% Cu showed a slight increase in the wear volume along the length of the ingots. An exceptionally high value was observed for the binary Al–2.5Cu alloy at the position corresponding to the TC2 thermocouple (30 mm) probably due to the presence of microporosity. In the alloys with 3.5%Cu, regardless of the percentage of Cr, the wear volume showed similar values in all positions. In the 4.5% Cu alloys, a tendency to decrease the wear volume loss was observed as the percentage of Cr and the distance from the bottom of the ingot increased.

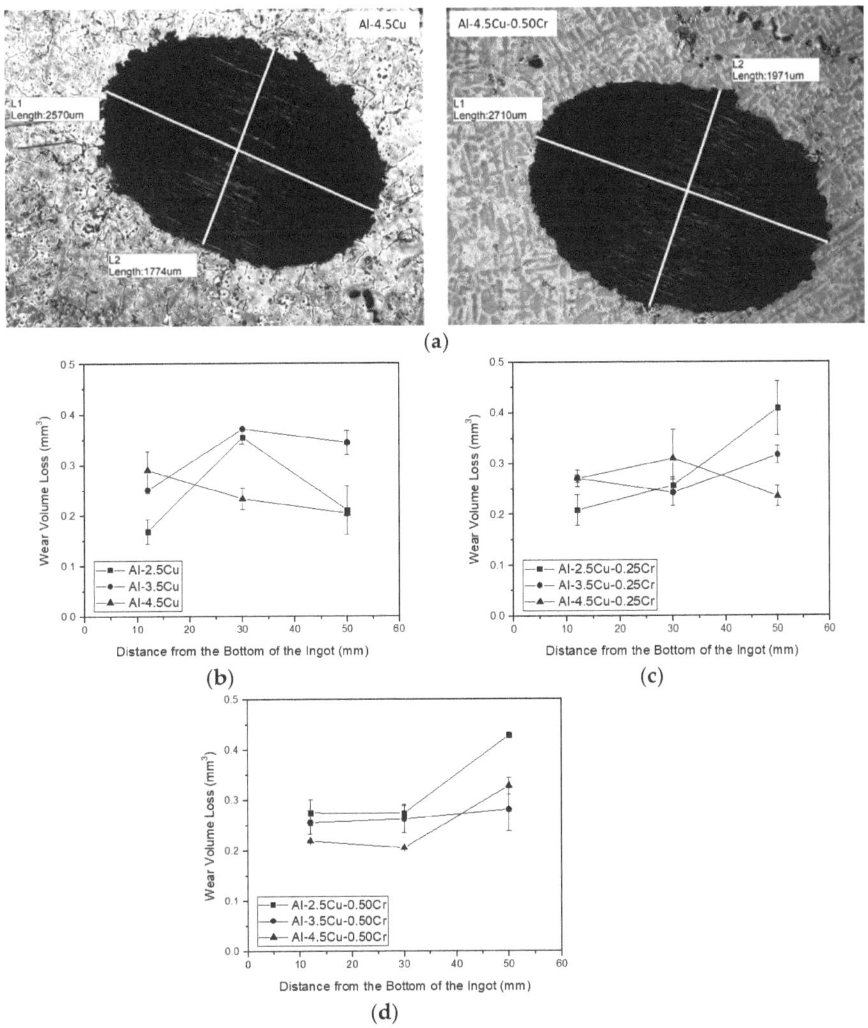

Figure 14. (**a**) Examples of wear track surfaces of the Al–4.5Cu and Al–4.5Cu–0.50Cr alloys; average wear volume loss as a function of the distance of the bottom of the ingots for all alloys: (**b**) Al–xCu, (**c**) Al–xCu–0.25Cr, and (**d**) Al–xCu–0.50Cr.

Since hypoeutectic Al–Cu alloys with a Cu content lower than 5.65 (wt.%) can be subjected to precipitation-hardening heat treatments (solutioning and natural or artificial aging steps), improved mechanical, corrosion, wear, and thermal properties can be achieved [40] with the use of small Cr additions to form new secondary phases or intermetallic precipitates depending on the Cu–Cr ratios. The present results provide an insight into the correlation between solidification conditions, microstructural formation and some mechanical properties, allowing for future investigations of the behavior after heat treatments.

4. Conclusions

The following main conclusions can be drawn based on the experimental results obtained in this study:

- Additions of Cu and Cr decreased the Liquidus temperatures (from 648.9 and 647.4 °C for the Al–2.5Cu alloys with 0.25 and 0.50%Cr, respectively, to 644.8 and 643.9 °C for the Al–4.5Cu alloys with 0.25 and 0.50%Cr, respectively) and the enthalpies of transformation of the alloys;
- The addition of Cu significantly affected cooling rates, decreasing as the Cu content increased, from 3.1 °C/s (Al–2.5Cu–0.25Cr alloy) to 2.3 °C/s (Al–4.5Cu–0.25Cr alloy), and from 3.3 °C/s (Al–2.5Cu–0.50Cr alloy) to 2.5 °C/s (Al–4.5Cu–0.50Cr alloy) at those positions close to the bottom of the ingots). On the other hand, higher Cr contents increased cooling rates;
- When comparing alloys with the same Cu content, Cr addition improved hardness, especially for lower Cu-containing alloys (≅10% at those positions near the bottom of the ingots). In all alloys, hardness decreased with increasing λ_1;
- Additions of Cu and Cr to the alloys resulted in a slight decrease in COF, with average values ranging from 0.30 to 0.20. Analyzing individually the effect of Cu and Cr additions on the wear volume loss, insignificant changes were observed, regardless of microstructure refinement. However, it is important to note that, when analyzed together, there is a tendency of decreasing wear volume loss with the coarsening of the microstructure.

Author Contributions: Conceptualization, C.A.d.S.; methodology, R.V.L., A.M.S.M. and T.V.P.; investigation, R.V.L., A.M.S.M., T.V.P., E.M.d.C. and C.A.d.S.; writing—original draft preparation, R.V.L., A.M.S.M., T.V.P., E.M.d.C. and C.A.d.S.; writing—review and editing, E.M.d.C. and C.A.d.S., funding acquisition, C.A.d.S. All authors have read and agreed to the published version of the manuscript.

Funding: This research was funded by CNPq (National Council for Scientific and Technological Development: grant: 403303/2016-8) and FINEP (Financiadora de Estudos e Projetos: grant: 0082/16).

Data Availability Statement: No new data were created or analyzed in this study. Data sharing is not applicable to this article.

Acknowledgments: The authors acknowledge the support provided by CNPq (National Council for Scientific and Technological Development: grant: 403303/2016-8), FINEP (Financiadora de Estudos e Projetos: grant: 0082/16), and PUCRS (Pontifícia Universidade Católica do Rio Grande do Sul). This study was also financed in part by the Coordenação de Aperfeiçoamento de Pessoal de Nível Superior—Brasil (CAPES)—Finance Code 001.

Conflicts of Interest: The authors declare no conflict of interest.

References

1. Mondolfo, L.F. *Aluminum Alloys: Structure and Properties*, 1st ed.; Butterworth-Heinemann: Oxfordshire, UK, 1976; 971p.
2. Kim, H.T.; Nam, S.W.; Hwang, S.H. Study on the solidification cracking behaviour of high strength aluminum alloy welds: Effects of alloying elements and solidification behaviours. *J. Mater. Sci.* **1996**, *31*, 2859–2864. [CrossRef]
3. Rana, R.S.; Purohit, R.; Das, S. Reviews on the Influences of Alloying elements on the Microstructure and Mechanical Properties of Aluminum Alloys and Aluminum Alloy Composites. *Int. J. Sci. Res. Publ.* **2012**, *2*, 1–7.
4. Razumovskiy, V.I.; Ruban, A.V.; Razumovskii, I.M.; Lozovoi, A.Y.; Butrim, V.N.; Vekilov, Y.K. The effect of alloying elements on grain boundary and bulk cohesion in aluminum alloys: An ab initio study. *Scr. Mater.* **2011**, *65*, 926–929. [CrossRef]
5. Zakharov, V.V. About alloying of aluminum alloys with transition metals. *Met. Sci. Heat Treat.* **2017**, *59*, 67–71. [CrossRef]
6. Dorin, T.; Ramajayam, M.; Lamb, J.; Langan, T. Effect of Sc and Zr additions on the microstructure/strength of Al-Cu binary alloys. *Mater. Sci. Eng. A* **2017**, *707*, 58–64. [CrossRef]

7. Kaufman, J.G.; Rooy, E.L. *Aluminum Alloy Castings: Properties, Processes, and Applications*, 1st ed.; ASM International: Novelty, OH, USA, 2004.
8. Kimura, T.; Nakamoto, T.; Ozaki, T.; Miki, T. Microstructures and mechanical properties of aluminum-transition metal binary alloys (Al-Fe, Al-Mn, and Al-Cr) processed by laser powder bed fusion. *J. Alloys Compd.* **2021**, *872*, 159680. [CrossRef]
9. Predel, B. Al-Cr (Aluminum-Chromium) Datasheet from Landolt-Börnstein-Group IV Physical Chemistry. In *"Ac-Ag... Au-Zr"-Phase Equilibria, Crystallographic and Thermodynamic Data of Binary Alloys*; Springer Materials: Cham, Switzerland, 2006; Volume 12A. [CrossRef]
10. Chen, H.; Wang, Q.; Wang, Y.; Qiang, J.; Dong, C. Composition rule for Al–transition metal binary quasicrystals. *Philos. Mag.* **2010**, *90*, 3935–3946. [CrossRef]
11. Grushko, B.; Przepiórzyński, B.; Pavlyuchkov, D.; Mi, S.; Kowalska-Strzęciwilk, E.; Surowiec, M. Complex intermetallics in Al–Cu–Cr system. *J. Alloys Compd.* **2007**, *442*, 114–116. [CrossRef]
12. Sugiyama, K.; Saito, H.; Hiraga, K. On the crystal structures of the Al–Cu–Cr alloy system. *J. Alloys Compd.* **2002**, *342*, 148–152. [CrossRef]
13. Grushko, B. A contribution to the Al–Cu–Cr phase diagram. *J. Alloys Compd.* **2017**, *729*, 426–437. [CrossRef]
14. Sviridova, T.A.; Schevchukov, A.P.; Shelekhov, E.V.; Diakonov, D.L.; Tcherdyntsev, V.V.; Kaloshkin, S.D. The quasicrystalline phaseformation in Al–Cu–Cr alloys produced by mechanicalalloying. *J. Alloys Compd.* **2011**, *509S*, S299–S303. [CrossRef]
15. Fu, Y.; Kang, N.; Liao, H.; Gao, Y.; Coddet, C. An investigation onselective laser melting of Al–Cu–Fe–Cr quasicrystal: Fromsingle layer to multilayers. *Intermetallics* **2017**, *86*, 51–88. [CrossRef]
16. Salimon, A.L.; Shevchukov, A.P.; Stepashkin, A.A.; Tcherdyntsev, V.V.; Olifirov, L.K. Mechanical alloying as a solid state route forfabrication of Al–Cu–M(=Fe, Cr) quasicrystalline phases. *J. Alloys Compd.* **2017**, *707*, 315–320. [CrossRef]
17. El-Nasser, G.A.; Samy, S.M.; Nassef, A. Effect of chromium additions on formability and mechanical behavior of Al-Cu alloy. In Proceedings of the 10th International Mining, Petroleum, and Metallurgical Engineering Conference, Suez, Egypt, 6–8 March 2007; pp. 84–94.
18. Ravikumar, A.; Sellamuthu, R.; Saravanan, R. Effect of Cr addition on mechanical properties and wear rate of cast Al–Cu alloy. *Indian J. Sci. Technol.* **2016**, *9*, 34. [CrossRef]
19. Ribeiro, T.M.; Catellan, E.; Garcia, A.; Santos, C.A. The effects of Cr addition on microstructure, hardness and tensile properties of as-cast Al–3.8wt.%Cu–(Cr) alloys. *J. Mater. Res. Technol.* **2020**, *9*, 6620–6631. [CrossRef]
20. Amer, S.M.; Glavatskikh, M.V.; Barkov, R.Y.; Loginova, I.S.; Pozdniakov, A.V. Effect of Cr on the Microstructure and Mechanical Properties of the Al-Cu-Y-Zr Alloy. *Metals* **2023**, *13*, 349. [CrossRef]
21. Monti, C.; Turani, M.; Papis, K.; Bambach, M. A new Al-Cu alloy for LPBF developed via ultrasonic atomization. *Mater. Des.* **2023**, *229*, 111907. [CrossRef]
22. Ache, C.T.; Lopes, M.M.; Reis, B.P.; Garcia, A.; Santos, C.A. Dendritic Spacing/Columnar Grain Diameter of Al–2Mg–Zn Alloys Affecting Hardness, Tensile Properties, and Dry Sliding Wear in the As-Cast/Heat-Treated Conditions. *Adv. Eng. Technol.* **2020**, *22*, 1901145. [CrossRef]
23. Rosso, E.; Santos, C.A.; Garcia, A. Microstructure, Hardness, Tensile Strength, and Sliding Wear of Hypoeutectic Al–Si Cast Alloys with Small Cr Additions and Fe-Impurity Content. *Adv. Eng. Mater.* **2021**, *24*, 2001552. [CrossRef]
24. Quaresma, J.M.V.; Santos, C.A.; Garcia, A. Correlation between unsteady-state solidification conditions, dendrite spacings, and mechanical properties of Al-Cu alloys. *Metall. Mater. Trans. A* **2000**, *31*, 3167–3178. [CrossRef]
25. Reis, B.P.; Lopes, M.M.; Garcia, A.; Santos, C.A. The correlation of microstructure features, dry sliding wear behavior, hardness and tensile properties of Al-2wt%Mg-Zn alloys. *J. Alloys Compd.* **2018**, *764*, 267–278. [CrossRef]
26. *ASTM E3*; Standard Guide for Preparation of Metallographic Specimens. ASTM International: West Conshohocken, PA, USA, 2017. [CrossRef]
27. *ASTM E407*; Standard Practice for Microetching Metals and Alloys. ASTM International: West Conshohocken, PA, USA, 2015. [CrossRef]
28. *ASTM E10*; Standard Test Method for Brinell Hardness of Metallic Materials. ASTM International: West Conshohocken, PA, USA, 2018. [CrossRef]
29. *ASTM G133*; Standard Test Method for Linearly Reciprocating Ball-on-Flat Sliding Wear. ASTM International: West Conshohocken, PA, USA, 2010. [CrossRef]
30. Jaradeh, M.; Carlberg, T. Differential thermal analysis and differential scanning calorimetry studies of aluminum 3003 alloys with Zn and Cu additions. *Metall. Mater. Trans. A* **2007**, *38*, 2138–2147. [CrossRef]
31. Liu, B.; Yang, X.G.; Huang, H.J.; Guo, Z. The effect of alloying elements on the microstructure of Al–5Fe alloys. *J. Miner. Metals Mater. Soc.* **2012**, *64*, 316–322. [CrossRef]
32. Santos, C.A.; Quaresma, J.M.V.; Garcia, A. Determination of transient interfacial heat transfer coefficients in chill mold castings. *J. Alloys Compd.* **2001**, *319*, 174–186. [CrossRef]
33. Santos, C.A.; Siqueira, C.A.; Garcia, A.; Quaresma, J.M.V.; Spim, J.A. Metal–mold heat transfer coefficients during horizontal and vertical Unsteady-State solidification of Al–Cu and Sn–Pb Alloys. *Inv. Probl. Sci. Eng.* **2004**, *12*, 279–296. [CrossRef]
34. Souza, E.N.; Cheung, N.; Santos, C.A.; Garcia, A. Factors affecting solidification thermal variables along the cross-section of horizontal cylindrical ingots, *Mater. Sci. Eng. A* **2005**, *397*, 239–248. [CrossRef]

35. Alfaia, M.Â.O.; Oliveira, R.; Lima, T.S.; Mariani, F.E.; Casteletti, L.C.; Cheung, N.; Garcia, A. Effects of cooling rate and microstructure scale on wear resistance of unidirectionally solidified Al-3.2wt.%Bi-(1; 3) wt.%Pb alloys. *Mater. Today Comm.* **2020**, *25*, 101659. [CrossRef]
36. He, C.; Yu, W.; Li, Y.; Wang, Z.; Wu, D.; Xu, G. Relationship between cooling rate, microstructure evolution, and performance improvement of an Al–Cu alloy prepared using different methods. *Mater. Res. Express* **2020**, *7*, 116501. [CrossRef]
37. Júnior, P.F.; de Olivé Ferreira, L.; Garção, W.J.L.; de Paula Almeida, R.; Melo, C.M.; Ferreira, A.F. Heat-flow parameters affecting microstructure and mechanical properties of Al–Cu and Al–Ni alloys in directional solidification: An experimental comparative study. *Int. J. Mater. Res.* **2022**, *113*, 181–193. [CrossRef]
38. Stapóri, S.; Górny, M.; Kawalec, M.; Gracz, B. Effect of manganese content on microstructure of Al-Cu alloys. *Arch. Metall. Mater.* **2020**, *65*, 1377–1383. [CrossRef]
39. Wang, J.; Yang, Z.; Ma, Z.; Bai, Y.; Duan, H.; Tao, D.; Shi, G.; Zhang, J.; Wang, Z.; Li, J. Influence of cerium content on the microstructure and thermal expansion properties of suction cast Al-Cu-Fe alloys. *R. Soc. Open Sci.* **2021**, *8*, 210584. [CrossRef] [PubMed]
40. Czerwinski, F. Thermal stability of aluminum alloys. *Materials* **2020**, *13*, 3441. [CrossRef]

metals

MDPI

Article

Novel Cast and Wrought Al-3Zn-3Mg-3Cu-Zr-Y(Er) Alloys with Improved Heat Resistance

Maria V. Glavatskikh, Ruslan Yu. Barkov, Leonid E. Gorlov, Maxim G. Khomutov and Andrey V. Pozdniakov *

Department of Physical Metallurgy of Non-Ferrous Metals, NUST MISIS, 119049 Moscow, Russia; glavatskikh@edu.misis.ru (M.V.G.); barkov@misis.ru (R.Y.B.); m1803226@edu.misis.ru (L.E.G.); khomutov@misis.ru (M.G.K.)

* Correspondence: pozdniakov@misis.ru

Abstract: The main weaknesses of commercial high-strength Al-Zn-Mg-Cu-based alloys are the low casting properties, corrosion and heat resistance. Al-Zn-Mg-Cu-based alloys with Zn/Mg ratio equal to 1 combine good strength, corrosion and heat resistance. Al alloys with atomic ratio Cu/Y(Er) equal to 4 have a narrow solidification range and high solidus temperature. Two basic principles were taken into consideration to develop novel heat-resistant Al-Zn-Mg-Cu-based alloys with improved casting properties: 1—mass ratio of Zn/Mg = 1, and 2—atomic ratio of Cu/Y(Er) = 4. The microstructure, phase transformation and tensile properties of the novel cast and wrought Al–3Zn–3Mg–3Cu–0.2Zr–Y(Er) alloys were investigated. The structure and phase composition were investigated via thermodynamic calculation, optical and scanning electron microscopy and X-ray diffraction methods. A two-step solution treatment with higher temperature in the second step provides a microstructure with better elongation, making possible to increase the hot rolling temperature of the Y or Er-containing alloys. The yield strength (YS) of the alloys decreased insignificantly from 270 to 290 MPa at room temperature to 225 to 260 MPa at 200 °C after casting, solution treatment, water quenching and aging. A better combination of the YS = 291–345 MPa and elongation (El.) (11–14.8%) was achieved in the Al3Zn3Mg3CuY and Al3Zn3Mg3CuEr alloys after solution treatment, rolling, recrystallization annealing, water quenching and aging compared with the Al3Zn3Mg3Cu alloy with YS = 245–340 MPa and El. = 6.8–12.5%.

Keywords: aluminum alloys; microstructure; mechanical properties; precipitates; erbium; yttrium

Citation: Glavatskikh, M.V.; Barkov, R.Y.; Gorlov, L.E.; Khomutov, M.G.; Pozdniakov, A.V. Novel Cast and Wrought Al-3Zn-3Mg-3Cu-Zr-Y(Er) Alloys with Improved Heat Resistance. *Metals* **2023**, *13*, 909. https://doi.org/10.3390/met13050909

Academic Editors: Frank Czerwinski and Noé Cheung

Received: 11 April 2023
Revised: 4 May 2023
Accepted: 5 May 2023
Published: 8 May 2023

1. Introduction

Al-Zn-Mg-Cu-based compositions are part of the high-strength wrought Al alloys. The compositions of this wrought Al alloy group were developed to achieve the highest strength and impact strength. The basic weaknesses of the Al-Zn-Mg-Cu-based alloys are the low casting properties, corrosion and heat resistance [1–4]. The cast Al-Zn-Mg-Cu alloys of the 7xxx series have the same flaws [5,6]. The main way to improve the casting and high-temperature properties is alloying with eutectic forming elements [7–14]. On the other hand, Al-Zn-Mg-Cu-based alloys with a Zn/Mg ratio near 1 combine good strength, corrosion and heat resistance [2–5,13,14]. Trace amounts of rare earth scandium or zirconium improve the room- and high-temperature strength and casting properties due to $L1_2$-precipitate formation during annealing and grain refining during solidification [14–18]. However, Sc is a very high-cost additive in Al alloys. In this case, the search for an adequate substitution of scandium in Al alloys is an important task. Titanium, yttrium and erbium in combination with zirconium provide the same effect in strengthening Al due to nucleation of the $L1_2$-precipitates [19–29]. Titanium [19–22], erbium [23–28] and yttrium [29] substitute the Zr atoms in the $Al_3(Zr,Me)$ (Me=Ti,Y, or Er) precipitates. Titanium addition does not improve the coarsening resistance of the ternary alloy at temperatures of 375–425 °C [20]. However, Ti reduces the incubation time for nucleation and also increases the peak hardness achieved during aging at temperatures of 450–600 °C [21]. The $Al_3(Zr,Er)$ precipitates demonstrate a high coarsening resistance [24,25] and improve the strength and recrystallization behavior [26–28]. The effect of Y with Zr in the structure and properties has not been investigated much [29]. Last year's investigations demonstrate the complex effect of Y and Er in Al alloys [30–39]. Yttrium or erbium with copper in Al provide the eutectic $Al_8Cu_4Y(Er)$ phase formation [30–35]. The alloys with an atomic ratio Cu/Y(Er) equal to 4 have a narrow solidification interval with a high melting point [30–37]. In this case, the first role of the Y and Er is the eutectic forming elements. The $Al_3(Zr,Y)$ or $Al_3(Zr,Er)$ precipitates nucleate in a high-temperature homogenization treatment for Al-Cu-Y- and Al-Cu-Er-based alloys and improve the room- and high-temperature strength and recrystallization behavior [32–39]. This is the second role of Y and Er precipitates forming elements. Based on these two principles, novel heat-resistant cast and wrought Al alloys were developed [38,39]. Trace additives of Er or Y in Al–Zn–Mg–Cubased alloys affect the aging strengthening and corrosion resistance [40–45]. As a result, two basic principles should be taken into consideration to develop novel heat-resistant Al-Zn-Mg-Cu-based alloys with improved casting properties: 1—mass ratio of Zn/Mg = 1, and 2—atomic ratio of Cu/Y(Er) = 4. Based on these theses, the effect of yttrium and erbium in the phase composition, casting properties, room- and high-temperature mechanical properties of the Al–4.5Zn–4.5Mg–2.5Cu–0.2Zr alloy was investigated [46]. Additional alloying with Mn decreases the aging effect due to the formation of complex Mn-containing phases of a solidification origin [47]. A very high content in the alloying elements provides low deformation behavior. The main conclusion from a previous study is that the content of the Zn and Mg should be decreased and the Mn removed to improve the plasticity. The aim of the present investigation is to determine the effect of Y or Er alloying on the phase content, tensile properties and heat resistance of the Al–3Zn–3Mg–3Cu–0.2Zr alloy with Fe and Si impurities to develop novel cast and wrought alloys. The novel alloys should combine the high casting properties and strength at room and elevated temperatures.

2. Materials and Methods

2.1. Alloy Preparation

The Al3Zn3Mg3Cu, Al3Zn3Mg3CuY and Al3Zn3Mg3CuEr alloys (Table 1) were melted from pure Al (99.7%), Zn (99.7%), Mg (99.5%), Cu (99.5%) and Al-10Y, Al-10Er, Al-5Zr and Al-5Ti-1B master alloys in the resistance furnace. The melting and pouring temperatures were 800–810 °C. Pure copper and master alloys, rather than pure Zn and Mg, were introduced into the Al melt separately at 800–810 °C. Casting was carried out into a copper water-cooling mold (CM) and steel mold (SM). The ingot dimensions were 120 mm × 40 mm × 20 mm for a CM. The ingots from a copper water-cooling mold were used for rolling. The cylinders from SM with diameter of 24 mm and length of 290 mm were used for tensile test sample of cast alloys. The hot tearing of the alloys was determined via hot cracking index (HCI) using "pencil probe" [5,14].

Table 1. The chemical composition of the investigated alloys in wt.%.

Alloy	Al	Zn	Mg	Cu	Zr	Ti	Fe and Si	Y or Er
Al3Zn3Mg3Cu	bal.	2.8	2.7	3.0	0.2	0.1	0.3	-
Al3Zn3Mg3CuY	bal.	3.1	2.9	2.9	0.2	0.1	0.3	0.6
Al3Zn3Mg3CuEr	bal.	2.9	2.8	3.0	0.2	0.1	0.3	1.4

2.2. Microstructure and Phase Composition Analyses

The grain structure of the ingots and rolled sheets was analyzed with optical microscope (OM) Zeiss (Carl Zeiss, Oberkochen, Germany) and electron backscattered diffraction (EBSD) detector in scanning electron microscope (SEM) TESCAN VEGA 3LMH (Tescan, Brno, Czech Republic). OM was applied to grain structure analysis under polarized light. Microstructure and phase composition of the alloy were investigated in detail using SEM and transmission electron microscope (TEM) JEOL—2100 EX (Jeol Ltd., Tokyo, Japan). SEM images were scanned in back-scattered electron (BSE) detector at ×3000 magnification at voltage of 20 kV. Phase identification in SEM was obtained using the electron diffraction X-ray (EDX) detector X-max 80. Additional phase analysis was performed using X-ray diffraction (XRD) with Cu-Kα radiation on a Bruker D8 Advance diffractometer (Bruker, Billerica, MA, USA). The XRD scanning was applied with $2\theta = 0.05°$ step and dwell time of 5 s in a range of $2\theta = 20–90°$.

2.3. Preparation of the Specimens for Microstructure Investigation

Specimens for OM and SEM investigation were mechanically ground and polished. Struers Labopol equipment was used for mechanical grinding and polishing. Grinding disk sheets with numbers of #800, #1200, #2000 and #4000 and OP-S suspension were used for specimen preparation. The grain structure for OM was electrochemically etched (15–25 V, 0–5 °C) using Barker's reagent (46 mL of HBF_4, 7 g of HBO_3 and 970 mL of H_2O). The average grain size value was measured using the random secant method in three images. The specimens for the TEM were prepared using the A2 electrolyte on Struers Tenupol-5 equipment. The TEM specimens were mechanically thinned to a 0.25 mm thickness before electrochemical procedure. The XRD investigation was conducted using alloy powder prepared through mechanical grinding.

2.4. Heat Treatment and Rolling Processing

The solidus and phase transformation temperatures of the alloys were determined via a Labsys Setaram differential scanning calorimeter (SETARAM Instrumentation, Caluire, France) (DSC). The solution treatment at 480–520 °C for 3 and 6 h was carried out in the resistant furnace with accuracy about 1 °C. The aging after solution treatment and water quenching was carried out at 120–250 °C in the resistant furnace with accuracy about 3 °C. The ingot after solution treatment at 480–520 °C and water quenching was hot rolled at 440 °C (Al3Zn3Mg3Cu) and 500 °C (Al3Zn3Mg3CuY and Al3Zn3Mg3CuEr) from thickness of 20 mm. The Al3Zn3Mg3Cu alloy ingot was rolled at 440 °C for 2 mm and at room temperature to 1 mm thickness sheet. The Al3Zn3Mg3CuY and Al3Zn3Mg3CuEr alloy ingots were rolled at 500 °C for 5 mm and at room temperature to 1 mm thickness sheets. Samples from rolled sheets were annealed at 150–450 °C for different times to investigate the grain structure, hardness and tensile properties evaluations. The rolled sheets were recrystallized at 480–520 °C for 15 min, water quenched and aged at 120–250 °C to determine the tensile properties.

2.5. Thermodynamic Calculations

Thermodynamic computations of the multicomponent phase equilibrium and non-equilibrium solidification (Sheil model) were carried out in the Thermo-Calc software in the TTAl5 database.

2.6. Mechanical Properties Measurements and Calculations

The Vickers method under 5 kg load was used for hardness measurements. The tensile samples were stretched on a Zwick/Roell Z250 Allround (Zwick/Roell, Kennesaw, GA, USA) test machine with an extensometer. The tensile samples with gage diameter of 6 mm and gage length of 42 mm were cut out from heat-treated ingots obtained in the steel mold. The tensile samples with gage width of 6 mm and gage length of 22 mm were cut out from 1 mm thickness sheets. Minimum 3 samples were tested per state, and average value was calculated. The rupture stress was determined on an Instron Creep M3 (Instron, Grove City, PA, USA) test machine.

3. Results and Discussion

3.1. Thermodynamic Calculation of the Phase Composition

Figure 1 illustrates the calculated phase equilibria in the Al–3Zn–3Mg–3Cu–0.1Ti–0.15Si–0.15Fe–(0–0.4)Zr system and non-equilibrium cooling curve of the Al3Zn3Mg3Cu alloy. The increase in Zr content in the Al–3Zn–3Mg–3Cu–0.1Ti–0.15Si–0.15Fe alloy provides an increase in the liquidus temperature from about 680 °C in the alloy without Zr to 800 °C in the alloy with 0.27%Zr (Figure 1a). The melting of the Zn and Mg content alloy should be proceeded at temperatures lower than 800 °C to avoid the losses of these elements. In this case, the maximum content of Zr should be 0.2%. The solidification of the Al3Zn3Mg3Cu alloy starts from Al_3(Zr,Ti) phase nucleation due to the calculated cooling curve (Figure 1b). The Sheil model includes the assumption that the distributive and equalizing diffusion in a liquid was completely processed. However the formation of the Al_3(Zr,Ti) usually suppresses, and the clusters of the Al_3(Zr,Ti) have a role as a substrate for primary aluminum solid solution (Al). Al_3Fe, Mg_2Si, Al_7Cu_2Fe, S(Al_2CuMg), T (Al,Zn,Mg,Cu) and $MgZn_2$ phases solidify during non-equilibrium solidification (Figure 1b).

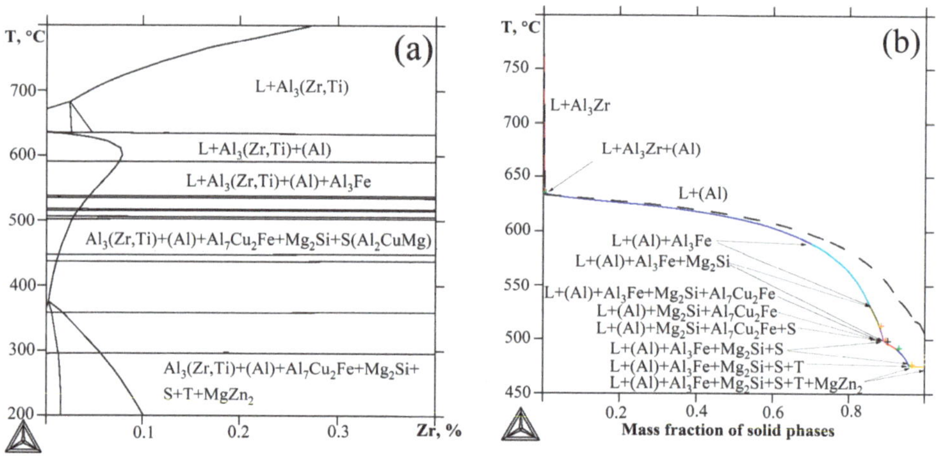

Figure 1. Calculated (**a**) polythermal section Al–3Zn–3Mg–3Cu–0.1Ti–0.15Si–0.15Fe– (0-0.4)Zr and (**b**) non-equilibrium cooling curve of the Al3Zn3Mg3Cu alloy (dash line—equilibrium cooling curve).

The non-equilibrium cooling curve should be used to calculate the effective solidification range (ESR) and the HCI of the alloys [14]. The ESR of the Al–Zn–Mg–Cu-based alloy should be calculated as a difference between temperature in the 65% of solid phase formation and non-equilibrium solidus [14]. The linear empirical equation was used to calculate the HCI using the calculated value of the ESR (Table 2). The HCI = 14 mm is suitable for the Al-Si-Cu-Mg-based alloys with good casting properties [48]. Additional alloying of the Al3Zn3Mg3Cu alloy with Y or Er should improve the casting properties due to the eutectic fraction increasing. The experimental HCI of the Al3Zn3Mg3Cu is 14–16 mm and 14 mm for the Al3Zn3Mg3CuY and Al3Zn3Mg3CuEr alloys.

Table 2. Calculated critical temperatures, ESR and HCI of the Al3Zn3Mg3Cu alloy.

T_L, °C	T_S, °C	$T_{65\%}$, °C	T_{NS}, °C	ESR, °C	HCI_c, mm
761	505	595	475	120	14

3.2. As-Cast Microstructure and Phase Composition

The investigated alloys were poured into SM and CM. The cooling rate in the steel and copper molds was different, which may affect the microstructure. Figures 2 and 3 demonstrate the grain and dendritic cell structure of the as-cast alloys. Differences in the cooling rate do not affect the grain structure of the alloys (compare a–c and d–e images in Figure 2). The average grain size of the Al3Zn3Mg3Cu alloy is 100 ± 15 μm (Figure 2a,d). Yttrium and erbium may substitute the Zr or Ti atom in the Al_3(Zr,Ti) phase clusters and increase the volume fraction of the substrates for primary (Al). As a result, a slightly fine grain structure with a size of 80 ± 10 μm was obtained in the ingots of the Al3Zn3Mg3CuY alloy (Figure 2b,e). Erbium provides a significantly higher refining effect, as indicated in the literature [38–43]. An average grain size of 45 ± 10 μm was obtained in the Al3Zn3Mg3CuEr ingots (Figure 2c,f).

Figure 2. (**a**–**c**) As-cast grain structure of the (**a**,**d**) Al3Zn3Mg3Cu, (**b**,**e**) Al3Zn3Mg3CuY and (**c**,**f**) Al3Zn3Mg3CuEr alloys poured into (**a**–**c**) CM and (**d**,**e**) SM.

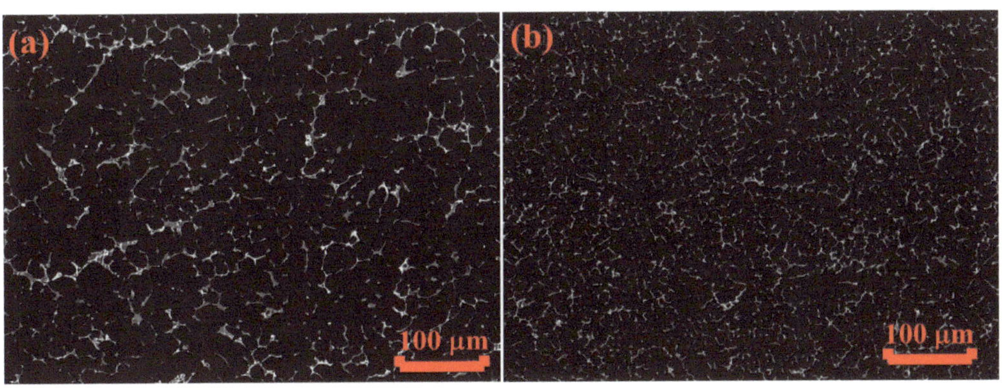

Figure 3. As-cast microstructure of the Al3Zn3Mg3CuY alloy poured into (**a**) SM and (**b**) CM.

The cooling rate in the copper mold is higher than in the steel one, which provides a finer dendritic cell structure in the ingots poured into the coper mold (Figure 3). The size of the secondary dendritic arm spacing (SDAS) in the SM and CM ingots is 22 ± 4 μm and 12 ± 3 μm, respectively (Figure 3). This difference in the SDAS size is not significant for mechanical properties, as demonstrated by the authors in [49].

The as-cast microstructure and XRD patterns of the Al3Zn3Mg3Cu, Al3Zn3Mg3CuY and Al3Zn3Mg3CuEr alloys are presented in Figure 4. Three phase particles are presented in the microstructure of the Al3Zn3Mg3Cu alloy: Zn-, Cu- and Mg-rich particles, Mg- and Si-rich particles and Cu- and Fe-rich particles correspond to T, Mg_2Si and Al_7Cu_2Fe

phases (Figure 4a). These results are in good agreement with calculation (Figure 1) and XRD (Figure 4d) data. The main peaks at angles of about 27, 32, 35 and 41° are marked as S phase (XRD card no. 03-065-2504). The peaks at 37 and 40.5° correspond to the T phase (XRD card no. 00-043-1442 (Q)). The main peaks from (Al) cover some peaks from S and T phases. A very low fraction of S phase particles was also identified in the microstructure. Point analyses demonstrate the presence of 2–4%Zn in these particles. Zn atoms may substitute Al in the phase. Al_8Cu_4Y and A_8Cu_4Er phases formed in the Al3Zn3Mg3CuY and Al3Zn3Mg3CuEr alloys, respectively, in addition to the T and Mg_2Si phases (Figure 4b–d). The main two peaks at 41 and 41.5° are indicated for Al_8Cu_4Y and A_8Cu_4Er phases (Figure 4d, XRD card no. 01-070-9419 and 00-033-0006 (I), respectively). Fe impurity dissolved in the Al_8Cu_4Y and A_8Cu_4Er phases (see the distribution of the alloying elements between phases) at 1–2% content. The same results were obtained in an earlier investigation [38]. Separate particles of the $(Al,Cu)_{11}Y_3$ (XRD card no. 01-070-9423) and Al_3Er (XRD card no. 03-065-9825) with dissolved Zn were also identified in the Al3Zn3Mg3CuY and Al3Zn3Mg3CuEr alloys, respectively. Zirconium completely dissolved in the (Al) in all investigated alloys, without coarse particle formation. Yttrium and erbium content in the (Al) was 0.2 and 0.3%, respectively.

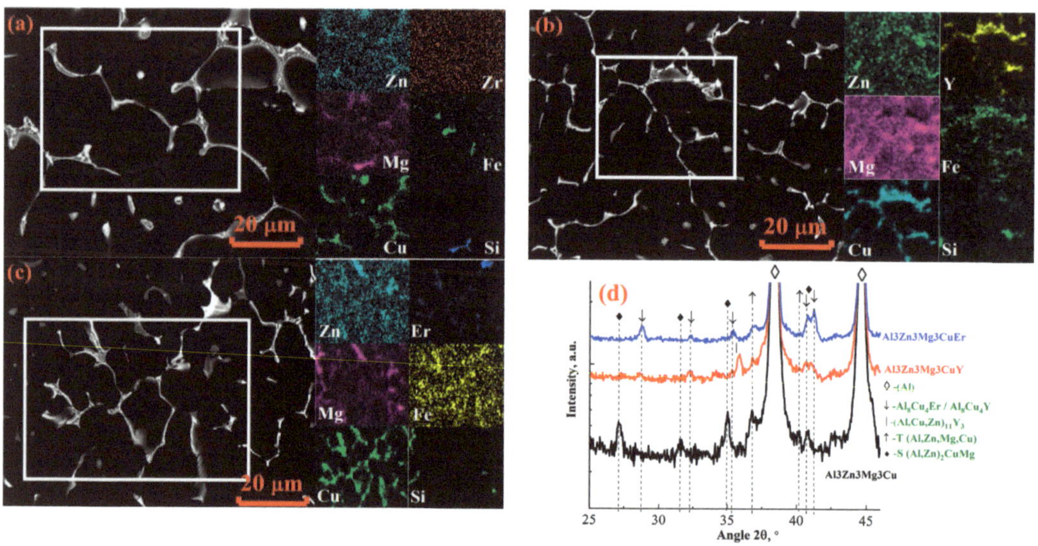

Figure 4. As-cast microstructure of the (**a**) Al3Zn3Mg3Cu, (**b**) Al3Zn3Mg3CuY and (**c**) Al3Zn3Mg3CuEr alloys and (**d**) XRD patterns (distribution of alloying elements into white rectangle in (**a**–**c**)).

3.3. *Evaluation of the Microstructure under Solution Treatment*

The DSC curves of the investigated alloys are presented in Figure 5. Three melting peaks of the T, Al_7Cu_2Fe and Mg_2Si at temperatures of 493, 525 and 540 °C were identified on the heating DSC curves of the investigated alloys. The Y- and Er-containing phases in the Al3Zn3Mg3CuY and Al3Zn3Mg3CuEr alloys melted in a temperature range of 540–580 °C. The small peak at 477 °C was found in the Al3Zn3Mg3CuEr alloy, which corresponded to the $MgZn_2$ phase in accordance with the calculation (Figure 1b). The investigated alloys were solution-treated at 480 °C for 3 h. The non-equilibrium fraction of T, S and $MgZn_2$ phases dissolved during solution treatment. These phases dissolved completely in the Al3Zn3Mg3CuY and Al3Zn3Mg3CuEr alloys. The solidus temperature of the Al3Zn3Mg3CuY and Al3Zn3Mg3CuEr alloys increased to 533 °C (Figure 5b).

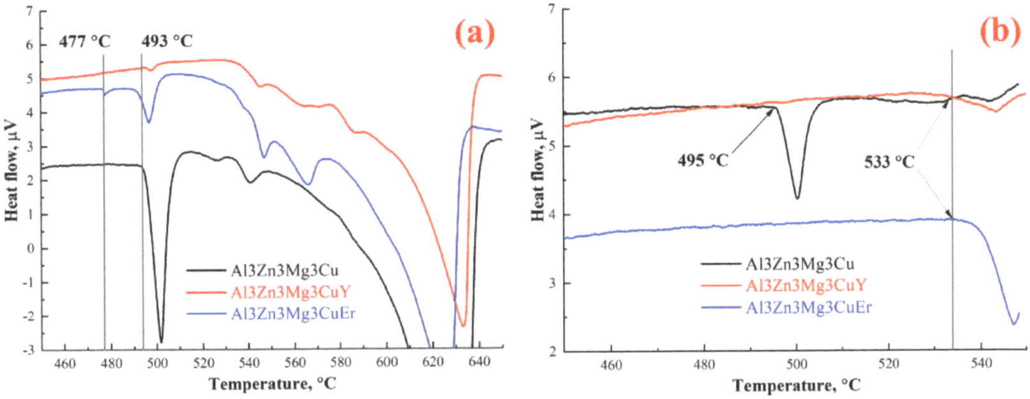

Figure 5. DSC curves of the (**a**) as-cast and (**b**) solution treated at 480 °C for 3 h alloys.

Figure 6 demonstrates the microstructure changes. The T phase transforms to S phase and spheroidizes in the Al3Zn3Mg3Cu alloy (Figure 6a). These results are in good agreement with the calculations. The S, Mg_2Si, Al_7Cu_2Fe and $Al_3(Zr,Ti)$ should be in equilibrium with (Al) in accordance with the calculation (Table 3). The $Al_3(Zr,Ti)$ phase must nucleate from the supersaturated (Al) during solution treatment. The Y- and Er-containing phases in the Al3Zn3Mg3CuY and Al3Zn3Mg3CuEr alloys (Figure 6b,c) do not significantly change the morphology compared to the as-cast state (Figure 4b,c). This fact confirms a high thermal stability of Y- and Er-containing phases.

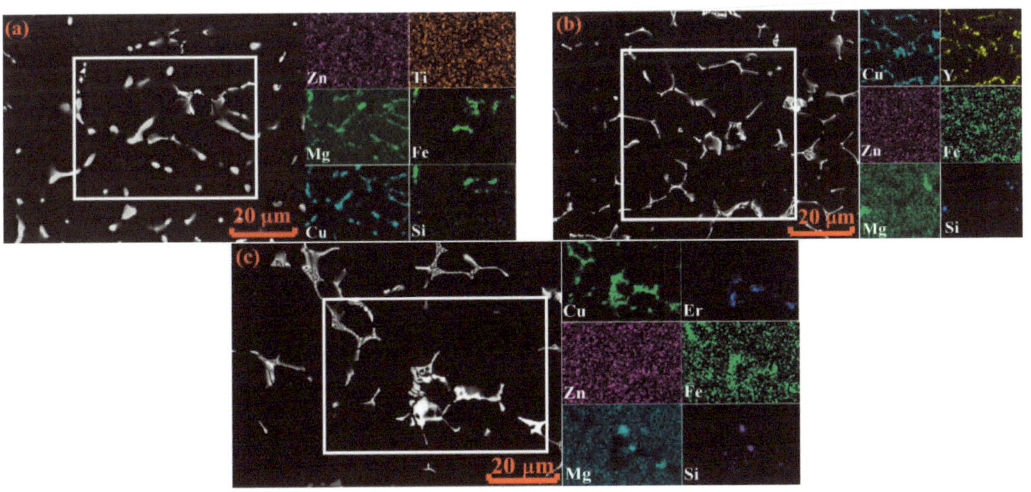

Figure 6. Microstructure of the of the (**a**) Al3Zn3Mg3Cu, (**b**) Al3Zn3Mg3CuY and (**c**) Al3Zn3Mg3CuEr alloys after solution treatment at 480 °C for 3 h.

Table 3. Calculated mass fraction of phases and composition of (Al) at 480 °C.

(Al)	S	Al_7Cu_2Fe	Mg_2Si	$Al_3(ZrTi)$
bal. (3.1Zn-2.6Mg-2.1Cu)	1.5	1	0.37	0.4

The second step of the solution treatment at 520 °C for 6 h was applied to the Al3Zn3Mg3CuY and Al3Zn3Mg3CuEr alloys. The Y- and Er-containing phase particles fragmentized and sheroidized (Figure 7). The microstructure with more compact particles of the solidification origin should provide better plasticity. The experimentally measured composition of the (Al) is presented in Table 4. The Mg content in the (Al) of the Al3Zn3Mg3Cu alloy is lower than in the Y- and Er-containing alloys due to the presence of the S phase. However, the Cu content in the Al3Zn3Mg3CuY and Al3Zn3Mg3CuEr alloys is lower than in the Al3Zn3Mg3Cu alloy due to the presence of Y- and Er-containing phases. In addition, the Y and Er solute atoms in the (Al) during solidification should provide a higher fraction of the precipitates after solution treatment in the Al3Zn3Mg3CuY and Al3Zn3Mg3CuEr alloys.

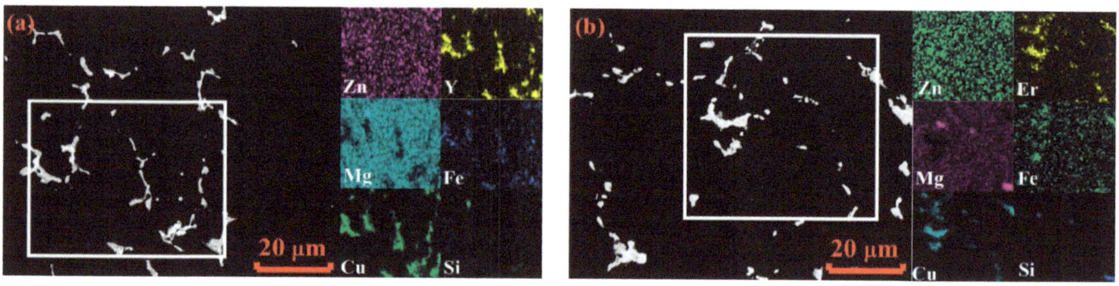

Figure 7. Microstructure of the (**a**) Al3Zn3Mg3CuY and (**b**) Al3Zn3Mg3CuEr alloys after solution treatment at 480 °C for 3 h + 520 °C for 6 h.

Table 4. Composition of the Al solid solution in mass.% (EDX SEM).

Alloy	As-Cast State			480 °C, 3 h			480 °C, 3 h + 520 °C, 6 h		
	Zn	Mg	Cu	Zn	Mg	Cu	Zn	Mg	Cu
Al3Zn3Mg3Cu	2.7	2.0	0.6	3.2	2.5	1.6	-	-	-
Al3Zn3Mg3CuY	2.2	2.2	1.0	3.0	3.0	1.3	3.1	3.1	1.3
Al3Zn3Mg3CuEr	2.2	1.9	1.0	3.0	3.0	1.2	3.0	3.0	1.2

3.4. Hardening under Aging Treatment and Mechanical Properties

The metastable $MgZn_2$(1.4–3 mass.%), S (0.8–1.3 mass.%) and T (5.7–6.2 mass.%) phases should provide strengthening during aging at 120–210 °C for the Al3Zn3Mg3Cu alloy in accordance with the measured composition of (Al) (Table 4) and calculations. The same calculation for the Al3Zn3Mg3CuY and Al3Zn3Mg3CuEr alloys shows that the T (8.7–9.9 mass.%) phase should be nucleated during aging. Figure 8 illustrates the HV curves at different aging temperatures of the alloys quenched at 480 °C and 520 °C (for the Al3Zn3Mg3CuY and Al3Zn3Mg3CuEr alloys). The highest strengthening was achieved in the Al3Zn3Mg3Cu alloy due to the formation of $MgZn_2$ and S phases. Slightly higher strengthening was achieved after a two-step solution treatment. The Er- or Y-containing precipitates that formed during the solution treatment may affect the aging process [40–45]. The size and distribution of the precipitates after a one- and two-step solution treatment are different. A detailed investigation of the precipitation strengthening will be presented in a future paper.

The hardness of 120–140 HV was achieved after 40–50 h of annealing at 120–150 °C in all investigated alloys. Slightly lower strengthening was achieved after aging at 180–210 °C. The aging temperature of the heat-resistant alloys should cover the operating temperatures. In this case, aging at 210 °C for 3 h was applied for investigated alloys before tensile tests at room and elevated temperatures.

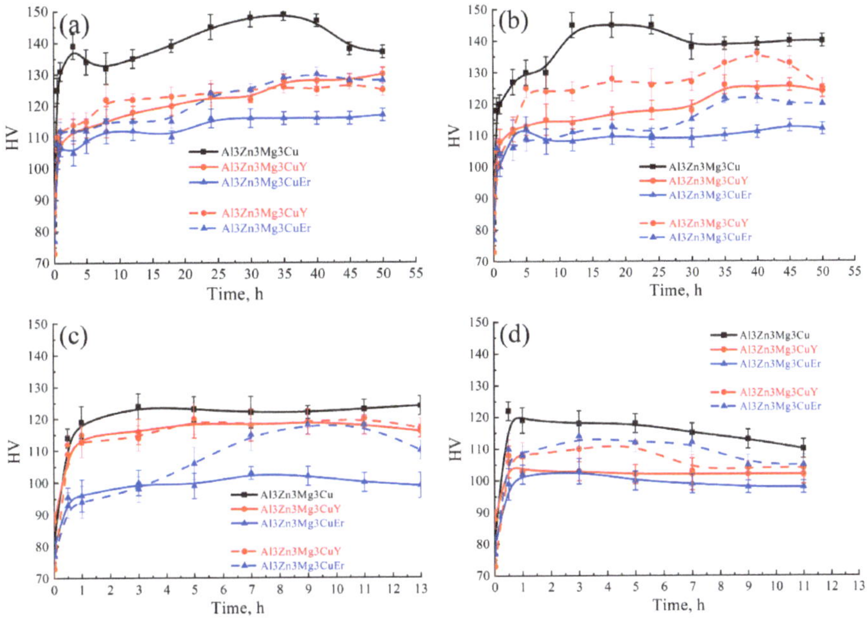

Figure 8. HV curves at different aging temperatures (**a**) 120 °C, (**b**) 150 °C, (**c**) 180 °C and (**d**) 210 °C (solid lines—quenched at 480 °C, dash lines—quenched at 520 °C).

The tensile test results after tension at room and elevated temperatures are presented in Table 5. The Al3Zn3Mg3Cu alloy was solution-treated at 480 °C for 3 h and water quenched. A two-step solution treatment with water quenching was applied for the Al3Zn3Mg3CuY and Al3Zn3Mg3CuEr alloys: 480 °C for 3 h and 520 °C for 6 h. The investigated alloys demonstrate a good YS of 270–290 MPa at room temperature. The two-step solution treatment provides a higher elongation of 2–3% for the Al3Zn3Mg3CuY and Al3Zn3Mg3CuEr alloys. The YS decreased insignificantly from 270–290 MPa to 225–260 MPa after tension at 200 °C. The commercial cast 771.0 (7Zn–0.9Mg–0.13Cr) alloy had a substantially lower YS = 185 MPa and the same elongation of 3% after aging at 205 °C for 6 h [6].

Table 5. Tensile properties of the alloys aged at 210 °C for 3 h at different temperatures.

Alloy	YS, MPa	UTS, MPa	El., %
	20 °C		
Al3Zn3Mg3Cu	290 ± 10	340 ± 20	1.4 ± 0.4
Al3Zn3Mg3CuY	280 ± 8	340 ± 20	3 ± 1
Al3Zn3Mg3CuEr	270 ± 4	330 ± 10	2.0 ± 0.5
	200 °C		
Al3Zn3Mg3Cu	260 ± 8	270 ± 6	3.0 ± 0.5
Al3Zn3Mg3CuY	230 ± 6	250 ± 5	5.2 ± 0.2
Al3Zn3Mg3CuEr	225 ± 4	237 ± 6	3.5 ± 0.6

3.5. Microstructure and Hardness Evaluation after Rolling and Annealing

The high content of the alloying elements and high fraction of the solidification origin phases may decrease the manufacturability during rolling. An increase in the rolling temperature provides rolled sheets without cracks and defects. The higher solidus temperature of the Al3Zn3Mg3CuY and Al3Zn3Mg3CuEr alloys after a two-step solution treatment may possibly increase the rolling temperature. Al3Zn3Mg3CuY and Al3Zn3Mg3CuEr alloy sheets were obtained with a higher fraction of cold deformation than the Al3Zn3Mg3Cu alloy sheet. As a result, higher hardness was achieved in the Y- and Er-containing alloys (Figure 9). A higher deformation strengthening provides a faster softening during annealing of the rolled sheets at temperatures lower than 250 °C. Deformed grain structure is maintained up to 300 °C. The recrystallized grains were found after 1 h of annealing at 350 °C (Figure 10a–c). A significantly finer grain structure was achieved in the Al3Zn3Mg3CuY and Al3Zn3Mg3CuEr alloys due to a higher content of rare earth elements, the precipitates of which suppress recrystallization.

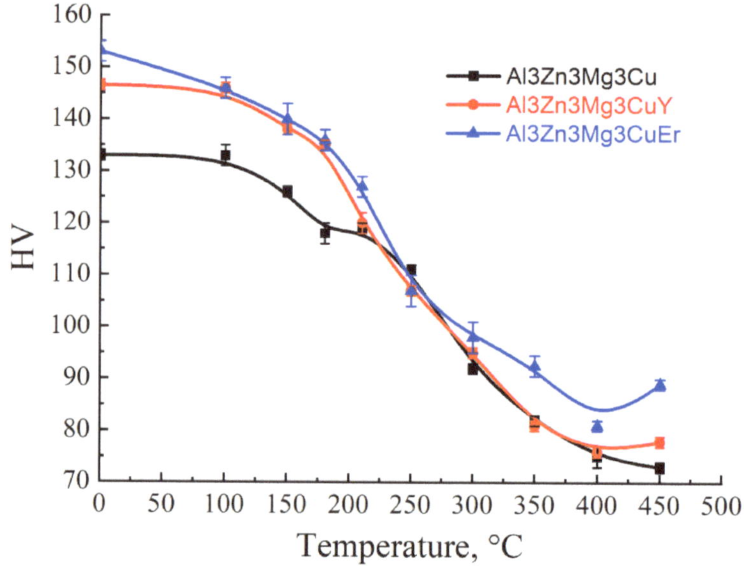

Figure 9. HV–temperature curves after 1 h of annealing of the alloy sheets.

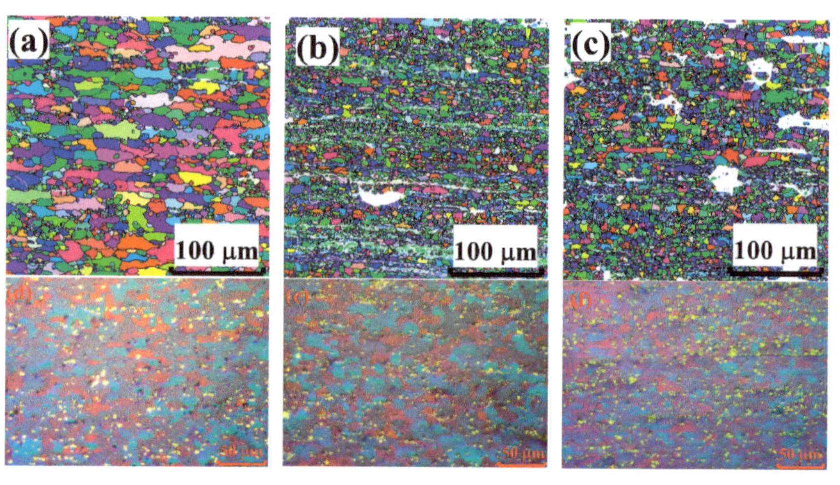

Figure 10. Grain structure of the (**a**,**d**) Al3Zn3Mg3Cu, (**b**,**e**) Al3Zn3Mg3CuY and (**c**,**f**) Al3Zn3Mg3CuEr alloys after 1 h of annealing at (**a**–**c**) 350 °C (EBSD) and after 15 min of annealing at (**d**) 480 °C (OM) and (**e**,**f**) 520 °C (OM).

The Al3Zn3Mg3Cu alloy was recrystallized at 480 °C for 15 min and water quenched. The Al3Zn3Mg3CuY and Al3Zn3Mg3CuEr alloys were recrystallized at 520 °C for 15 min and water quenched. The uniform recrystallized grain structure with a size of 8–10 μm was formed in the investigated alloys (Figure 10d–f).

The aging strengthening of the rolled sheet (Figure 11) is similar to the aging of the cast, solution-treated and water-quenched alloys (Figure 7). The tensile test results of the aged alloys sheets at different temperatures are presented in Table 6. A better combination of the YS = 291–345 MPa and elongation (11–14.8%) was achieved in the Al3Zn3Mg3CuY and Al3Zn3Mg3CuEr alloys in comparison to the Al3Zn3Mg3Cu alloy with YS = 245–340 MPa and El. = 6.8–12.5%.

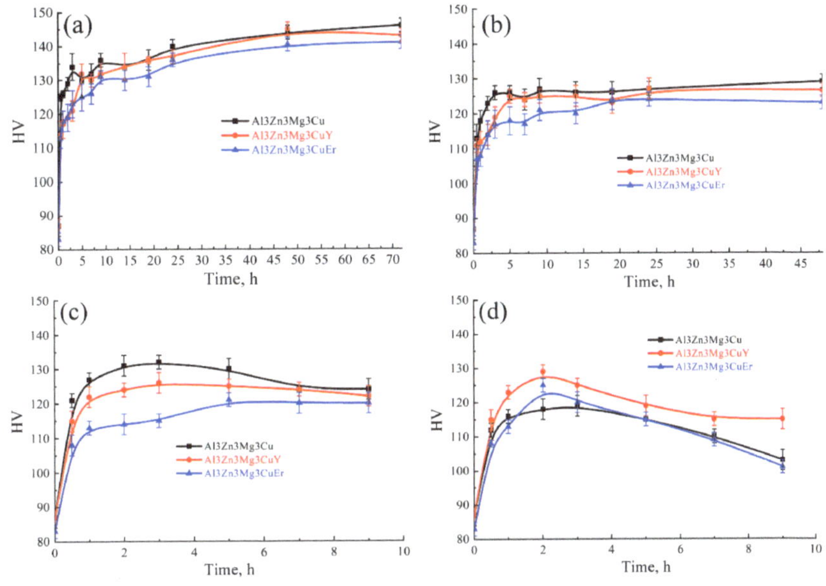

Figure 11. HV curves at different aging temperatures (**a**) 120 °C, (**b**) 150 °C, (**c**) 180 °C and (**d**) 210 °C (Al3Zn3Mg3Cu alloy quenched at 480 °C after 15 min, Al3Zn3Mg3CuY and Al3Zn3Mg3CuEr alloys quenched at 520 °C after 15 min).

Table 6. Tensile properties of the rolled, quenched and aged alloy sheets.

State	YS, MPa	UTS, MPa	El., %
	Al3Zn3Mg3Cu		
As rolled	407 ± 2	440 ± 1	4.0 ± 0.4
480 °C for 15 min/120 °C for 48 h	340 ± 10	440 ± 10	8.0 ± 1.0
480 °C for 15 min/150 °C for 32 h	290 ± 7	435 ± 15	9 ± 3
480 °C for 15 min/180 °C for 5 h	285 ± 5	425 ± 5	12.5 ± 0.5
480 °C for 15 min/210 °C for 2 h	245 ± 10	370 ± 5	6.8 ± 0.3
	Al3Zn3Mg3CuY		
As rolled	455 ± 8	477 ± 3	2 ± 1.5
520 °C for 15 min/120 °C for 48 h	345 ± 2	471 ± 2	12.8 ± 1.2
520 °C for 15 min/150 °C for 32 h	315 ± 1	460 ± 1	13.6 ± 0.2
520 °C for 15 min/180 °C for 5 h	291 ± 2	449 ± 1	14.8 ± 1.5
520 °C for 15 min/210 °C for 2 h	316 ± 1	430 ± 1	10.3 ± 1.5
	Al3Zn3Mg3CuEr		
As rolled	450 ± 2	474 ± 3	3.7 ± 0.5
520 °C for 15 min/120 °C for 48 h	334 ± 1	446 ± 1	14.8 ± 0.1
520 °C for 15 min/150 °C for 32 h	300 ± 2	434 ± 3	12.5 ± 2
520 °C for 15 min/180 °C for 5 h	295 ± 13	445 ± 5	14.8 ± 0.5
520 °C for 15 min/210 °C for 2 h	303 ± 1	413 ± 1	11 ± 2

4. Conclusions

The microstructure, phase transformation and mechanical properties of the novel cast and wrought Al–3Zn–3Mg–3Cu–0.2Zr–Y(Er) alloys were investigated. Thermodynamic calculation was calculated with optical and scanning electron microscopy, and the X-ray diffraction methods were used for phase identification.

The Y and Er addition in the Al3Zn3Mg3Cu alloy increased the solidus temperature from 493 to 533 °C. A two-step solution treatment (480 °C, 3 h + 520 °C, 6 h) with higher temperature in the second step provided the microstructure with better elongation and made it possible to increase the temperature of the hot rolling. The YS of the cast, solution-treated, water-quenched and aged alloys decreased insignificantly from 270–290 MPa at room temperature to 225–260 MPa after tension at 200 °C. A significantly finer-grain structure was achieved in the Al3Zn3Mg3CuY and Al3Zn3Mg3CuEr alloys after rolling and annealing due to a higher content of the rare earth elements precipitating to suppress recrystallization. A better combination of the YS=291-345 MPa and elongation (11–14.8%) was achieved in the Al3Zn3Mg3CuY and Al3Zn3Mg3CuEr alloys after solution treatment, rolling (hot and cold), recrystallization annealing (520 °C, 15 min), water quenching and aging (120–210 °C) in comparison with the Al3Zn3Mg3Cu alloy with YS = 245–340 MPa and El. = 6.8–12.5%.

Author Contributions: Conceptualization, A.V.P.; methodology, M.V.G. and L.E.G.; formal analysis, M.G.K. and A.V.P.; investigation, R.Y.B. and A.V.P.; data curation, M.G.K., R.Y.B.; writing—original draft preparation, A.V.P.; writing—review and editing, A.V.P.; visualization, M.V.G. and A.V.P.; supervision, A.V.P.; funding acquisition, M.G.K. All authors have read and agreed to the published version of the manuscript.

Funding: The work was supported by the Russian Science Foundation (Project No. 22-79-10142), https://rscf.ru/project/22-79-10142/, accessed on 4 May 2023.

Conflicts of Interest: The authors declare no conflict of interest.

References

1. Gerchikova, N.S.; Fridlyander, I.N.; Zaitseva, N.I.; Kirkina, N.N. Change in the structure and properties of Al-Zn-Mg alloys. *Met. Sci. Heat Treat.* **1972**, *14*, 233–236. [CrossRef]
2. Zolotorevsky, V.S. Microstructure and Mechanical Properties of Cast Aluminum Alloys. Ph.D. Thesis, MISIS, Moscow, Russia, 1979. (In Russian)
3. Zou, Y.; Wu, X.; Tang, S.; Zhu, Q.; Song, H.; Guo, M.; Cao, L. Investigation on microstructure and mechanical properties of Al-Zn-Mg-Cu alloys with various Zn/Mg ratios. *J. Mater. Sci. Technol.* **2021**, *85*, 106–117. [CrossRef]
4. Novikov, I.I. *Hot-Shortness of Non-Ferrous Metals and Alloys*; Nauka: Moscow, Russia, 1966. (In Russian)
5. Pan, Y.; Zhang, D.; Liu, H.; Zhuang, L.; Zhang, J. Precipitation hardening and intergranular corrosion behavior of novel Al–Mg–Zn(-Cu) alloys. *J. Alloys Compd.* **2021**, *853*, 157199. [CrossRef]
6. ASM International Handbook Committee. *ASM Handbook Volume 2: Properties and Selection—Nonferrous Alloys and Special-Purpose Materials*; ASM: Almere, The Netherlands, 2001; ISBN 0871700077.
7. Belov, N.A.; Alabin, A.N.; Eskin, D.G.; Istomin-Kastrovskii, V.V. Optimization of hardening of Al–Zr–Sc cast alloys. *J. Mater. Sci.* **2006**, *41*, 5890–5899. [CrossRef]
8. Shurkin, P.K.; Akopyan, T.K.; Galkin, S.P.; Aleshchenko, A.S. Effect of Radial Shear Rolling on the Structure and Mechanical Properties of a New-Generation High-Strength Aluminum Alloy Based on the Al-Zn-Mg-Ni-Fe System. *Met. Sci. Heat Treat.* **2019**, *60*, 764–769. [CrossRef]
9. Akopyan, T.K.; Belov, N.A.; Alabin, A.N.; Zlobin, G.S. Calculation-experimental study of the aging of casting high-strength Al-Zn-Mg-(Cu)-Ni-Fe aluminum alloys. *Russ. Metall.* **2014**, *2014*, 60–65. [CrossRef]
10. David, M.D.; Foley, R.D.; Griffin, J.A.; Monroe, C.A. Microstructural Characterization and Thermodynamic Simulation of Cast Al-Zn-Mg-Cu Alloys. *Int. J. Met.* **2016**, *10*, 2–20. [CrossRef]
11. Petrova, A.N.; Brodova, I.G.; Razorenov, S.V.; Shorokhov, E.V.; Akopyan, T.K. Mechanical Properties of the Al–Zn–Mg–Fe–Ni Alloy of Eutectic Type at Different Strain Rates. *Phys. Met. Metallogr.* **2019**, *120*, 1221–1227. [CrossRef]
12. Brodova, I.G.; Shirinkina, I.G.; Rasposienko, D.Y.; Akopyan, T.K. Structural Evolution in the Quenched Al–Zn–Mg–Fe–Ni Alloy during Severe Plastic Deformation and Annealing. *Phys. Met. Metallogr.* **2020**, *121*, 899–905. [CrossRef]
13. Zolotorevskii, V.S.; Pozdnyakov, A.V.; Churyumov, A.Y. Search for promising compositions for developing new multiphase casting alloys based on Al-Zn-Mg matrix using thermodynamic calculations and mathematic simulation. *Phys. Met. Metallogr.* **2014**, *115*, 286–294. [CrossRef]

14. Pozdniakov, A.V.; Zolotorevskiy, V.S.; Mamzurina, O.I. Determining hot cracking index of Al–Mg–Zn casting alloys calculated using effective solidification range. *Int. J. Cast Met. Res.* **2015**, *28*, 318–321. [CrossRef]
15. Guangyu, Y.; Shaojun, L.; Wanqi, J. Effects of minor Sc addition on the microstructures and mechanical properties of Al-Zn-Mg-Cu casting aluminum alloy. In *The Minerals, Metals & Materials Series*; Elsevier: Amsterdam, The Netherlands, 2016; pp. 463–467.
16. Nadella, R.; Eskin, D.; Katgerman, L. Effect of grain refining on defect formation in DC Cast Al-Zn-Mg-Cu alloy billet. In *Essential Readings in Light Metals*; Springer International Publishing: Cham, Switzerland, 2016; pp. 842–847.
17. Senkov, O.N.; Bhat, R.B.; Senkova, S.V.; Schloz, J.D. Microstructure and properties of cast ingots of Al-Zn-Mg-Cu alloys modified with Sc and Zr. *Metall. Mater. Trans. A* **2005**, *36*, 2115–2126. [CrossRef]
18. He, Y.; Zhang, X.; You, J. Effect of minor Sc and Zr on microstructure and mechanical properties of Al-Zn-Mg-Cu alloy. *Trans. Nonferrous Met. Soc. China* **2006**, *16*, 1228–1235. [CrossRef]
19. Srinivasan, S.; Desch, P.B.; Schwarz, R.B. Metastable phases in the Al3X (X = Ti, Zr, and Hf) intermetallic system. *Scr. Metall. Mater.* **1991**, *25*, 2513–2516. [CrossRef]
20. Knipling, K.E.; Dunand, D.C.; Seidman, D.N. Precipitation evolution in Al–Zr and Al–Zr–Ti alloys during isothermal aging at 375–425 °C. *Acta Mater.* **2008**, *56*, 114–127. [CrossRef]
21. Gao, H.; Feng, W.; Wang, Y.; Gu, J.; Zhang, Y.; Wang, J.; Sun, B. Structural and compositional evolution of Al3(Zr,Y) precipitates in Al-Zr-Y alloy. *Mater. Charact.* **2016**, *121*, 195–198. [CrossRef]
22. Knipling, K.E.; Dunand, D.C.; Seidman, D.N. Precipitation evolution in Al-Zr and Al-Zr-Ti alloys during aging at 450–600 °C. *Acta Mater.* **2008**, *56*, 1182–1195. [CrossRef]
23. Knipling, K.; Dunand, D. Creep resistance of cast and aged Al–0.1Zr and Al–0.1Zr–0.1Ti (at.%) alloys at 300–400 °C. *Scr. Mater.* **2008**, *59*, 387–390. [CrossRef]
24. Wen, S.P.; Gao, K.Y.; Li, Y.; Huang, H.; Nie, Z.R. Synergetic effect of Er and Zr on the precipitation hardening of Al–Er–Zr alloy. *Scr. Mater.* **2011**, *65*, 592–595. [CrossRef]
25. Li, H.; Bin, J.; Liu, J.; Gao, Z.; Lu, X. Precipitation evolution and coarsening resistance at 400 °C of Al microalloyed with Zr and Er. *Scr. Mater.* **2012**, *67*, 73–76. [CrossRef]
26. Wen, S.P.; Gao, K.Y.; Huang, H.; Wang, W.; Nie, Z.R. Precipitation evolution in Al-Er-Zr alloys during aging at elevated temperature. *J. Alloys Compd.* **2013**, *574*, 92–97. [CrossRef]
27. Gao, Z.; Li, H.; Lai, Y.; Ou, Y.; Li, D. Effects of minor Zr and Er on microstructure and mechanical properties of pure aluminum. *Mater. Sci. Eng. A* **2013**, *580*, 92–98. [CrossRef]
28. Li, H.; Gao, Z.; Yin, H.; Jiang, H.; Su, X.; Bin, J. Effects of Er and Zr additions on precipitation and recrystallization of pure aluminum. *Scr. Mater.* **2013**, *68*, 59–62. [CrossRef]
29. Huang, H.; Wen, S.P.; Gao, K.Y.; Wang, W.; Nie, Z.R. Age hardening behavior and corresponding microstructure of dilute Al-Er-Zr alloys. *Metall. Mater. Trans. A Phys. Metall. Mater. Sci.* **2013**, *44*, 2849–2856. [CrossRef]
30. Pozdniakov, A.V.; Barkov, R.Y. Microstructure and materials characterisation of the novel Al–Cu–Y alloy. *Mater. Sci. Technol.* **2018**, *34*, 1489–1496. [CrossRef]
31. Amer, S.M.; Barkov, R.Y.; Yakovtseva, O.A.; Pozdniakov, A.V. Comparative Analysis of Structure and Properties of Quasibinary Al–6.5Cu–2.3Y and Al–6Cu–4.05Er Alloys. *Phys. Met. Metallogr.* **2020**, *121*, 476–482. [CrossRef]
32. Pozdnyakov, A.V.; Barkov, R.Y.; Sarsenbaev, Z.; Amer, S.M.; Prosviryakov, A.S. Evolution of Microstructure and Mechanical Properties of a New Al–Cu–Er Wrought Alloy. *Phys. Met. Metallogr.* **2019**, *120*, 614–619. [CrossRef]
33. Pozdniakov, A.V.; Barkov, R.Y.Y.; Amer, S.M.; Levchenko, V.S.; Kotov, A.D.; Mikhaylovskaya, A.V. Microstructure, mechanical properties and superplasticity of the Al–Cu–Y–Zr alloy. *Mater. Sci. Eng. A* **2019**, *758*, 28–35. [CrossRef]
34. Amer, S.M.; Barkov, Y.R.; Yakovtseva, O.A.; Loginova, I.S.; Pozdniakov, A.V. Effect of Zr on microstructure and mechanical properties of the Al–Cu–Er alloy. *Mater. Sci. Technol.* **2020**, *36*, 453–459. [CrossRef]
35. Amer, S.M.M.; Mikhaylovskaya, A.V.V.; Barkov, R.Y.Y.; Kotov, A.D.D.; Mochugovskiy, A.G.G.; Yakovtseva, O.A.A.; Glavatskikh, M.V.V.; Loginova, I.S.S.; Medvedeva, S.V.V.; Pozdniakov, A.V.V. Effect of Homogenization Treatment Regime on Microstructure, Recrystallization Behavior, Mechanical Properties, and Superplasticity of Al-Cu-Er-Zr Alloy. *JOM* **2021**, *73*, 3092–3101. [CrossRef]
36. Amer, S.M.; Barkov, R.Y.; Pozdniakov, A.V. Effect of Mn on the Phase Composition and Properties of Al–Cu–Y–Zr Alloy. *Phys. Met. Metallogr.* **2020**, *121*, 1227–1232. [CrossRef]
37. Amer, S.; Yakovtseva, O.; Loginova, I.; Medvedeva, S.; Prosviryakov, A.; Bazlov, A.; Barkov, R.; Pozdniakov, A. The Phase Composition and Mechanical Properties of the Novel Precipitation-Strengthening Al-Cu-Er-Mn-Zr Alloy. *Appl. Sci.* **2020**, *10*, 5345. [CrossRef]
38. Amer, S.M.; Barkov, R.Y.; Prosviryakov, A.S.; Pozdniakov, A.V. Structure and Properties of New Heat-Resistant Cast Alloys Based on the Al–Cu–Y and Al–Cu–Er Systems. *Phys. Met. Metallogr.* **2021**, *122*, 908–914. [CrossRef]
39. Amer, S.M.; Barkov, R.Y.; Prosviryakov, A.S.; Pozdniakov, A. V Structure and Properties of New Wrought Al-Cu-Y- and Al-Cu-Er-Based Alloys. *Phys. Met. Metallogr.* **2021**, *122*, 915–922. [CrossRef]
40. Zhong, H.; Li, S.; Wu, J.; Deng, H.; Chen, J.; Yan, N.; Chen, Z.; Duan, L. Effects of retrogression and re-aging treatment on precipitation behavior, mechanical and corrosion properties of a Zr+Er modified Al-Zn-Mg-Cu alloy. *Mater. Charact.* **2022**, *183*, 111617. [CrossRef]
41. Zhang, Z.; Li, D.; Li, S.; Deng, H.; Zhang, S.; Fang, J.; Yuan, H.; Deng, B.; Qi, L. Effect of direct aging treatment on microstructure, mechanical and corrosion properties of a Si-Zr-Er modified Al-Zn-Mg-Cu alloy prepared by selective laser melting technology. *Mater. Charact.* **2022**, *194*, 112459. [CrossRef]

42. Wang, Y.; Wu, X.; Cao, L.; Tong, X.; Couper, M.J.; Liu, Q. Effect of trace Er on the microstructure and properties of Al–Zn–Mg–Cu–Zr alloys during heat treatments. *Mater. Sci. Eng. A* **2020**, *792*, 139807. [CrossRef]
43. Lu, J.-T.; Huang, H.; Wu, H.; Wen, S.-P.; Gao, K.-Y.; Wu, X.-L.; Nie, Z.-R. Mechanical properties and corrosion behavior of a new RRA-treated Al–Zn–Mg–Cu–Er–Zr alloy. *Rare Met.* **2023**, *42*, 672–679. [CrossRef]
44. Huang, Y.; Zhang, C.; Ma, Y.; Liu, Y. Effects of homogenization on the dissolution and precipitation behaviors of intermetallic phase for a Zr and Er containing Al–Zn–Mg–Cu alloy. *Prog. Nat. Sci. Mater. Int.* **2020**, *30*, 47–53. [CrossRef]
45. Li, J.; Zhang, Y.; Li, M.; Hu, Y.; Zeng, Q.; Zhang, P. Effect of combined addition of Zr, Ti and Y on microstructure and tensile properties of an Al-Zn-Mg-Cu alloy. *Mater. Des.* **2022**, *223*, 111129. [CrossRef]
46. Glavatskikh, M.V.; Barkov, R.Y.; Khomutov, M.G.; Pozdniakov, A.V. The Effects of Yttrium and Erbium on the Phase Composition and Aging of the Al–Zn–Mg–Cu–Zr Alloy with a High Copper Content. *Phys. Met. Metallogr.* **2022**, *123*, 617–623. [CrossRef]
47. Glavatskikh, M.V.; Barkov, R.Y.; Khomutov, M.G.P.A.V. Structure and properties of the Al-Zn-Mg-Cu-Zr-Y(Er) alloys sheets doped by manganese. *Phys. Met. Metallogr.* **2023**, *in press*.
48. Pozdniakov, A.V.; Zolotorevskiy, V.S. Determining hot cracking index of Al–Si–Cu–Mg casting alloys calculated using effective solidification range. *Int. J. Cast Met. Res.* **2014**, *27*, 193–198. [CrossRef]
49. Liu, X.; Wang, B.; Li, Q.; Wang, J.; Zhang, C.; Xue, C.; Yang, X.; Tian, G.; Liu, X.; Tang, H. Quantifying the Effects of Grain Refiners Al-Ti-B and La on the Microstructure and Mechanical Properties of W319 Alloy. *Metals* **2022**, *12*, 627. [CrossRef]

metals MDPI

Article

Influence of Minor Additions of Be on the Eutectic Modification of an Al-33wt.%Cu Alloy Solidified under Transient Conditions

Adilson Vitor Rodrigues [1,2,*], Rafael Kakitani [1], Cássio Silva [1], Leonardo Giovanetti [1], Marcelino Dias [3], Hani Henein [3], Amauri Garcia [1] and Noé Cheung [1]

1 Faculty of Mechanical Engineering, University of Campinas, Campinas 13083-860, Brazil
2 Federal Institute of Education, Science and Technology of São Paulo, IFSP, Bragança Paulista 12903-000, Brazil
3 Chemical and Materials Engineering Department, University of Alberta, Edmonton, AB T6G 2R3, Canada
* Correspondence: adilsonrodrigues@ifsp.edu.br

Abstract: Al-based alloys are often selected for use in various engineering applications as well as in the aircraft and aerospace industry. The improvement of their performances under severe conditions have required the use of alloying elements. In the present work, Be is added to the eutectic Al-33 wt.%Cu alloy with a view to investigating the resulting effects on microstructural features and hardness. A directional solidification technique is used, yielding a wide range of solidification cooling rates. This permits microstructural features to be investigated with emphasis on the role of Be on the eutectic morphology. The directionally solidified microstructures are formed by eutectic colonies for the Al-33 wt.%Cu alloy, however, with additions of both 0.05 and 0.5 wt.% Be, the original cell-like morphology is transformed into a trefoil-like morphology. This together with the α-Al dendritic pattern characterize the microstructure of the Al-33Cu-Be alloys examined. Solidification growth laws are experimentally derived relating dendritic and eutectic colony spacings to solidification cooling rates and growth rates. The length scale of such spacings is shown not to affect the Vickers hardness of the Al-33Cu-Be alloys examined; however, the additions of Be are shown to improve the brittle behavior of the Al-33 wt.%Cu alloy.

Keywords: Al-33Cu-Be alloys; directional solidification; microstructure; hardness

Citation: Rodrigues, A.V.; Kakitani, R.; Silva, C.; Giovanetti, L.; Dias, M.; Henein, H.; Garcia, A.; Cheung, N. Influence of Minor Additions of Be on the Eutectic Modification of an Al-33wt.%Cu Alloy Solidified under Transient Conditions. *Metals* **2023**, *13*, 94. https://doi.org/10.3390/met13010094

Academic Editor: Giulio Timelli

Received: 23 November 2022
Revised: 26 December 2022
Accepted: 30 December 2022
Published: 2 January 2023

1. Introduction

Aluminum alloys are often a preferred choice in the aeronautical and aerospace industries, as well as for various engineering applications due to their lightweight properties and high strength-to-weight ratio when compared to steels [1,2]. Such outstanding mechanical and physical characteristics have required the increased use of alloying elements in order to improve the performance of components under extreme conditions for their use in strategic fields [3]. On the other hand, Deschamps and collaborators [4] emphasize that the challenges in working with multicomponent alloys concern the formation of different phases and the identification of useful alloy compositions. Moreover, the complex compositions of multiple elements offer many challenges to optimize the design of alloys. Among the latest developments of aluminum-based alloys for aeronautic applications, the 2xxx series—(Al-Cu) based ones, particularly those containing Mg—have attracted attention due to the precipitation strengthening of the Al_2Cu and Al_2CuMg phases and good resistances to damage and to fatigue crack growth when compared to other Al-alloys series. However, improvements are required concerning the yield strength for applications under high load conditions and corrosion resistance due to the anodic behavior of the Al_2CuMg phase. On the other hand, controlled addition of alloying elements such as Sn, In, Cd and Ag is reported to be beneficial to mechanical properties, e.g., hardness, yield strength and ultimate tensile strength [5].

In terms of castings, not only the alloying content affects the final properties but also the cooling conditions during solidification [6]. In this regard, in recent years our research group has emphasized the study of the simultaneous effect of different cooling rates and alloying elements during solidification of several alloys systems on the resulting properties. For example: the addition of vanadium to the Al-Si-Fe alloy system [7], of zinc to the Al-Si system [8], of bismuth to the Zn-Al system [9]. We believe that such an approach can contribute to the understanding of the formation of phases and intermetallic compounds in more complex systems, such as multicomponent and high entropy alloys, when some common alloying elements are involved.

The present work aims to analyze the influence of additions of beryllium (0.05 and 0.5 wt.%) on the microstructure of the binary Al-Cu eutectic alloy (Al-33 wt.%Cu) during transient solidification conditions. To the best of the present authors' knowledge, such study has not yet been performed. As a first goal, it is intended to support the ongoing work analyzing the solidification of Al-Cu hypoeutectic alloys with additions of beryllium, since such binary alloy has an Al-rich primary phase immersed in a eutectic matrix. The main goal is to generate knowledge on the Al-Cu-Be system in order to understand the effects of the additions of Ag, Ni, Si and Sn since in previous studies the following systems have been analyzed by our research group: Al-Ag-Cu [10], Al-Cu-Ni [11], Al-Si-Cu [12], Al-Sn-Cu [13], which can be used for a comparative study.

Works concerning the addition of Be to Al-Cu alloys are very scarce. A potential application in the aerospace field has been reported by Houska [14] highlighting the formation of a BeO protective surface layer when 0.1–0.3 wt.%Be is added to an Al-based alloy. This oxide layer exhibits excellent resistance to thermal oxidation corrosion. The author also mentions that Be accelerates the age hardening process in Al-Cu alloys, which is also interesting for aeronautical and structural applications. The rate of precipitation of the θ phase is significantly increased by the addition of beryllium, and a faster transformation rate is shown to occur in the earlier θ ′ ~ θ transition [15]. On the other hand, the as-cast microstructure influences the aging process, such as the eutectic morphology, as emphasized by Yuan [16]. Thus, it is necessary to conduct a detailed study on the formation of the as-cast microstructure before exploring the alloy aging. In the present work, a comparative analysis of the solidification of Al-33 wt.%Cu-X wt.%Be alloys (X = 0, 0.05 and 0.5—the last one representing an increase of ten times in Be content when compared to the first addition and such Be contents aim to cover two regions of the Al-Cu-Be phase diagram) under a wide range of cooling rates is performed, with a view to analyzing the increasing influence of beryllium on the experimental microstructural growth laws and on the modification of the eutectic morphology.

2. Materials and Methods

The investigated Al-based alloys, Al-33 wt.%Cu, Al-33 wt.%Cu-0.05 wt.%Be and Al-33 wt.%Cu-0.5 wt.%Be, were prepared using commercially pure Al, Cu and a Cu-Be master alloy, for which the respective compositions are summarized in Table 1. At first, to produce each alloy, 800 g of Al were placed in a silicon carbide crucible, previously coated with alumina, and melted in a muffle furnace at 800 °C. Subsequently, the required amounts of Cu and Be were incorporated into the molten Al. For the eutectic Al-Cu alloy, only chips of commercial-purity Cu were added. For the ternary alloys, additional chips of a Cu-10 wt.%Be master alloy were used. The molten alloy was mechanically stirred using an alumina-coated stainless-steel bar and the resulting mixture was again placed into the furnace for about 45 min. This mechanical homogenization procedure was repeated three more times. In order to eliminate possible gases trapped in the liquid, ultrapure Argon (UN1006) was injected into the molten mixture for approximately 2 min and at about 1atm. The molten alloy was poured into a 60 mm diameter and 150 mm height stainless-steel cylindrical split mold, previously arranged inside the casting chamber of the directional solidification apparatus, as schematically represented in Figure 1. It is worth noting that in order to minimize the radial heat losses and facilitate the removal of the casting, the mold was previously coated internally with a highly refractory alumina-based material. It is worth mentioning that the bottom of the mold, made of AISI 1020 carbon steel sheet, is directly in contact with the liquid metal, i.e., it is not covered with alumina, being the region responsible for heat extraction. The alloy was remelted and when a temperature of about 5% above the alloy liquidus temperatures (only considering Al, Cu, Fe and Be as elements in the alloy composition) was reached, the electric heaters were switched off and the water flow (20 L/min—controlled by a rotameter), at the bottom of the mold started the vertical upward directional solidification. It is worth noting that in this work, solidification occurs under transient heat flow conditions in which both growth and cooling rates are free to vary over time, unlike the steady-state regime, in which these parameters are set at constant values in a Bridgman-type solidification furnace [17]. During the cooling process, 8 (eight) K-type thermocouples continuously monitored the temperatures at different positions along the length of the casting while the thermal profiles (temperature–time) were recorded through a data logger system (LynxADS1000, São Paulo, Brazil), connected to a computer, at a frequency of 20 Hz. Subsequently, these data were used to determine the evolution of both the growth rates (V) and cooling rates ($\dot{T}$) over the length of the casting.

Table 1. Chemical compositions (wt.%) of the elements and of the Cu-Be master alloy used in the preparation of the alloys.

Element/Master Alloy	Al	Si	Mg	Cu	Fe	Pb	Zn	Ni	Sn	Mn	Co	Be
Al	Bal.	0.006	0.0011	0.01	0.073	0.006	0.005	-	-	-	-	-
Cu	-	-	-	Bal.	-	-	-	0.008	0.009	0.008	-	-
Cu-Be	0.029	0.097	-	Bal.	0.053	0.0028	-	0.01	-	-	0.014	9.83

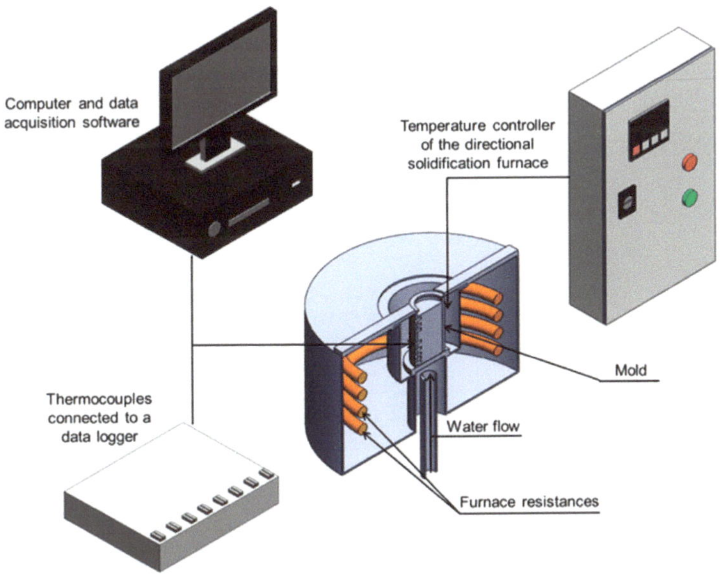

Figure 1. Schematic representation of the water-cooled upward directional solidification apparatus.

As schematically represented in Figure 2, longitudinal and transverse samples were extracted from the directionally solidified (DS) castings for microstructural characterization. The samples were ground with silicon carbide papers from 100 up to 1200 mesh and subsequently polished with diamond paste (1 and 3 μm). Microstructural spacings measurements were performed on images acquired using an Olympus Inverted Metallurgical Microscope (model 41GX, Olympus, Tokyo, Japan). The eutectic colony/primary dendritic arm spacings (λ_{EC}/λ_1) were measured on transverse samples using the triangle method, while the lamellar (λ_L) and secondary dendritic arm (λ_2) spacings were quantified using the linear intercept method, λ_L in transverse sections and λ_2 in longitudinal sections.

Both methods are depicted in Figure 2 [18]. A scanning electron microscope (SEM) FEI Inspect F5 (Thermo Fisher Scientific, Waltham, MA, USA) with an energy dispersive X-ray spectrometer (EDS) was used to verify the Cu content along the length of the DS castings and to quantify the distribution of the elements (Al, Cu and Fe) in the microstructure. The analysis of Be content was performed by using the inductively coupled plasma optical emission spectrometry technique. Samples with similar $\dot{T}$ values were extracted from the DS alloys castings and subjected to X-ray diffraction (XRD) analysis. The analyzed samples were extracted from different positions from the cooled surface of the DS castings (P): P = 5, 15, 30 and 70 mm, for Al-33 wt.%Cu and Al-33 wt.%Cu-0.05 wt.%Be alloys, and P = 5, 10, 30 and 70 mm, for the Al-33 wt.%Cu-0.5 wt.%Be alloy. A X'Pert-MDP (Philips Analytical X Ray, Malvern Panalytical, Cambridge, UK) diffractometer, employing CuK$_\alpha$ radiation with a wavelength of 0.15406 nm in the 2θ range from 20° to 80°, was used to determine the phases. Vickers hardness tests were performed using a Shimadzu HMV-2 model hardness tester using a load of 0.5 kgf and a dwell time of 15 s. The average of at least 20 measurements was the hardness value adopted for each representative sample.

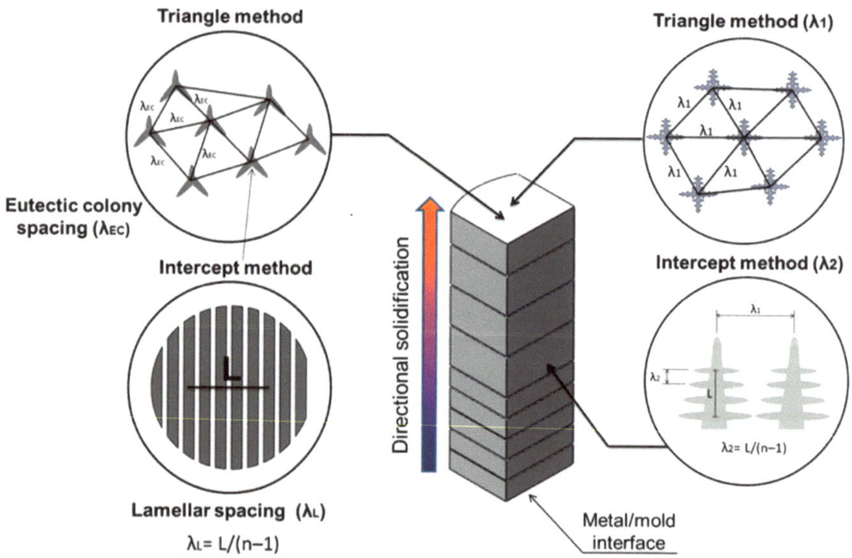

Figure 2. Schematic representation of samples extraction and of measurement methods used to quantify the length scale of the dendritic microstructure: λ_1 is primary dendritic arm spacing, λ_2 is secondary dendritic arm spacing, λ_{EC} is eutectic colony spacing, λ_L is lamellar spacing, 'L' is the length of the line and 'n' is the number of intercepted phases.

3. Results and Discussion

3.1. *Thermo-Calc Simulations*

The partial pseudo-binary Al-33 wt.%Cu-0.2 wt.%Fe-xBe (Be up to 0.1 wt.%) phase diagram is shown in Figure 3a, with a red dashed line indicating 0.05 wt.%Be (Al-33 wt.%Cu-0.2 wt.%Fe-0.05 wt.%Be alloy). In Figure 3b, a blue dashed line indicates 0.5 wt.%Be (Al-33 wt.%Cu-0.2 wt.%Fe-0.5 wt.%Be alloy). The 0.5 wt.%Be represents an increase of ten times when compared to the first addition and also, the chosen Be contents are to involve characteristic alloys of each region of the phase diagram: to the left and right of the 0.23 wt% Be point. At the right zone, it was intended to obtain the α-Be phase. The 0.2 wt.% Fe amount, shown later in this section, is a result from the alloy macrosegregation analysis, i.e., as an impurity present in aluminum and in the Cu-Be master alloy (Table 1), as well as from a result of diffusion from the cooled steel mold at elevated temperatures. Both diagrams were calculated under equilibrium conditions with Thermo-Calc software version

2021b using the TCAL8 database. The partial pseudo-binary Al-33 wt.%Cu-0.2 wt.%Fe phase diagram is not shown since the difference concerns only to the absence of α-Be (pure beryllium) when compared to the Al-33 wt.%Cu-0.2 wt.%Fe-xBe diagram. Hereafter, the alloys are referred to as Al-33 wt.%Cu, Al-33 wt.%Cu-0.05 wt.%Be and Al-33 wt.%Cu-0.5 wt.%Be, as shortened forms. During the solidification process, it can be noted that the α-Al and α-Be phases as well as the θ-Al_2Cu and Al_7Cu_2Fe intermetallic compounds (IMC) are common to both alloys. The precipitation sequence predicted by the phase diagram, in the case of the alloy with the lowest Be content (0.05 wt.%), is: primary phases, Al_7Cu_2Fe and α-Al, binary eutectic reaction L → L + α-Al + θ-Al_2Cu and ternary eutectic reaction L → α-Be + θ-Al_2Cu + α-Be. In the case of the alloy with 0.5 wt.%Be the sequence is: primary phases, Al_7Cu_2Fe and α-Be, binary eutectic reaction L → L + α-Be + θ-Al_2Cu and ternary eutectic reaction L → α-Be + θ-Al_2Cu + α-Al. The α-Al and α-Be phases are FCC and HCP based structures respectively.

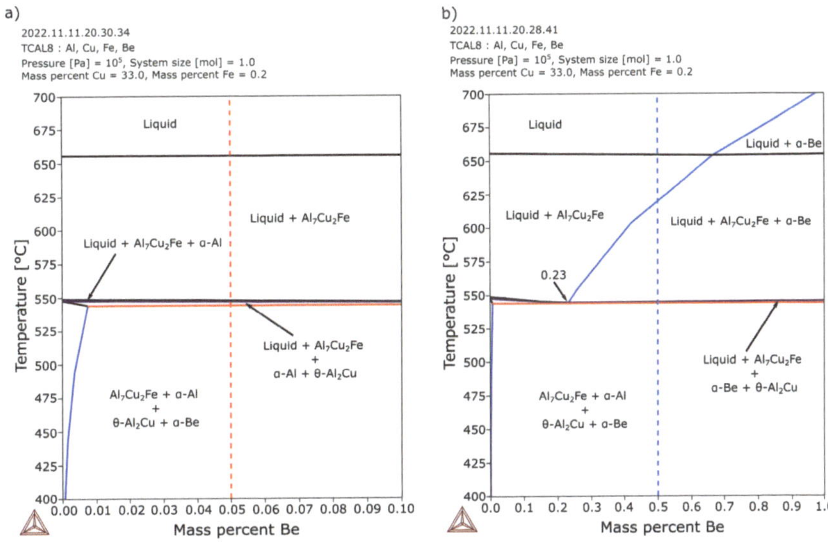

Figure 3. Partial pseudo-binary phase diagram Al-33 wt.%Cu-0.2 wt.%Fe−xBe with dashed lines indicating the compositions of the analyzed alloys (**a**) Al-33 wt.%Cu-0.2 wt.%Fe−0.05 wt.%Be and (**b**) Al-33 wt.%Cu-0.2 wt.%Fe−0.5 wt.%Be (Thermo-Calc 2021b, TCAL8 database).

3.2. Solidification Thermal Parameters

It is worth mentioning that some results on the Al-33 wt.%Cu eutectic alloy were previously published by Kakitani and coauthors [19]. In order to analyze the effects of Be addition to the Al-Cu eutectic alloy, these results, whenever necessary, will be presented and referenced throughout the work. The cooling rate ($\dot{T}$) and growth rate (V) profiles are shown in Figure 4a,b, respectively. These solidification thermal parameters were calculated from the temperature profiles obtained along the length of the DS castings through thermocouples readings during the experiments.

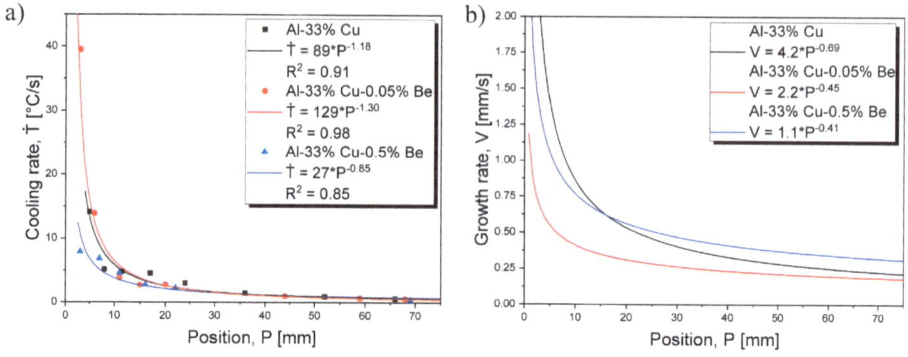

Figure 4. (**a**) Cooling rate and (**b**) growth rate as functions of position from metal/mold interface for Al-33 wt.%Cu date from [19], Al-33 wt.%Cu-0.05 wt.%Be and Al-33 wt.%Cu-0.5 wt.%Be alloys.

The cooling rate profiles were obtained from the first time-derivative of the function T = f(t), in which t represents the time corresponding to the isotherm of eutectic transformation passing by each thermocouple position. The growth rates were obtained from the first time-derivative of the equation P = f(t) that represents the position (P) of each thermocouple, from the cooled bottom of the casting, as a function of t_L. As expected, higher values of $\dot{T}$ and V are predominant at positions near to the bottom of the cooled mold. Two factors contribute to the decrease in $\dot{T}$ and V toward the top of the casting. First, the advance of the solidification front leads to increase in the thermal resistance between the cooled mold and the remaining molten alloy. Second, the increase in the metal/mold interfacial thermal resistance due to the gap evolving between the inner wall of the mold and the casting bottom surface. When compared, Al-33 wt.%Cu and Al-33 wt.%Cu-0.05 wt.%Be alloys present similar cooling rate profiles. In this case, the amount of 0.05 wt.%Be seems to have little influence on this thermal parameter. However, increasing the amount of Be to 0.5%, the $\dot{T}$ values decrease specifically at positions close to metal/mold interface (up to P $\approx$ 20 mm). From this position, almost the same cooling rates can be observed for different positions in the DS alloys castings studied in this work. Although the Al-33 wt.%Cu and Al-33 wt.%Cu-0.05 wt.%Be alloys presented mostly the same ($\dot{T}$) profiles, this trend did not occur for the (V) profiles, because the alloys differed from each other by having distinct thermal gradients (G). This can be explained by the definition of the cooling rate, which is provided by the product of thermal gradient and growth rate, i.e., ($\dot{T}$) = G.V.

3.3. Microstructural Analysis

Transverse sections images obtained through scanning electron microscopy (SEM) are presented in Figure 5. Similar $\dot{T}$ values (3.8 °C/s and 0.6 °C/s) were adopted as reference to establish a comparison among the microstructures of the alloys studied in this work. First, eutectic colonies can be noted in the Al-33 wt.%Cu alloy, and dendrites with a trefoil morphology, in the Al-33 wt.%Cu-0.05 wt.%Be and Al-33 wt.%Cu-0.5 wt.%Be alloys. Additionally, the intermetallic compounds (IMCs) are finer for positions close to the cooled bottom of the castings (higher cooling rates) for any alloy examined. Second, beryllium seems to contribute not only to the microstructural coarsening, but also to the microstructural morphology modification. Figure 6 shows longitudinal images that provide another view of the microstructure being influenced by the increase in beryllium. It can be noted that the increase in beryllium gradually transforms the cellular morphology of the eutectic mixture into a dendritic one. The red line contour in Figure 6, for the alloy with 0.05 wt.%Be, indicates the secondary dendritic arms with the eutectic mixture inside, which is very curious. It seems that a stage of the transition from a cellular structure to a dendritic one has occurred. The alloy with 0.5 wt.%Be shows a characteristic dendritic morphology, although with a different nature, i.e., Al2Cu instead of the eutectic mixture (Al2Cu + α-Al)—

which will be further shown in the elemental SEM-EDS analysis. How this mechanism leading to the Al_2Cu prevalence occurs over the eutectic mixture, induced by the addition of beryllium is a study that deserves to be carried out. Jafari and Amiryavari [20] investigated the influence of zirconium and beryllium additions on the microstructure, mechanical properties and corrosion behavior of an as-cast AZ63 (Mg-Zn-Al) alloy. The AZ63 alloy containing 0.0001 and 0.001 wt.%Be exhibited microstructure coarsening effect, while a morphological change (from sixfold symmetrical to irregular shape) was observed to occur for the AZ63 alloy containing 0.01 and 0.1 wt.%Be. Longitudinal images, obtained by optical microscopy, are shown in Figure 6. The secondary dendritic arms are evident in alloys containing beryllium, but they are not observed in the binary Al-Cu eutectic alloy.

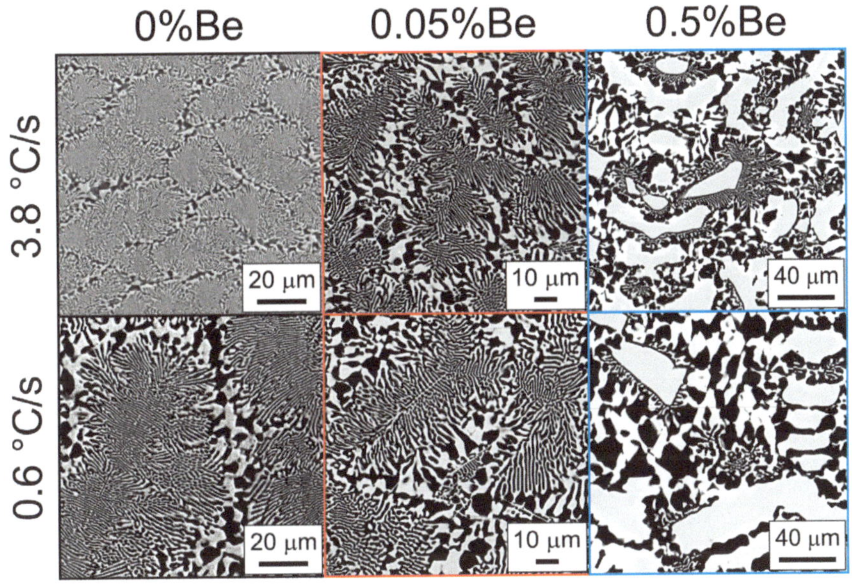

Figure 5. Transverse SEM images of Al-33 wt.%Cu, Al-33 wt.%Cu-0.05 wt.%Be and Al-33 wt.%Cu-0.5 wt.%Be alloys under different cooling rates (3.8 °C/s and 0.6 °C/s).

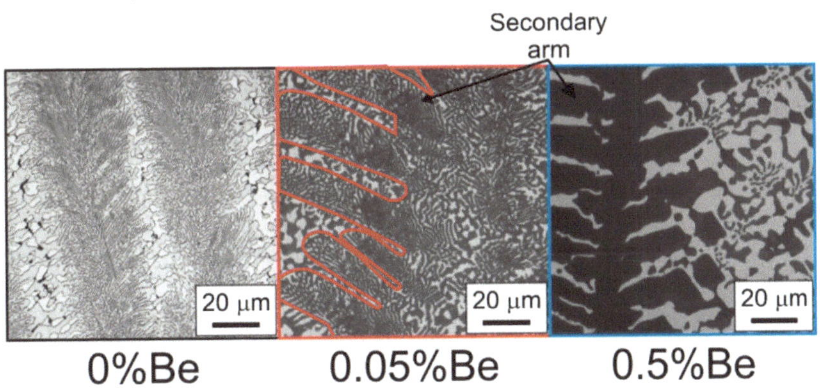

Figure 6. Longitudinal images obtained by optical microscopy for Al-33 wt.%Cu, Al-33 wt.%Cu-0.05 wt.%Be and Al-33 wt.%Cu-0.5 wt.%Be alloys solidified under a cooling rate of 1.6 °C/s. The red line contour delimitates secondary arms of a dendritic morphology.

3.4. Chemical Composition

Figure 7 shows the Cu distribution profiles and the residual Fe profiles along the length of the DS castings. While Fe remains close to 0.2 wt.% along the length of all the alloys castings with Be, a small inverse Cu macrosegregation can be seen at positions closer to the water-cooled bottom of the Be-containing alloys. Considering the samples of Al-33 wt.%Cu, the Fe content remained at about 0.15 wt.% along the casting [19]. This behavior was observed to occur for other ternary alloys as Al-Cu-Si [12] and Al-Cu-Ni [11,21] alloys. The inverse macrosegregation was described by Grandfield et al. [22] as a result of movement of the interdendritic fluid in the opposite direction with respect to the advance of the solidification front. The increase in Cu content promotes microstructural refinement [23]. This fact, in addition to the high cooling rates imposed at positions close to the cooled bottom of the DS castings, may also have contributed to the microstructural refinement in this region. In order to verify the amount of Be in alloys and in view of the difficulty of detecting beryllium using EDS (the use of EDS is feasible when the phase contain a very high Be content) [24], the plasma optical emission spectrometry method was applied, and the results are presented in Table 2. As can be seen, the effective amount of Be is in good agreement with the nominal composition of the ternary alloys.

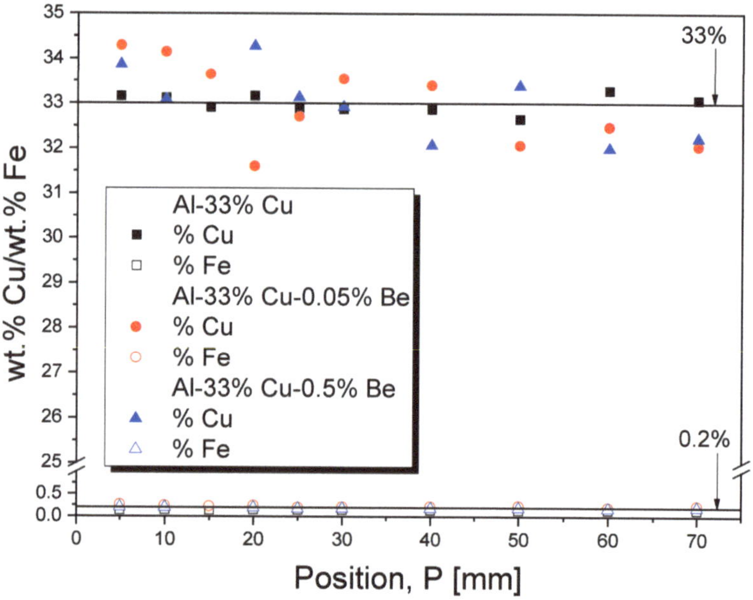

Figure 7. Cu and Fe (residual) distribution along the length of Al-33 wt.%Cu data from [19], Al-33 wt.%Cu-0.05 wt.%Be and Al-33 wt.%Cu-0.5 wt.%Be DS castings.

Table 2. Beryllium content (wt.%) in Al-33 wt.%Cu-0.05 wt.%Be and Al-33 wt.%Cu-0.5 wt.%Be alloys.

Alloy	%Be
Al-33wt.%Cu-0.05wt.%Be	0.055 ± 0.02
Al-33wt.%Cu-0.5wt.%Be	0.60 ± 0.01

3.5. Microstructural Growth Laws

The values of primary dendritic arm spacing (λ_1), eutectic colony spacing (λ_{EC}), lamellar spacing (λ_L) and secondary dendritic arm spacing (λ_2) are correlated with the thermal parameters $\dot{T}$ and V in Figure 8. λ_{EC} and λ_1 decrease with increasing $\dot{T}$ and V and the following experimental growth laws can be derived: $\lambda_{EC} = 132\ \dot{T}^{-0.25}$ and $\lambda_{EC} = 80\ V^{-0.50}$, for the Al-33 wt.%Cu-0.05 wt.%Be alloy, and $\lambda_1 = 132\ \dot{T}^{-0.55}$ and $\lambda_1 = 24\ V^{-1.10}$, for the Al-33 wt.%Cu-0.5 wt.%Be alloy. When compared with the binary DS Al-33 wt.%Cu alloy, studied by Kakitani and coauthors [19], the analyzed alloys in this work show higher λ_1 values. It can be noted that, close to metal/mold interface, the amount of beryllium exerts a greater influence on the microstructure refinement as compared to that exerted by the cooling rates, since the alloy containing 0.5 wt.%Be has a higher degree of refinement, even being subjected to lower cooling rates when compared to the Al-33 wt.%Cu-0.05 wt.%Be alloy, as previously presented. However, as the distance from the cooled bottom of the DS casting increases, where both alloys are almost under the same cooling rates, the microstructure of the alloy with 0.5 wt.%Be becomes slightly coarser than that observed for the alloy with 0.05 wt.%Be. The use of −0.55 and −1.1 exponents in experimental laws correlating the primary dendritic arm spacing with cooling rate and growth rate, respectively, has been demonstrated to be effective for several ternary Al-based alloys solidified under unsteady-state conditions [25–27].

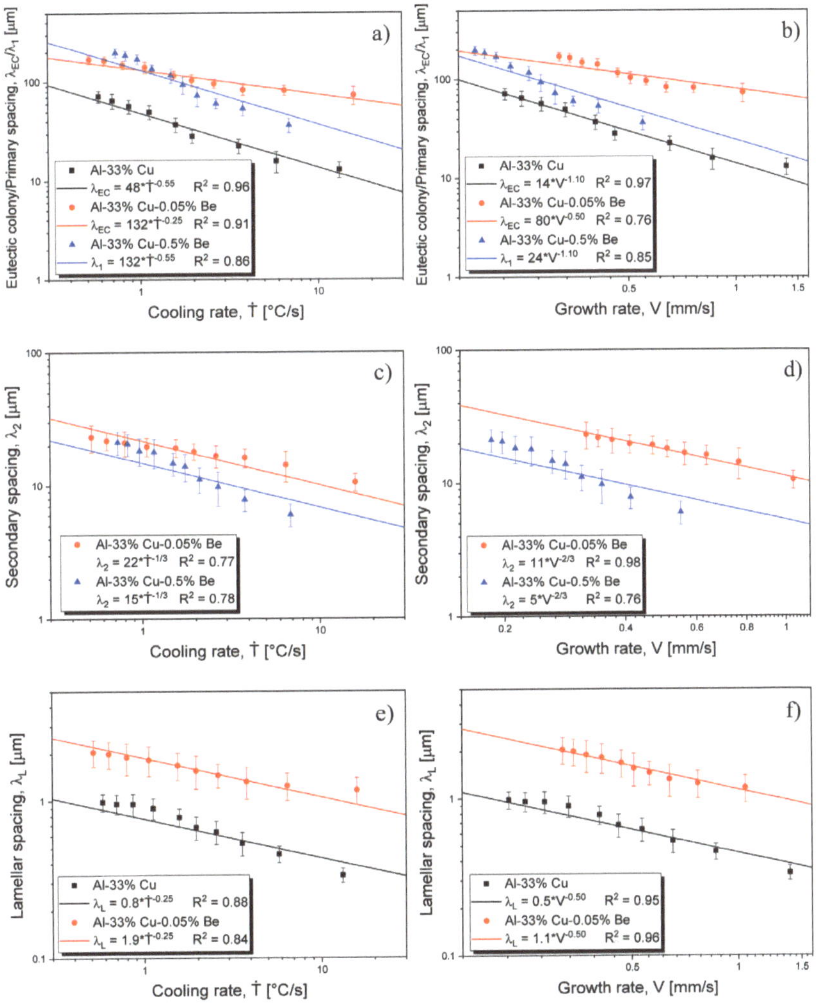

Figure 8. Eutectic colony/primary dendritic spacing, λ_{EC}/λ_1, as a function of (**a**) $\dot{T}$ and (**b**) V, lamellar spacing, λ_L, as a function of (**c**) $\dot{T}$ and (**d**) V and secondary dendritic spacing, λ_2, as a function of (**e**) $\dot{T}$ and (**f**) V for Al-33 wt.%Cu data from [19], Al-33 wt.%Cu-0.05 wt.%Be and Al-33 wt.%Cu-0.5 wt.%Be alloys.

Despite that, in the case of the Al-33 wt.%Cu-0.05 wt.%Be alloy, the use −0.25 and −0.50 exponents promoted better adjustments in the experimental growth laws correlating λ_{EC} to $\dot{T}$ and V, respectively. When compared to the binary Al-33 wt.%Cu alloy, the addition of 0.05wt.%Be provided coarsening of the lamellae of eutectic colonies and, consequently, higher λ_L values were found. The experimental growth laws for lamellar spacing are $\lambda_L = 1.9\ \dot{T}^{-0.25}$ and $\lambda_L = 1.1\ V^{-0.50}$. The exponents −0.25 and −0.5 used in experimental laws correlating the lamellar spacing with cooling rate and growth rate, respectively, were successfully applied for Al-Ni [28] and Al-Co [29] alloys. The exponential value of the growth rate was found to be 0.50 using the Jackson–Hunt theory of eutectic growth [30]. The use of the 0.50 exponent proved to also be satisfactory in studies with ternary alloys, e.g., alloys from the Al-Cu-Ag system [31]. Finally, the increase in Be content to 0.5wt.% promoted the refinement of secondary dendritic arm spacings, and the following experimental growth laws are proposed: $\lambda_2 = 22\ \dot{T}^{-1/3}$ and $\lambda_2 = 11\ V^{-2/3}$, for the Al-33 wt.%Cu-0.05 wt.%Be alloy, and $\lambda_2 = 15\ \dot{T}^{-1/3}$ and $\lambda_2 = 5\ V^{-2/3}$, for the Al-33 wt.%Cu-0.5 wt.%Be alloy. The use of $-1/3$ and $-2/3$ exponents to correlate secondary dendritic arm spacing with cooling rate and growth rate, respectively, was shown to be effective for several binary and ternary alloys solidified under unsteady-state conditions [7,8,23]. These results suggest that the addition of Be to the eutectic Al-33 wt.%Cu alloy (Figure 8a,b) promotes coarsening of the microstructure. However, comparing both Be additions, the increase in Be content from 0.05 to 0.5 wt.%, as shown in Figure 8c,d, promotes microstructural refining.

3.6. Microstructural Phases

The XRD patterns of the samples extracted along the length of the DS Al-33 wt.%Cu, Al-33 wt.%Cu-0.05 wt.%Be and Al-33 wt.%Cu-0.5 wt.%Be alloys castings are shown in Figure 9a–c, respectively. Generally, the studied alloys had three phases in common: α-Al, Al_2Cu and Al_7Cu_2Fe. Additionally, alloys containing Be present characteristic peaks of the Al_4Cu_9 phase, which are not predicted by the Al-Cu-Be pseudo-binary phase diagram (Figure 3). All phases were identified by comparison based on the diffraction patterns available in the Inorganic Crystal Structure Database (ICSD). The α-Al peaks are related not only to the alloy matrix, but also to the α-Al + Al_2Cu eutectic mixture. Fe is the most common impurity found in Al-Cu alloys [32] and plate-like Al_7Cu_2Fe (referred to as β-CuFe) is one of the possible IMCs formed. Al_7Cu_2Fe is considered a deleterious constituent, since it may be the starting point of fatigue crack and localized corrosion of high-strength aluminum alloys [33,34]. When formed in large volume, Al_7Cu_2Fe will consume atoms from the α-Al matrix, thus decreasing the strengthening effect of Cu-related precipitates such as Al_2Cu [35]. As previously presented (Figure 7), all the alloys investigated in this work show an average iron contamination of about 0.2 wt.%, which, associated with the peaks found by XRD analyses, suggests the formation of Al_7Cu_2Fe, which is in agreement with the prediction of the phase diagrams in Figure 3. Xu and coauthors [36] studied the evolution of Fe-rich IMCs and their effects on the mechanical properties of a 2219 Al-Cu alloy under different processing approaches. Alloys containing 0.03, 0.10, 0.15 and 0.20 wt.%Fe were analyzed. For Fe content less than 0.03 wt.%, Al_2Cu was the main IMC observed. When the Fe amount increased to 0.10 wt.%, a new needle-like IMC appeared, either Al_7Cu_2Fe or Al_7Cu_2(Fe, Mn). The authors emphasized that the increase in Fe content did not change the morphology of IMCs, but they became longer and wider.

The peaks that suggest the formation of the Al_4Cu_9 IMC occurred at 48.17°, in the Al-33 wt.%Cu-0.05 wt.%Be alloy, and at 38.73° and 66.82°, in the Al-33 wt.%Cu-0.5 wt.%Be alloy. It is worth noting that the peak that occurred at 48.17° in the Al-33 wt.%Cu-0.05 wt.%Be alloy is more intense than that observed for the α-Al matrix, which can be related to a possible texturing of the sample. Karov and Youdelis [37] determined that the degree of solubility of beryllium in Al_2Cu is in the range of 0.8–2.07 wt.%. These authors examined Al-Cu alloys with eutectic composition containing 0, 0.14, 0.3 and 0.5 wt.%Be. For alloys with 0.14 and 0.3 wt.%Be, only α-Al and Al_2Cu phases were found, indicating that all the Be was absorbed by the Al_2Cu IMC. The authors reported that the aluminum atoms were replaced with beryllium ones. However, the authors pointed out that the common fine lamellar microstructure of the eutectic became irregular and coarse with increasing beryllium content and a new phase within the Al_2Cu regions was observed to occur, specifically in the alloy containing 0.5 wt.%Be. Moreover, in this work, the authors observed diffusion diffraction lines of the Al_4Cu_9 phase, identified as the Cu-rich phase in the Al-3Cu-0.1Be alloy. These results showed that, in these alloys, Be is concentrated in the precipitated phases. In this work, the observed peaks, associated with the high solubility of Be in Cu-rich phases, suggest the formation of the Al_4Cu_9 phase.

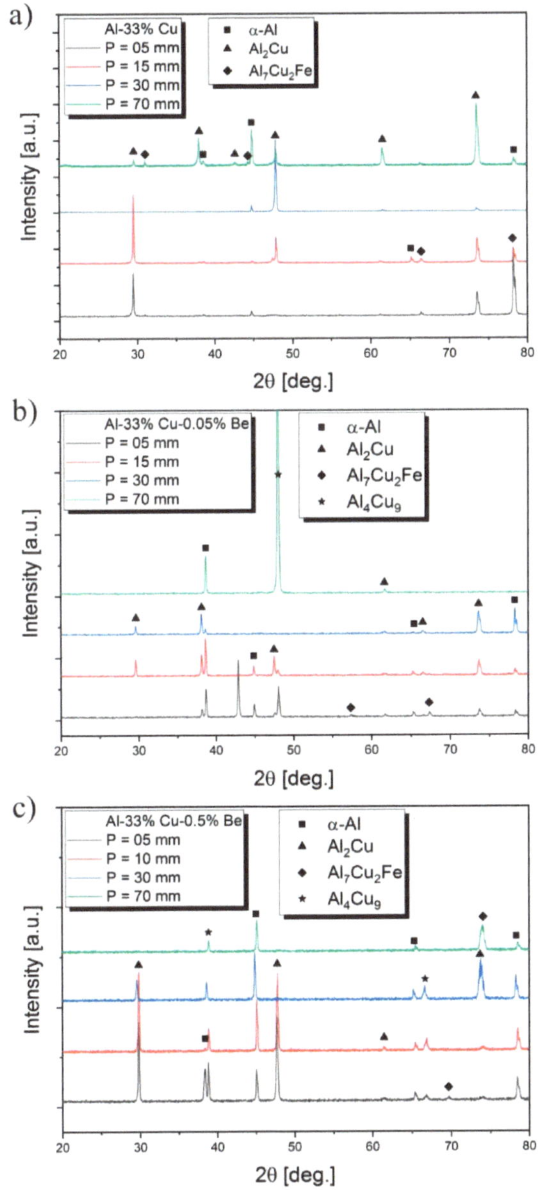

Figure 9. XRD patterns at four different positions (P), chosen by similarity of $\dot{T}$ values, in the (**a**) Al-33 wt.%Cu, (**b**) Al-33 wt.%Cu-0.05 wt.%Be and (**c**) Al-33 wt.%Cu-0.5 wt.%Be alloys.

It is worth noting that no peaks of IMC phases containing Be were observed, although expected in the pseudo-binary phase diagrams (Figure 3) and in the solidification paths based on the Scheil–Gulliver (S–G) model (Figure 10), which assumes nonequilibrium conditions that are complete diffusion in the liquid and no diffusion in the solidified phases. Amirkhanlou [38] reported that the distribution of a Be-rich phase in the Al matrix depends on the relative amounts of Be and on the processing method. Bodwen [39] found fine dispersions of Be phase in the Al matrix of samples processed by powder metallurgy, due to the high amount of Be in the Al-63%Be eutectic alloy. In this work, the α-Be phase is expected to occur in the alloy with 0.05%Be only at the end of solidification (~544 °C, Figure 10a), probably with the remaining liquid achieving eutectic transformation with a tiny amount of α-Be being part of the eutectic mixture. On the other hand, in the alloy with 0.5%Be, α-Be arises as the primary phase, i.e., at the beginning of solidification. In the alloy with 0.5% Be, the practically vertical green line of Figure 10b indicates a low amount of α-Be, similarly as Al_7Cu_2Fe, when compared to the other phases. It worth noting that increasing the beryllium content the solidification interval associated with the formation of Al_7Cu_2Fe decreases, as shown comparatively by the red line lengths in Figure 10a,b. This decrease in the solidification range can also be realized in Figure 3b, due to the existence of the two-phases zone (L+α-Be) in L+ Al_7Cu_2Fe that was absent in Figure 3a. Thus, in both alloys, α-Be appears in small amount, which may make its detection through XRD unfeasible. Although the S–G model considers nonequilibrium solidification, it is worth bearing in mind that the water-cooled mold can shift solidification to a condition much further than that considered by the S–G model, undermining or suppressing the α-Be phase. Next, a complementary discussion involving SEM-EDS analysis on the α-Be phase is conducted.

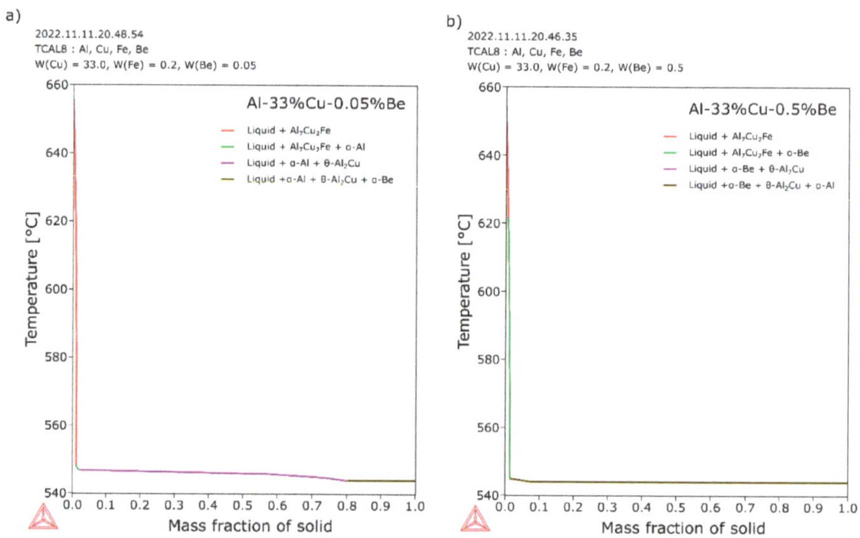

Figure 10. Solidification paths of (**a**) Al-33 wt.%Cu-0.05 wt.%Be and (**b**) Al-33 wt.%Cu-0.5 wt.%Be alloys simulated by the Thermo-Calc software based on the Scheil–Gulliver model.

SEM-EDS analyses on transverse sections of Al-33 wt.%Cu-0.05 wt.%Be and Al-33 wt.%Cu-0.5 wt.%Be alloys (P = 70 mm) are presented in Figure 11a,b, respectively. It can be seen that Be is not detected in any analyzed region. Some well-known characteristics of Be are its low atomic number (Z = 4), low relative atomic mass (A = 9.012) and low density (ρ = 1.85 g/cm^3). Beryllium presents a very low fluorescence yield, which makes the emission of Auger electrons more likely than the emission of characteristic X-rays, which can easily be absorbed by the surface layers and the contaminations on the surface.

Such attributes make the detection of Be, using the EDS technique, very difficult [40]. Characterizing Be-containing phases is reported as challenging in the literature [40], mainly for Be-diluted alloys and to the best of the present authors' knowledge, this task has not been successfully accomplished in related works. The EDS analysis for the alloy with 0.05wt.%Be shows in point #1, that the amount of Cu in solid solution is about the expected value for the Al-Cu binary system [41]. The same can be observed for point #2 (Al_2Cu) and point #3 (eutectic mixture). Considering the EDS analysis for the alloy with 0.5%wt.%Be, the composition of the α-Al matrix (point #4) remained mostly the same as that of point#1, as well as for the Al_2Cu phase (point #5). Point #6 represents the Al_4Cu_9 phase with a faceted morphology curiously located inside the Al_2Cu phase. The ratio between at.% Cu and Al at point#6 may not follow 9:4, probably due to interference from what is beneath the analyzed area, which is a common issue in the Energy Dispersive X-ray spectroscopy technique. Figure 12 shows the same SEM image of the Al-33 wt.%Cu-0.05 wt.%Be alloy (Figure 12a), but with its respective EDS elemental maps. The maps present the elemental distribution of Cu, Al, Fe and Be elements in the α-Al, eutectic and the Al_7Cu_2Fe phases. Beryllium seems homogeneously distributed throughout the sample and the iron distribution demonstrates that the Al_7Cu_2Fe phase has needle-like and non-elongated morphologies, although reported as plate-like by Zhao [32], which may cause deleterious effects on the mechanical properties, similarly as the β-AlFeSi in Fe contaminated Al-Si alloys [42,43].

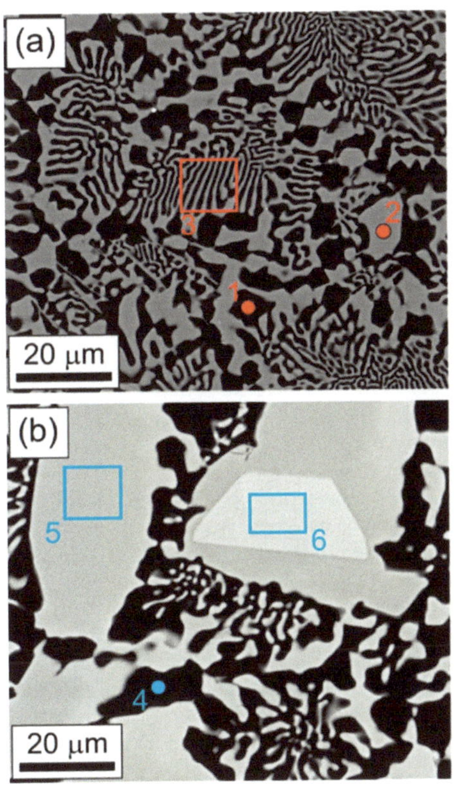

#1: α-Al matrix	wt.%	at.%
Al	95.71	98.13
Cu	4.29	1.87

#2: Al_2Cu	wt.%	at.%
Al	50.05	70.23
Cu	49.85	29.7
Fe	0.10	0.07

#3: Eutectic	wt.%	at.%
Al	68.36	83.57
Cu	31.56	16.38
Fe	0.08	0.05

#4: α-Al matrix	wt.%	at.%
Al	95.00	97.81
Cu	5.00	2.19

#5: Al_2Cu	wt.%	at.%
Al	52.48	72.23
Cu	47.51	27.73
Fe	0.01	0.01

#6: Al_4Cu_9	wt.%	at.%
Al	26.44	45.81
Cu	72.78	53.54
Fe	0.78	0.65

Figure 11. Elemental SEM-EDS analyses in transverse sections of the (**a**) Al-33 wt.%Cu-0.05 wt.%Be and (**b**) Al-33 wt.%Cu-0.5 wt.%Be alloys (P = 70 mm).

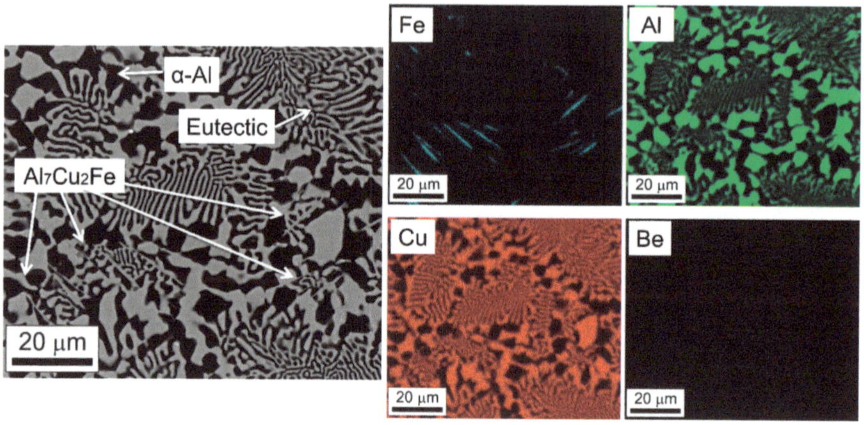

Figure 12. SEM image with EDS mapping detailing the phases present in the Al-33 wt.%Cu-0.05 wt.%Be alloy.

Figure 13 schematically suggests the nucleation and growth of the observed phases for the three analyzed alloys. Considering the eutectic binary alloy without Be addition, the transient solidification conditions favored the nucleation and growth of eutectic colonies surrounded by the same eutectic mixture rather coarser. The tiny addition of 0.05 wt.% Be seems to be responsible for modifying the anisotropic interface of the eutectic colonies, transforming the original cell-like morphology into a trefoil-like one. Observing the phase diagram (Figure 3a), the expected α-Al primary phase, as previously discussed, is not favored in the microstructure (Figures 5 and 6) instead of the eutectic phase, which may nucleate and grow preferentially. Thus, in the eutectic mixture, atoms of Al may have been replaced by Be, as previously mentioned by Karov and Youdelis [37]. In addition, considering that Be has a different crystal structure as compared to Al and Cu, i.e., HCP instead of FCC [44]. The gradual replacement of Al atoms by Be may have induced the growth of a trefoil morphology. In addition, considering the interfacial energy anisotropy on the phase morphology could be performed as an attempt to respond the influence of Be, as shown in the literature concerning the influence of Zn on the dendrite morphology of Al-Zn alloys [45]. With the increase in the addition of Be to 0.5 wt.%, the phase diagram (Figure 3b) predicts a primary α-Be phase, which was not found. Instead, the formation of primary phases of θ-Al_2Cu and Al_4Cu_9 occurred. It seems that the higher Be content, associated with nonequilibrium conditions, induced the formation of Al_xCu_y type primary phases. The trefoil-like morphology observed in both alloys, with 0.05 and 0.5 wt%. Be, seems to indicate that Be has been incorporated into the phases. Be constituting other IMCs is favored as compared to the formation of the α-Be phase, as reported in the literature, such as: Al_8Fe_2SiBe [46], $Al_{92}Mn_4Be_2Cu_2$ [40] and $Be_4Al(Mn,Cu)$ [24]. An exception is the formation of a fine eutectic, i.e., without a trefoil morphology around the primary phases that may have served as a nucleating agent, as can be seen in Figure 5 for the alloy with 0.5%Be. It was curious to observe that in some regions, as shown in Figure 10b, the Al_4Cu_9 IMC played the role of a nucleating agent for the Al_2Cu IMC. Additionally, the Al_7Cu_2Fe IMC was shown to have two morphologies (needle-like and non-elongated) in the alloys with additions of Be although it was reported in in a previous study [19] that the Al_7Cu_2Fe IMC has only the non-elongated morphology.

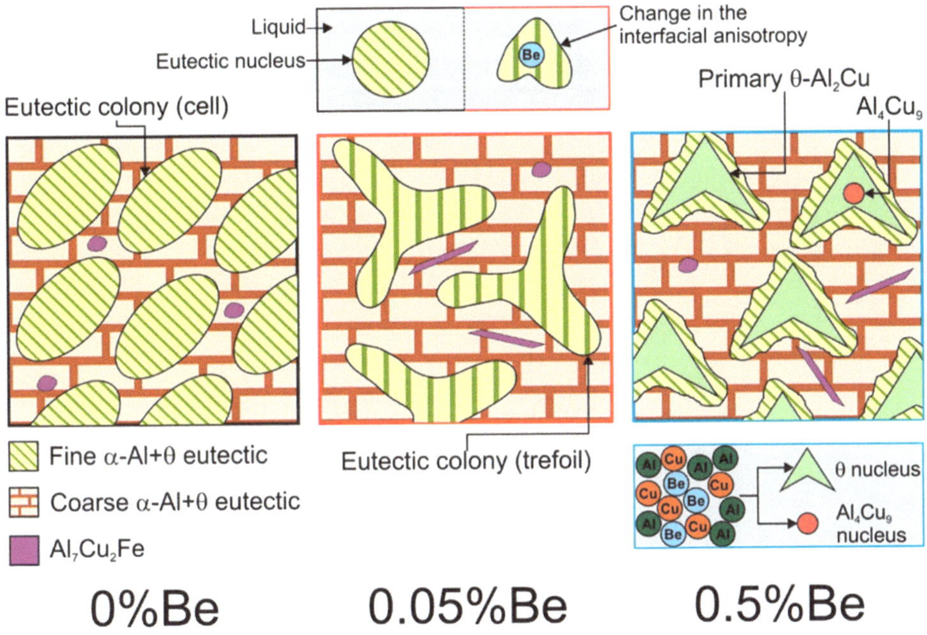

Figure 13. Schematic representation of the mechanism proposed to explain the microstructural changes promoted by beryllium in the Al-33 wt.%Cu alloy.

3.7. Hardness

Figure 14 presents the Vickers hardness (HV) of Al-33 wt.%Cu, Al-33 wt.%Cu-0.05 wt.%Be and Al-33 wt.%Cu-0.5 wt.%Be alloys as a function of λ_{EC}/λ_1. A Hall–Petch-type equation has been fitted to most of the experimental scatter only for the alloy without Be, in which HV increases along the lower range of cellular spacings due to the more homogeneous distribution of the eutectic mixture throughout the alloy microstructure. For the range of $\lambda_{EC}^{-1/2}/\lambda_1^{-1/2}$ values below 0.15 $\mu m^{-1/2}$, λ_{EC} and λ_1 are shown not to affect HV for any alloy examined. On the other hand, the decrease in HV indicates that the addition of Be has the beneficial effect of improving the brittle behavior of the Al-33%Cu alloy, even with the formation of the Al_4Cu_9 IMC, which has HV of about two times higher than that of the Al_2Cu IMC (Table 3). Both tensile strength and hardness are indicators of the mechanical strength of the metal to plastic deformation. When the strength (and hardness) increases, normally the ductility, which is a measure of the degree of plastic deformation developed by the material until fracture, is sacrificed resulting in brittle behavior, which involves very little or no plastic deformation. Thus, hardness is inversely related to ductility [44]. Hence, the decrease in HV provided by the addition of Be improves the brittle behavior of the Al-33%Cu alloy.

Table 3. Vickers hardness of θ-Al_2Cu and Al_4Cu_9 IMCs.

Phase	$HV_{0.05}$ (HV)
θ-Al_2Cu	397 ± 61
Al_4Cu_9	824 ± 46

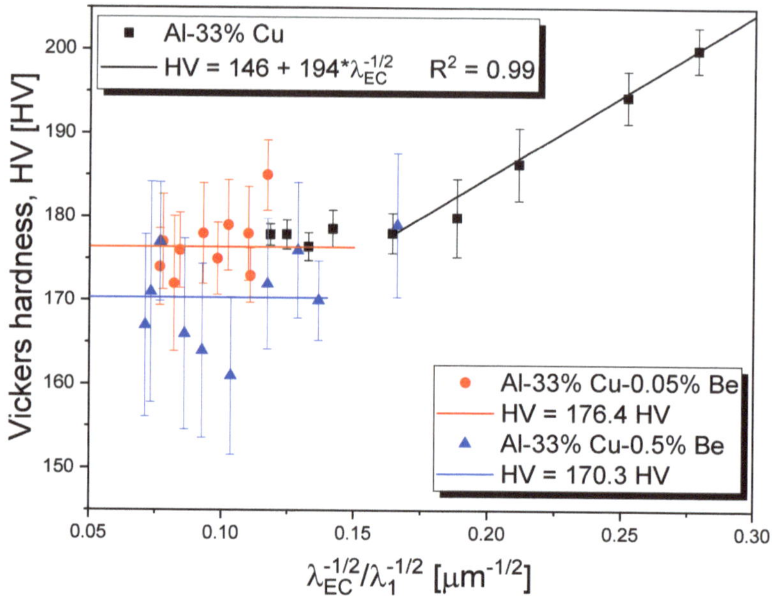

Figure 14. Vickers hardness of Al-33 wt.%Cu data from [19], Al-33 wt.%Cu-0.05 wt.%Be and Al-33 wt.%Cu-0.5 wt.%Be alloys as a function of $\lambda_{EC}^{-1/2}/\lambda_1^{-1/2}$.

4. Conclusions

The directionally solidified microstructures were shown to be formed by eutectic colonies for the Al-33 wt.%Cu alloy, and dendrites with a trefoil morphology, in Al-33 wt.%Cu-0.05 wt.%Be and Al-33 wt.%Cu-0.5 wt.%Be alloys. The secondary dendritic arms are evident in alloys containing beryllium, but they are not observed in the Al-Cu eutectic binary alloy.

The primary dendritic arm spacing (λ_1), the eutectic colony spacing (λ_{EC}), the lamellar spacing (λ_L) and the secondary dendritic arm spacing (λ_2) were correlated with the solidification cooling rate ($\dot{T}$), and the growth rate (V) and the following experimental growth laws were derived:

$$\lambda_{EC} = 132\ \dot{T}^{-0.25} \text{ and } \lambda_{EC} = 80\ V^{-0.50} \text{ for the Al-33 wt.\%Cu-0.05 wt.\%Be alloy}$$

$$\lambda_2 = 22\ \dot{T}^{-1/3} \text{ and } \lambda_2 = 11\ V^{-2/3}$$

$$\lambda_1 = 132\ \dot{T}^{-0.55} \text{ and } \lambda_1 = 24\ V^{-1.10}, \text{ for the Al-33 wt.\%Cu-0.5 wt.\%Be alloy}$$

$$\lambda_2 = 15\ \dot{T}^{-1/3} \text{ and } \lambda_2 = 5\ V^{-2/3}$$

where $\lambda_{EC}/\ \lambda_1/\ \lambda_2$ (μm); $\dot{T}$ (K/s) and V (mm/s).

For the Al-33 wt.%Cu alloy, the transient solidification conditions favored the nucleation and growth of eutectic colonies surrounded by a coarser eutectic mixture. The tiny addition of 0.05 wt.%Be was shown to be responsible for modifying the anisotropic interface of the eutectic colonies, transforming the original cell-like morphology into a trefoil-like one. Such trefoil-like morphology was shown to occur for both alloys, with 0.05 and 0.5 wt% Be, seeming to indicate that Be has been incorporated into the phases.

A Hall–Petch-type equation has been fitted to most of the experimental hardness scatter only for the alloy without Be, in which HV was shown to increase in the lower range of λ_{EC}. For the range of $\lambda_{EC}^{-1/2}/\ \lambda_1^{-1/2}$ values below 0.15 $\mu m^{-1/2}$, λ_{EC} and λ_1 were shown not to affect HV for any alloy examined. In contrast, the addition of Be was shown to have the beneficial effect of improving the brittle behavior of the Al-33%Cu alloy.

Author Contributions: Formal analysis, A.V.R., R.K., C.S., L.G., M.D. and N.C.; Methodology, A.V.R., R.K. and C.S.; Writing—original draft, A.V.R., R.K., C.S. and N.C.; Writing—review and editing, H.H., A.G. and N.C. All authors have read and agreed to the published version of the manuscript.

Funding: This study was financed in part by FAPESP-São Paulo Research Foundation, Brazil (Grant 2021/11439-0); CNPq-National Council for Scientific and Technological Development–CNPq, Brazil; CAPES-Coordenacão de Aperfeiçoamento de Pessoal de Nível Superior, Brazil (Finance Code 001); and FAEPEX/UNICAMP-Fundo de Apoio ao Ensino, à Pesquisa e à Extensão (grants 2162/21 and 2549/22).

Data Availability Statement: Not applicable.

Acknowledgments: This research used facilities of the Brazilian Nanotechnology National Laboratory (LNNano), part of the Brazilian Centre for Research in Energy and Materials (CNPEM), a private non-profit organization under the supervision of the Brazilian Ministry for Science, Technology, and Innovations (MCTI). The SEM–FIB staff is acknowledged for the assistance during the experiments (SEM–FIB–C1–20210595).

Conflicts of Interest: The authors declare no conflict of interest.

References

1. Cantor, B.; Assender, H.; Grant, P. *Aerospace Materials*; CRC Press: Boca Raton, FL, USA, 2001.
2. Kaufman, J.F.; Rooy, E.L. *Aluminum Alloy Castings: Properties, Processes, and Applications*; ASM International: Materials Park, OH, USA, 2004.
3. Chen, H.-L.; Chen, Q.; Engström, A. Development and applications of the TCAL aluminum alloy database. *Calphad* **2018**, *62*, 154–171. [CrossRef]
4. Deschamps, A.; Tancret, F.; Benrabah, I.-E.; De Geuser, F.; Van Landeghem, H.P. Combinatorial approaches for the design of metallic alloys. *C. R. Phys.* **2018**, *19*, 737–754. [CrossRef]
5. Gloria, A.; Montanari, R.; Richetta, M.; Varone, A. Alloys for Aeronautic Applications: State of the Art and Perspectives. *Metals* **2019**, *9*, 662. [CrossRef]
6. Dias, M.; Costa, T.; Rocha, O.; Spinelli, J.E.; Cheung, N.; Garcia, A. Interconnection of thermal parameters, microstructure and mechanical properties in directionally solidified Sn–Sb lead-free solder alloys. *Mater. Charac.* **2015**, *106*, 52–61. [CrossRef]
7. Silva, C.; Barros, A.; Rodrigues, A.V.; Kakitani, R.; Soares, T.; Garcia, A.; Cheung, N. Microstructure and Tensile Strength of an Al-Si-Fe-V Alloy: Vanadium and Solidification Thermal Parameters as Recycling Strategies. *Sustainability* **2022**, *14*, 13859. [CrossRef]
8. Kakitani, R.; Konno, C.; Garcia, A.; Cheung, N. The Effects of Solidification Cooling and Growth Rates on Microstructure and Hardness of Supersaturated Al-7%Si-x%Zn Alloys. *J. Mater. Eng. Perform.* **2022**, *31*, 1956–1970. [CrossRef]
9. Septimio, R.S.; Costa, T.A.; Silva, C.A.P.; Vida, T.A.; Damborenea, J.; Garcia, A.; Cheung, N. Correlation between unsteady-state solidification thermal parameters and microstructural growth of Zn–8 mass% Al and Zn–8 mass% Al–XBi tribological alloys. *J. Therm. Anal. Calorim.* **2020**, *139*, 1741–1761. [CrossRef]
10. Duarte, R.N.; Faria, J.D.; Brito, C.; Veríssimo, N.C.; Cheung, N.; Garcia, A. Length scale of the dendritic microstructure affecting tensile properties of Al–(Ag)–(Cu) alloys. *Int. J. Mod. Phys. B* **2016**, *30*, 1550261. [CrossRef]
11. Rodrigues, A.V.; Lima, T.S.; Vida, T.A.; Brito, C.; Garcia, A.; Cheung, N. Microstructure and Tensile/Corrosion Properties Relationships of Directionally Solidified Al-Cu-Ni Alloys. *Met. Mater. Int.* **2018**, *24*, 1058–1076. [CrossRef]
12. Ferreira, I.L.; Garcia, A.; Nestler, B. On Macrosegregation in Ternary Al-Cu-Si Alloys: Numerical and Experimental Analysis. *Scr. Mater.* **2004**, *50*, 407–411. [CrossRef]
13. Bertelli, F.; Freitas, E.S.; Cheung, N.; Arenas, M.A.; Conde, A.; Damborenea, J.; Garcia, A. Microstructure, tensile properties and wear resistance correlations on directionally solidified Al-Sn-(Cu; Si) alloys. *J. Alloys Compd.* **2017**, *695*, 3621–3631. [CrossRef]
14. Houska, C. Beryllium in Aluminum and Magnesium alloys. *Met. Mater.* **1988**, *4*, 100–104.
15. Karov, J.; Youdelis, W.V. Precipitation in Al–3Cu–0·1Be. *Mater. Sci. Technol.* **1987**, *3*, 1–6. [CrossRef]
16. Yuan, Z.; Guoa, Z.; Xiong, S.M. Effect of as-cast microstructure heterogeneity on aging behavior of a high-pressure die-cast A380 alloy. *Mater. Charact.* **2018**, *135*, 278–286. [CrossRef]
17. Çadırlı, E.; Nergiz, E.; Kaya, H.; Büyük, U.; Şahin, M.; Gündüz, M. Effect of growth velocity on microstructure and mechanical properties of directionally solidified 7075 alloy. *Int. J. Cast Metal. Res.* **2020**, *33*, 11–23. [CrossRef]
18. Gündüz, M.; Çadirli, E. Directional Solidification of Aluminium-Copper Alloys. *Mater. Sci. Eng. A* **2002**, *327*, 167–185. [CrossRef]
19. Kakitani, R.; Gouveia, G.L.; Garcia, A.; Cheung, N.; Spinelli, J.E. Thermal analysis during solidification of an Al–Cu eutectic alloy: Interrelation of thermal parameters, microstructure and hardness. *J. Therm. Anal. Calorim.* **2019**, *137*, 983–996. [CrossRef]
20. Jafari, H.; Amiryavari, P. The effects of zirconium and beryllium on microstructure evolution, mechanical properties and corrosion behaviour of as-cast AZ63 alloy. *Mater. Sci. Eng. A* **2016**, *A654*, 161–168. [CrossRef]
21. Rodrigues, A.V.; Kakitani, R.; Oliveira, R.; Barros, A.; Brito, C.; Garcia, A.; Cheung, N. Two-Phase Dendrite and Bimodal Structure in an Al-Cu-Ni Alloy: Their Roles in Hardness. *J. Mater. Eng. Perform.* **2022**, *31*, 3704–3715. [CrossRef]

22. Grandfield, J.; Eskin, D.G.; Bainbridge, I. *Direct-Chill Casting of Light Alloys: Science and Technology*; Wiley: New York, NY, USA, 2013; pp. 144–254.
23. Rocha, O.L.; Siqueira, C.A.; Garcia, A. Heat Flow Parameters Affecting Dendrite Spacings during Unsteady-State Solidification of Sn-Pb and Al-Cu Alloys. *Metall. Mater. Trans. A* **2003**, *34A*, 995–1006. [CrossRef]
24. Rozman, N.; Medved, J.; Zupanic, F. Microstructural evolution in Al-Mn-Cu-(Be) alloys. *Philos. Mag.* **2011**, *91*, 4230–4246. [CrossRef]
25. Barros, A.; Cruz, C.; Garcia, A.; Cheung, N. Corrosion behavior of an Al-Sn-Zn alloy: Effects of solidification microstructure characteristics. *J. Mater. Res. Technol.* **2021**, *12*, 257–263. [CrossRef]
26. Kakitani, R.; Cruz, C.B.; Lima, T.S.; Brito, C.; Garcia, A.; Cheung, N. Transient directional solidification of a eutectic Al-Si-Ni alloy: Macrostructure, microstructure, dendritic growth and hardness. *Materialia* **2019**, *7*, 100358. [CrossRef]
27. Brito, C.; Costa, T.A.; Vida, T.A.; Bertelli, F.; Cheung, N.; Spinelli, J.E.; Garcia, A. Characterization of Dendritic Microstructure, Intermetallic Phases, and Hardness of Directionally Solidified Al-Mg and Al-Mg-Si Alloys. *Metall. Mater. Trans. A* **2015**, *46A*, 3342–3355. [CrossRef]
28. Kakitani, R.; Reyes, R.V.; Garcia, A.; Spinelli, J.E.; Cheung, N. Relationship between spacing of eutectic colonies and tensile properties of transient directionally solidified Al-Ni eutectic alloy. *J. Alloys Compd.* **2018**, *733*, 59–68. [CrossRef]
29. Silva, C.A.P.; Kakitani, R.; Canté, M.V.; Brito, C.; Garcia, A.; Spinelli, J.E.; Cheung, N. Microstructure, phase morphology, eutectic coupled zone and hardness of Al-Co alloys. *Mater. Charact.* **2020**, *169*, 110617. [CrossRef]
30. Jackson, K.A.; Hunt, J.D. Binary Eutectic Solidification. *Trans. Met. Soc. AIME* **1966**, *6*, 843–852.
31. Wilde, J.D.; Froyen, L. Two-phase planar and regular lamellar coupled growth along the univariant eutectic reaction in ternary alloys: An analytical approach and application to the Al–Cu–Ag system. *J. Appl. Phys.* **2005**, *97*, 113515. [CrossRef]
32. Zhao, Y.; Zhang, W.; Yang, C.; Zhang, D.; Wang, Z. Effect of Si on Fe-rich intermetallic formation and mechanical properties of heat-treated Al-Cu-Mn-Fe alloys. *J. Mater. Res.* **2018**, *33*, 898–911. [CrossRef]
33. Payne, J.; Welsh, G.; Christ, R.J., Jr.; Nardiello, J.; Papazian, J.M. Observations of fatigue crack initiation in 7075-T651. *Int. J. Fatigue* **2010**, *32*, 247–255. [CrossRef]
34. Birbilis, N.; Cavanaugh, M.K.; Buchheit, R.G. Electrochemical behavior and localized corrosion associated with Al7Cu2Fe particles in aluminum alloy 7075-T651. *Corros. Sci.* **2006**, *48*, 4202–4215. [CrossRef]
35. Jiang, N.; Gao, X.; Zheng, Z. Microstructure evolution of aluminum-lithium alloy 2195 undergoing commercial production. *Trans. Nonferrous Met. Soc. China* **2010**, *20*, 740–745. [CrossRef]
36. Xu, D.; Zhu, C.; Xu, C.; Chen, K. Microstrutures and Tensile Fracture Behavior of 2219 Wrought Al-Cu Alloys with Different Impurity of Fe. *Metals* **2021**, *11*, 174. [CrossRef]
37. Karov, J.; Youdelis, W.V. Solubility of beryllium in $CuAl_2$. *Mater. Sci. Technol.* **1987**, *3*, 394–395. [CrossRef]
38. Amirkhanlou, S.; Ji, S. Casting lightweight stiff aluminum alloys: A review. *Crit. Rev. Solid State Mater. Sci.* **2019**, *45*, 171–186. [CrossRef]
39. Bowden, D.; Pokross, C.; Kaczynski, D.; London, G.; Starke, E.A.; Sanders, T.H.; Cassada, W.A. Characterization of aluminium-beryllium alloy sheet. *Mater. Sci. Forum* **2000**, *331–337*, 901–906. [CrossRef]
40. Zupanič, F. Extracting electron backscattering coefficients from backscattered electron micrographs. *Mater. Charact.* **2010**, *61*, 1335–1341. [CrossRef]
41. Okamoto, H. Supplemental Literature Review of Binary Phase Diagrams: Ag-Ni, Al-Cu, Al-Sc, C-Cr, Cr-Ir, Cu-Sc, Eu-Pb, H-V, Hf-Sn, Lu-Pb, Sb-Yb, and Sn-Y. *J. Phase Equilib. Diffus.* **2013**, *34*, 493–505. [CrossRef]
42. Kaiser, M.S.; Sabbir, S.H.; Kabir, M.S.; Soummo, M.R.; Al Nur, M. Study of Mechanical and Wear Behaviour of Hyper-Eutectic Al-Si Automotive Alloy Through Fe, Ni and Cr Addition. *Mater. Res.* **2018**, 21. [CrossRef]
43. Silva, C.; Barros, A.; Vida, T.; Garcia, A.; Cheung, N.; Reis, D.A.P.; Brito, C. Assessing Microstructure Tensile Properties Relationships in Al-7Si-Mg Alloys via Multiple Regression. *Metals* **2022**, *12*, 1040. [CrossRef]
44. Callister, W.D.; Rethwisch, D.G. *Materials Science and Engineering: An Introduction*, 8th ed.; Wiley: New York, NY, USA, 2009.
45. Zhao, Y.; Liu, K.; Hou, H.; Chen, L.-Q. Role of interfacial energy anisotropy in dendrite orientation in Al-Zn alloys: A phase field study. *Mater. Des.* **2022**, *216*, 110555. [CrossRef]
46. Wang, Y.; Xiong, Y. Effects of beryllium in Al-Si-Mg-Ti cast alloy. *Mater. Sci. Eng. A* **2000**, *280*, 124–127. [CrossRef]

metals

MDPI

Review

Technological Insights into the Evolution of Bronze Bell Metal Casting on the Korean Peninsula

Chun-Soo Won [1], Jae Pil Jung [1,*], Kwang-Sik Won [2] and Ashutosh Sharma [3,*]

1 Department of Material Science and Engineering, University of Seoul, Seoul 02504, Korea
2 National Intangible Cultural Heritage No. 112, Seupji-gil 35, Deoksan-eup, Jincheon-gun 27815, Korea
3 Department of Materials Science and Engineering, Ajou University, Suwon 16499, Korea
* Correspondence: jpjung@uos.ac.kr (J.P.J.); ashu.materials@gmail.com (A.S.); Tel.: +82-10-95345040 (A.S.)

Abstract: Bronze cast bells have been designed and developed for hundreds of years, with the worldwide spread of several faiths and religions such as Buddhism, Catholicism, and Protestantism. The exceptional ringtones of bronze bell metals have scientific healing and cultural importance. In this review article, we highlight the evolution of bronze bell metal over the decades, its composition, and the complex fabrication technologies used to date. Furthermore, we overview ancient and modern casting alloy technology, especially bronze bell castings in Asia. The bell shape, materials, and alloy casting technology have undergone dramatic change over the years. For comparison, we include different bronze cast bells and their characteristics produced from the Middle Ages to the present times. Based on the data obtained from the bell casting technology surrounding the Korean Peninsula, the major trends in the evolution of bronze bell castings and long-standing traditions of mold materials and alloys are described. In the present review, the effects of different elements on bell materials are qualitatively overviewed, with an assessment of the material and casting properties, service life, and bell sound. We also highlight the challenges of conventional bronze casting and possible solutions for future investment castings and rapid prototyping of bronze bells.

Keywords: cast alloys; bells; silicon; bronze; 3D printing; wax casting

Citation: Won, C.-S.; Jung, J.P.; Won, K.-S.; Sharma, A. Technological Insights into the Evolution of Bronze Bell Metal Casting on the Korean Peninsula. *Metals* **2022**, *12*, 1776. https://doi.org/10.3390/met12111776

Academic Editors: Carlos Alexandre Dos Santos and Srecko Stopic

Received: 22 August 2022
Accepted: 14 October 2022
Published: 22 October 2022

1. Introduction

With the progress of humanity, bronze cast bells have occupied nearly all facets of human life. Bronze bells have been an important component of the life of societies irrespective of the historical traditions and customs they represent. The historical bronze bells stem from the 19th century BCE. These bells were believed to be a signature of religious customs, morning prayers, Buddhism, and pieces of art [1]. In this journey, humans learned the process of casting materials into definite shapes at high temperatures, which is metal forming, which is still used today. Casting technology has been known to man for many centuries and employed to shape tools for daily life. Cast weapons were mainly used in war and in protection against dangerous animals, as temple bells, as household products, and as art [2,3].

Bells are used either singly, doubly, or in combination with several bells as instruments, which are usually hit by a wooden block or shaken by hand or in a clock-operated system. In small cities, temple bells are hung in a church tower to wake the citizens for morning prayers. Bells were also part of various ceremonies and weddings or used in ships as a sign of imminent danger [4]. Figure 1 shows a typical temple bell structure with different body parts.

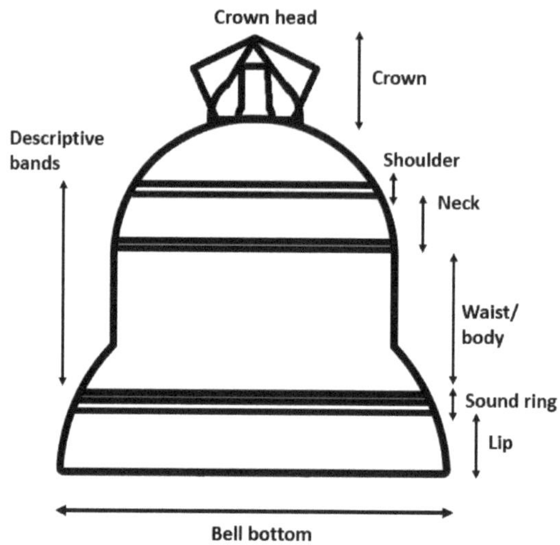

Figure 1. Schematic description of the bell structure.

The various parts of a temple bell include the bell bottom, which has different shapes: round, smooth, scalloped, etc. Lip and sound rings are located below the body. The upper regions consist of the neck, shoulder, and crowns of the bell. Several descriptive bands and scriptures can be written over the body, which are related to the specific culture or history of that religion. The crown heads consist of hanging holes or some divine creatures such as a dragon as seen in most Southeast Asian bells [4,5].

2. Historical Evolution of Casting Bell Technology in Asia

2.1. Appearance of Bells in Asia

From the beginning of the human era, bells have been an important part of the everyday ritual practice of Buddhist monks and monasteries in South Asia. Both the Buddhist and Chinese Daoist customs promoted the casting of copper and bronze alloys. Consequently, further spread of this technology occurred in the inner Asian countries, including Korea and Japan. The beginning of the Bronze Age occurred at different times in different geographical regions. This evolution began in the 7th century and passed to Tibetan bells in the 8th century. Bronze casting technology evolved in Southeast Asia in about the 10th century BCE [5]. The beginning of the Bronze Age took place around the 10th century BCE on the Korean Peninsula and the 15th century BCE in Manchuria (a region in northeast China) [6].

Written inscriptions such as epigraphy were either engraved or cast into bells, which proclaim the representative religion to serve narrative purposes. Without any doubt, many examples that once existed in the last millennium have either been spoiled, damaged, or burnt during the past centuries. For example, the Nazis confiscated over 175,000 bells throughout Europe for copper and tin. Of these, 150,000 were destroyed [7]. The Tsar Bell in Moscow was damaged during a fire in 1737 [8] and the Chinese Xi'an Bell Tower built during 1384 was bombed by the Japanese in 1939 [9]. Most Korean bells were lost during Japanese colonial rule (1910–1945), which were later restored by Kwang-Sik Won [10]. Thus, most of the precious data related to the initial form of bronze cast bells, their size and shapes, and the casting technology is mostly found in Buddhist Asia [4,5].

2.2. Types, Geographical Distribution, and Characteristics of Bronze Cast Bells

Bronze temple bells are largely divided into Eastern bells and Western bells. Eastern bells refer to temple bells and Western bells refer to worship bells in churches or cathedrals.

There are various types of bronze bells around the globe that require our attention due to their historical importance, their production, and the specific roles they played. The history of bells passes from generation to generation because bells have a longer life and are used in numerous events, outlasting humans. In Asia, common bronze cast temple bells include both flat and scallop-edged bottoms [1,11]. The important temple bells found in Asia include the Chinese bell of the Zhou dynasty (770–256 BCE), the Japanese temple bell, the Kanzeon-ji bell of 698 BCE, and the Korean Sangwonsa Temple bell, founded in 725 BCE [12–14]. Common bells located in Europe differ in the shape of the bell bottom. These include Tuba Dei, the Royal Sigismund Bell, Maria Bogurodzica, Cracow bell, etc., which are mostly used in churches for morning prayers [1,4,5]. Japanese bells have a convex-type bottom while concave types are common in Tibetan temple bells. Other East Asian temple bells are flat-bottomed, an uncommon observation in Tibetan bells [5]. In Korea, bronze bells were first made in the Bronze Age (3300–1200 BCE). However, these bells did not produce a good sound and were not musical and were often used in horse trappings and shamanist ceremonies as a symbol of authority [5]. Apart from the composition, the sound of the bell mostly depends on its shape. For example, a circular shape allows a wave to travel around the bell's perimeter. The standing waves generated around the circumference of the bell are responsible for the special tone of the ringing [15]. The first musical bells were made in China during 1116 BCE and were played in the *aak* orchestra and accompanied Confucian rites and ceremonies. The most famous Korean bells are big, shaped casting bells (*pomjong*) used in monasteries and Buddhist temples, which were introduced from China in the Tang dynasty [1,5]. However, utilizing other technology, such as papermaking and ceramics, the Koreans outperformed their teachers and developed exceptionally large and finely decorated bells in this era [5]. Papermaking of parts is generally used in metal castings to prevent melt leakage from the mold and prevent heat shrinkage [16]. Korean papermaking from mulberry has been used for printing, preparation of walls, and in ondol floors. The main sources of cellulosic fiber evolved as the ancient craft migrated from its origin in China to Korea and Japan, and then spread to Western countries [17]. Potteries such as ceramics were used on the bell surface to create engraved parts, which were filled with white and black clay, and various aesthetic decorations were produced over the bell [18]. Since Buddhist temple bells have different styles, varying from one country to another, they can be divided into Chinese bells, Korean bells, Japanese bells, Vietnamese bells, and Southeast Asian bells. According to the Gyeongju National Museum's report (a world heritage site in Korea), the Korean bell is rated as having the most beautiful sounds and patterns. There are various reasons why Korean bells make such a beautiful sound. The first reason is its unique shape and structure, which filters out undesired noise. This bell sits lower than most of the other bells. The caved-in bottom creates greater resonance. The top of the bell is almost half the diameter of the bottom, which also helps to improve the sound clarity [19]. The Korean bell is considered as a masterpiece among other bells around the world due to its unique and beautiful shape and clear and long-lasting sound. In simple words, the bell rings and reverberates like a "Maengnoli phenomenon", which is the sound of a heartbeat [20,21]. All these attributes led to its own scientific name "Korean bell", demonstrating the excellent aesthetic sense and workmanship of our ancestors [22,23]. The sound of the Chinese bell is relatively noisy and has short resonance. Other bells such as the Japanese bell are similar to the Chinese bell and cannot produce beautiful sounds, causing fluctuation from deep inside [18–21]. Some of the popular bell types on the Korean Peninsula are summarized in Table 1.

Table 1. Some common temple bells that evolved on the Korean Peninsula, data from [5,24,25].

Bell Name	Period	Remarks
Silla bells	57 BCE to 935 CE	Silla art, gold crowns, tombs, Buddhist temples
Sangwonsa bell	725 CE	Mount Odaesan near Pyeongchang
Songdok bell	771 CE	Gyeongju National Museum, Korea
Hwangnyong-sa bell	553-645 CE	Gyeonju, Korea
Goryeo bell	963 CE	Toksu Palace Museum of Fine Arts, Seoul, Korea
Kasagi bell	1196 CE	Kyoto, Japan

The largest temple bells in Korea were produced during the Silla kingdom and are shown in Figure 2a. The unified Silla period witnessed significant development in arts and crafts, and large bronze bells were hung in Buddhist temples. The other oldest surviving bells include Sangwonsa Temple bell on Mount Odaesan near Pyeongchang and Songdok bell, also known as the Emille Bell, at the end of the Silla period (57 BCE–935 CE). The Songdok bell, made using a lost wax casting, weighs 19 tons (Figure 2b). This bell is housed at Pongdok Temple in Gyeongju and is thus called the Gyeongju bell of Korea [26].

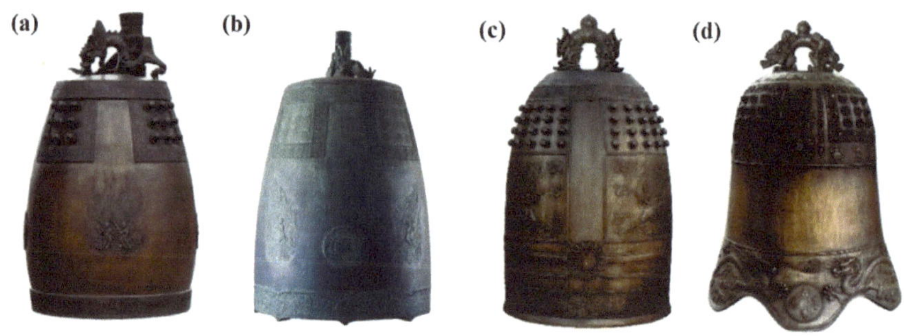

Figure 2. Important temple bells that evolved on Korean Peninsula. (**a**) Sangwonsa bell, (**b**) Songdok bell (Emille Bell), (**c**) Japanese bell, and (**d**) Chinese bell. Korean Buddhist 'Emille' Bell [26]. World History Encyclopedia. Last modified October 05, 2016. Retrieved from https://www.worldhistory.org/image/5838/korean-buddhist-emille-bell/ (accessed on 1 February 2022).

The height of the Songdok bell is about 3.5 m with a diameter of 2.27 m and is decorated with lotus and heavenly creatures suspended through a signature dragon loop. Some of the bells that were larger than Songdok bells were, unfortunately, short-lived. In addition, the massive Hwangnyongsa bell, which had a record weight of 300 tons, disappeared. According to the historical records, these bells were either buried or disappeared with the temples due to heavy floods in early 1400. Some of the largest bells of Silla were taken away during the Mongol invasion [27,28]. After the Silla period, the Goryeo bell dates to the Goryeo dynasty (918 to 1392 CE), in which Buddhism continued to be the dominant religion of the state. Many examples of such bells exist in Japan today, being donated or captured during the Hideyoshi invasion in the 16th century CE. Goryeo bells are smaller in size than Silla bells, which were bronze cast and mostly decorated with dragons and heavenly creatures. Examples include the Naesco Temple bell in southwest Korea (1222 CE), which has lotus petals over the upper rim of the bell, a wider border around the nine-nodule squares, and four spheres accommodating the dragon suspension loop [29].

The above discussion makes it clear that primitive Buddhist temple bells were mostly founded in East Asia, specifically in China, Korea, and Japan. In contrast to Korean bells, Japanese bells are flat-mouthed (Figure 2c), with a few rare exceptions of the temple bell from Kasagi Temple, Kyoto, cast in 1196, which had a different form of scalloping, which has exactly the opposite rim shape to that of the Songdok bell [5,11,22,29]. A major difference between Chinese and Korean bells lies in the shape of the bottom, being scalloped in Chinese bells (Figure 2d).

3. Bronze Alloys and Temple Bells

As discussed earlier, bells have been used in musical idiophones and have been known to humankind since the primitive ages [3]. The materials used for casting bells cover a wide range of elements. Primitive bells were cast using copper (Cu) or iron (Fe) sheets. In Southeast Asia, especially some parts of China, the production of iron occurred as early as 600 BCE. After the Iron Age, it gradually spread across the Korean Peninsula, where it is hard to locate the end of the Bronze Age, unlike other parts of the world. Bronze cast products continued to be made in large quantities even after people had adapted to working with iron. It was predicted that until the advent of the 2nd century CE, cast iron bells were more popular than bronze bells [30]. One such example of a cast iron bell is the iron bell of Jeondeungsa Temple (164 cm × Ø 100 cm) in Korea, which is similar to most Chinese bells. According to the inscriptions engraved over the bell, the bell belonged to China in around 1097 BCE [31]. At present, the major component alloy of temple bells is Cu-Sn alloy. A specific composition of 78 wt.%Cu and 22 wt.%Sn is widely known as bell metal [32,33]. Bell bronze is a structurally complex alloy, and the quality of bells mostly depends on the casting parameters and technology used for their fabrication [34,35]. The quality of cast bells is assessed using their acoustic properties, vibrations, and frequency analysis. The acoustic vibrations are also affected by the present porosity and structure of the cast alloy [36].

Copper alloys account for more than 90% of the materials used to make bells in both Eastern and Western countries and these materials are very limited to the extent that these bells are intermittently made of Fe and Cu-Ni. According to Strafford, it is believed that the production of cast iron (Fe-C) bells started in 1857 with the development of cast steel. However, Fe-C alloys were not suitable due to their poor sound and short life. Many ancient bronze castings included Ni as an impurity (<1%) associated with copper ores [37]. However, as Ni > 1%, this effect becomes noticeable, and it is believed that ancient men developed a certain type of copper bronze containing an appreciable amount of Ni [38]. Previous works pioneered by Audy et al. [34,37] showed that Ni < 2 wt.% is the recommended fraction for obtaining a balanced set of castability, strength, and sound quality properties of a bronze cast bell. This is due to the characteristics of bells that must make a sound. Among these materials, bronze is the most used material for bells because of its clear and resonant sound. The important characteristics of bronze are (a) the tone determination according to the speed of sound and loudness, (b) damping ability, and (c) toughness. Most world-famous bells, such as the American Liberty Bell and the Korean Emille Bell, are made of bronze [11,39]. The strength and sound of bronze depends on the content of tin. Eastern bells, characterized by their loudness and low notes, have a tin content of 12–18% while clear, high-pitched Western bells have a tin content of 20–24%. The tin content of Western bells is higher than that of Eastern bells [40,41].

3.1. Composition of Bronze Bells

The generated sound quality and functioning of bells depend on the cast alloy composition, their methods of fabrication, and inherent cast defects. In other words, the quality of the cast bell is highly dependent on the material type, shape, and methodology. Mostly, temple bells are composed of bronze castings. The major bell alloys include bronze, brass, cupronickel, and iron. Bronze accounts for more than 80%, and the chemical composition of bronze used for bells is reported to contain various metals such as Zn, Sb, Ni, and Pb in the Cu-(10–25)%Sn system. This trend does not differ much between the East and the West [34]. Table 2 shows the materials of the ancient Buddhist temple bells (Eastern bells). It can be seen that the tin contents of the same oriental bells, Japanese bells, are similar to those of Korean Buddhist temple bells.

Table 2. Chemical composition of some temple bells surrounding the Korean Peninsula [11].

Name of Bell	Chemical Composition (wt.%)						Impurity
	Cu	Sn	Pb	Au	Zn	Ag	S
Sangwonsa bell	83.87	13.26	2.12	-	0.32	-	-
Seonrimwon bell	80.2	12.2	-	-	2.2	-	0.1
Silsangsa bell	75.7	18.0	0.31	-	-	-	-
Bell of the Joseon dynasty	80.1	12.2	-	-	-	-	-
Japanese bell	82.1	9~13	-	-	1~1.7	-	-

The successful casting of bells is largely based on tin bronzes. Previous studies have shown wide variation in the composition of tin bronzes (Cu-Sn) used for the construction of temple bells. The compositions of the different bells shown in Table 2 were mostly analyzed for various alloying and impurity elements. The important bronze bells surrounding the Korean Peninsula are summarized in Table 2. Most of these bells contain Pb and Zn as alloying elements except Seonrimwon bell, which contain sulphur. The compositions that were not detected are not shown in Table 2. Traditionally, bell bronze consisted of numerous other alloying elements, for example, Pb, Ag, Zn, Sn, Bi, Sb, Fe, As, Ni, Si, S, and P [11,34,41,42]. Among these elements, Pb, Zn, Ni, Ag, Sb, and Fe are important alloying elements. In addition, Bi, As, Si, S, and P are impurities trapped within the bronze melt during melting with coke and charcoal [34].

In the search for a better bronze alloy composition, bell makers have experimented with a range of different alloying elements in various fractions. The alloy composition, however important, is not the sole factor used to assess a bell's properties. The problem of meeting the demands concerning the material properties becomes even more complicated as additional variables, thus far neglected or impossible to control, are considered. The requirements that need to be met by the bell alloy are as follows: (a) easy melting/casting, (b) high casting quality, (c) uniform and homogeneous bell surface properties, (d) durability, and (e) nice generated sound. Therefore, the casting technology used in bell preparation is a matter of secrecy in different bell foundries.

Bronze has a higher corrosion resistance than pure Fe, especially against harsh environmental conditions. The addition of Sn in Cu provides further strengthening of the Cu matrix and prevents mechanical damage to the bell when it is hit with a hammer, which in turn could degrade the sound. The alloying of Cu with Sn in moderation is preferred to avoid impact fracture upon prolonged striking of the bell. Additionally, Sn increases the sound amplitude; however, other elements such as Pb promote the castability but deteriorate the sound duration. A common Middle Age bronze bell composition falls within the range of 20–25 wt.% Sn; in recent times, lower amounts of Sn (10–15 wt.%) and Pb (1–3 wt.%) have been used [43]. The important bell compositions developed over the last decades are illustrated in Table 3. The ticks indicate the presence of a specific alloying element in the bronze cast bell.

Table 3 indicates that the material composition of bronze alloy has a significant impact on the mechanical and acoustic properties. According to the pioneering work of Audy et al. [34,37], the recommended concentration of Ni is less than 2 wt.% to achieve a high wear resistance, fatigue resistance, and castability of the melt. If Pb > 1.5 wt.%, the machinability, fluidity, and wear resistance are improved but at the cost of the sound quality. Additionally, if Sb > 1 wt.%, the brittleness increases, and the sound quality becomes poor. Other impurities such as $p = 0.01$ wt.% assist the deoxidizing properties of the melt during casting. When copper is melted, oxygen and other gaseous impurities trapped in the bell core cause further cracking after solidification. Thus, $p = 0.01$ wt.% is recommended for better bell casting. Based on these abovementioned factors, the most common bronze composition for optimum bell characteristics is recommended: ~20 wt.%Sn, <2 wt.%Ni, <1.5 wt.%Pb, <1 wt.%Sb, ~0.01 wt.%P, and Cu as a balance. There is large variation in different bells' composition in practice due to the different bell manufacturing practices [37,44–56]. Therefore, metallurgical analyses of bronze cast bells, including an evaluation of the mechanical properties, should be carried out in order to assess the quality of bells.

Table 3. Composition of bronze cast bells (Eastern and Western bells) and their possible effects on the mechanical properties. The tick mark (✓) indicates the presence of particular element in the casting.

Alloying Constituents (wt.%)											Properties	Ref.
Sn	**Pb**	**Zn**	**Bi**	**Ag**	**Sb**	**As**	**Ni**	**Fe**	**S**	**P**		
10–20	-	≈5	-	-	-	-	-	-	-	-	High tensile strength	[37,44–49]
10–20	✓	-	✓	-	-	✓	-	-	-	-	Reduced tensile strength	[44–47,49]
<23	-	<5	-	-	≈1	-	-	✓	✓	-	High hardness	[44–46,49]
>23	-	-	✓	-	>1	✓	-	-	-	-	High brittleness	[44–46,49]
-	✓	-	-	-	-	-	✓	✓	-	-	High wear resistance	[49]
-	-	<5	-	-	-	-	-	-	-	-	High elongation	[37,44,45,49]
<5	-	-	-	-	-	-	-	-	-	-	Reduced elongation	[43,44]
≈20	-	-	-	-	-	-	-	-	-	-	High elastic limit	[43,44,46,48]
>23	-	-	-	-	-	-	-	-	-	-	Low elastic limit	[43,44,46,48]
-	-	-	-	-	-	-	-	✓	-	-	Abrasion resistant	[50]
✓	-	-	-	-	-	-	-	-	-	-	High toughness	[21,22,24]
-	-	-	-	-	-	-	-	✓	-	-	Low toughness	[43,44,46]
-	-	-	-	-	-	-	✓	-	-	-	Fatigue resistant	[51]
✓	✓	-	-	-	-	-	-	-	-	-	High fluidity	[11,22,43]
-	-	-	-	-	-	-	>2	-	-	-	Low fluidity	[5,43,49]
-	-	-	-	-	≈1	-	-	-	-	0.01	Deoxidizing	[43]
-	-	-	-	-	-	-	✓	-	-	-	Reduced deformability	[52]
-	-	-	-	-	-	-	-	✓	-	-	Poor rusting resistance	[44,45,53]
>15	-	-	-	-	-	-	-	-	-	-	Poor crack resistance	[43,44]
✓	-	-	-	-	-	-	-	-	-	-	Change in cast color	[33,43–45]
-	-	✓	-	-	-	-	-	-	-	-	Reduced crystallization gap	[42–44,49,54]
-	-	-	✓	-	-	✓	-	-	-	-	Poor castability	[43–45,49,54]
✓	✓	-	-	-	-	-	-	-	✓	-	High machinability	[43–46,49,54]
-	-	-	-	-	-	-	-	✓	-	-	Poor machinability	[3,5,11,44]
<10	-	-	-	-	-	-	-	-	-	-	Highly porous	[2,44,45,54]
-	-	-	-	-	-	-	0.4–5	-	-	-	Reduced porosity	[55,56]
-	-	-	-	-	-	-	0.5–2	-	-	-	Softening increased	[45,55]
-	-	-	-	-	-	-	2–4	-	-	-	Structure stabilizes	[45,55,56]
20–23	-	-	-	-	≈1	-	-	-	-	-	Good sound quality	[37,44–46]
-	>1.5	>1.5	-	✓	-	-	-	-	-	-	Poor sound quality	[37,44–46]
✓	✓	✓	-	-	-	-	-	-	-	-	Reduced melting point	[37,43–45,47,48]
✓	-	-	-	✓	-	-	-	-	-	-	Expensive	[37,43–46]
-	✓	-	-	-	-	-	-	-	-	-	Cost-effective	[37,43–46]

3.2. *Microstructure of Bronze Bells*

The microstructure of bronze consists of an alpha (α) solid solution phase and an alpha + delta ($\alpha + \delta$) eutectoid phase, which is known to be affected by the Sn content and mold material. Here, the δ phase is $Cu_{31}Sn_8$, which is a brittle intermetallic compound. In the phase diagram shown in Figure 3 [42,57], it can be seen that the Cu_6Sn_5, Cu_6Sn_5, and Cu_3Sn intermetallic compound (IMC) phases can also be generated. Since these IMCs can cause cracks when the bell is used for a long time, it is necessary to suppress their formation. For this purpose, the use of a lower content of Sn or an increase in the cooling rate is recommended to reduce the size of IMCs or suppress their formation by supersaturating Sn [37]. Annealing procedures have also been suggested to remove the brittle δ-phase of the Cu-Sn system [58,59].

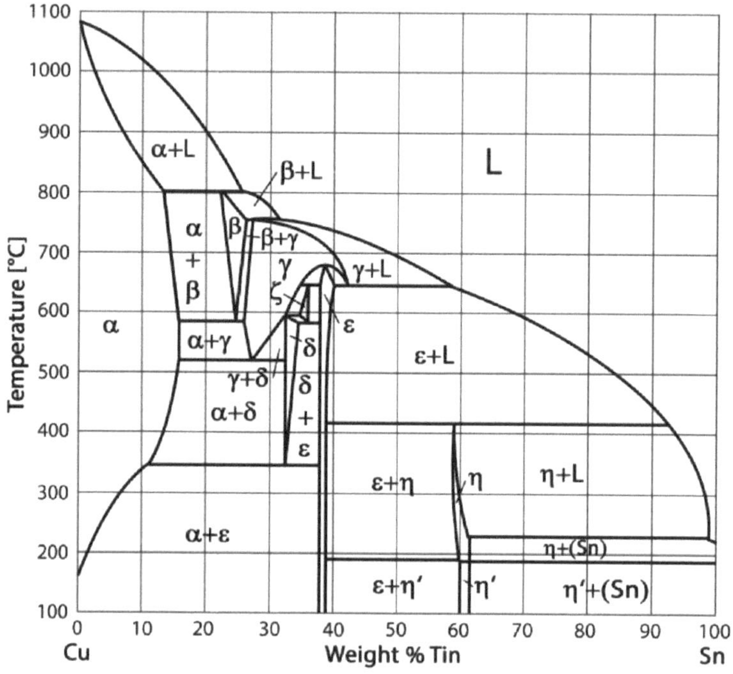

Figure 3. Cu-Sn phase diagram [42].

An increase in the Sn content improves the technological properties of tin bronze, especially regarding the improved hardness, wear and friction properties [60].

Meanwhile, the mechanical characteristics of cast bronze have a maximum tensile strength at 17% Sn, elongation at 4~5% Sn, and maximum hardness at 32% Sn [59]. Regarding the material of actual bells, the higher the strength and hardness, the better. When the tin content is greater than 20%, the hardness increases but brittleness appears, so it is not only weak to fatigue after impact but segregation also easily occurs during solidification, so it is difficult to obtain uniformity in the material [61,62]. Therefore, the tin content of bronze Buddhist temple bells produced in modern times is around 15~17%, and the tin content is adjusted according to the size of the bells [63,64]. A typical bronze bell microstructure is composed of α-dendrites and α + δ eutectoid as shown in Figure 4a. The black dots correspond to the porosity in the matrix (Figure 4b,c). Others have also reported a similar microstructure in the past [65–67]. The role of the α + δ microstructure in the corrosion degradation mechanism of bronze artifacts was studied in the past using potential pH Pourbaix diagrams [67,68]. The results indicated that the porosity and brittle particles provide additional reaction sites and dissolution occurs through crevices and pores, which promotes the entire corrosion process.

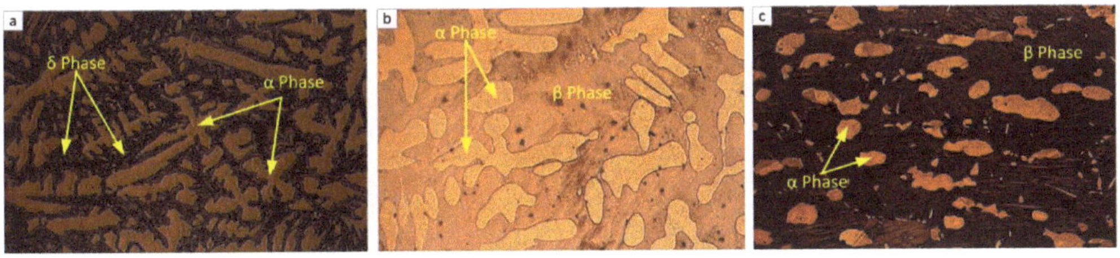

Figure 4. Optical micrograph of tin bronze (78 wt.%Cu-22 wt.%Sn). (**a**) As-cast, (**b**) quenched in oil, and (**c**) quenched in water. Reprinted with permission from Ref. [68]. Copyright 2021, Elsevier.

Recently, Sarkar et al. studied the effect of quenching media on the morphology of the brittle phases in as-cast bell metal (78wt.%Cu-22 wt.%Sn). The brittle δ-phase showed a needle-like structure. After quenching in oil and water media at 700 °C, the inter-dendritic region of all the samples differed. After comparison with the phase diagram, it was found that the dendritic structure of all three samples was in the α-phase while the interdendritic region of the as-cast sample was in the δ-phase; the β-phase was shown by the oil- and water-quenched samples [68]. The mechanical properties of the oil- and water-quenched samples were increased by 8% and 25%, respectively, after increasing the quenching temperature by 50 °C to above 700 °C due to the transformation of α- to β-phases by fragmentation of the dendrites. These results are consistent with several previous studies [69–72].

The binary phase diagram of tin bronze shows that the δ-phase solidifies first in the remaining Sn-rich liquid and other solute elements [73]. The interdendritic phases undergo eutectoid reaction to form an α + δ mixture, where δ is the brittle $Cu_{31}Sn_8$ compound. Rapid cooling prevents the generation of the brittle δ-phase and retains the α-solid solution. If the content of Sn is higher, i.e., 9–10%, the δ-phase inevitably forms, which impacts the mechanical strength and reduces the ductility of bronze. Some researchers have used high-temperature homogenization at 700 °C to improve the diffusion of brittle compounds into the α-phase, causing a single-phase state after fast cooling [74–76].

The composition of a bronze (80 wt.%Cu-20 wt.%Sn) bell was analyzed by Cekus et al. in a recent study [77] as given in Figure 5a–d. The compositional analysis showed that the test area A and B had α-fractions of 60% and 40%, respectively. Here, A and B refers to the "top" and "bottom" sections of the bell as shown in Figure 5a. The as-cast bronze microstructure consists of primary dendrites covered by α + δ eutectoid (Figure 5b–d), which is responsible for the sound quality of the bell. The eutectoid fraction was 40% and 60%, respectively, without taking into account the porosity. The microhardness of the α-phase in samples A and B was 132 and 135 HV. The microhardness of the δ-phase in samples A and B was 347 and 319 HV, respectively. The reduced mechanical properties along the thickest cross-section of the bell bottom did not affect the sound quality but did cause a decrease in the Sn content in the α-phase. The authors reported that this occurred to the differences in the melting, pouring, and solidification procedures of the alloys, i.e., a higher solidification time at the bell bottom creates a difference in the mechanical characteristics of the bell [77].

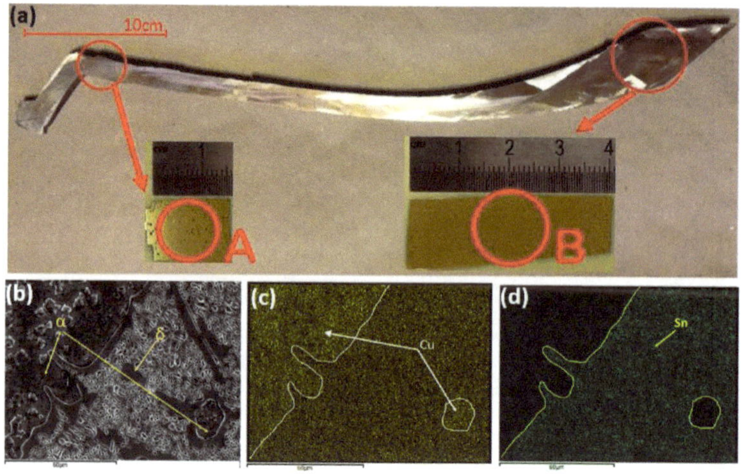

Figure 5. Microstructural analysis of an as-cast bronze bell. (**a**) The cut view areas of the test points A (top) and B (bottom) of the bronze cast bell. (**b–d**) Scanning electron microscopy and the elemental distribution in areas with increased Cu and Sn contents, respectively [77].

Low-tin bronze alloy depicts a single-phase microstructure and has better machinability, which is used to make thin sheets and artifacts [78]. Previous researchers have added Pb to compensate for the difference in the Sn content from the top to the bottom core of the bell. The addition of Pb improves the melt fluidity but remains undissolved and segregates in the solid state. The presence of Pb-spheroids and their size distribution contributes to the machinability of Cu-based alloys [78]. Pb is found as spherical aggregates along with the α and δ phases in the eutectoid. The microhardness of bronze bells varies between 160 and 200 HV0.5 due to the non-homogeneous microstructure ("harder" $\alpha + \delta$, α and Pb-globules, and porosity) [67].

Similar to tin bronze, aluminum bronze has also been used to enhance the acoustic properties of bronze bells and the feasibility of a substitution for dedicated Cu-Sn bronze. Al-bronze is characterized by an excellent hardness and strength. Previous authors have shown that castings made of $CuSn_{20}$ and $CuAl_{10}Fe_3Mn_2$ have similar acoustic vibrations, which can be used to develop cheaper and high-strength bell templates that can facilitate the desired curvature shape and acoustic characteristics without further additional tuning [36,62,79].

4. Metal Casting Technology

According to the metallurgy knowledgebase, casting is a material fabrication process in which molten material is introduced into a mold, allowing it to solidify within the mold, and then the mold is detached or broken to obtain the fabricated part. Castings are mainly used to make complex shapes that would be otherwise difficult or uneconomical to prepare using other methods, such as cutting, turning, or joining a solid [80]. In this section, we highlight the different methods used in casting technology.

Two major categories of metal castings include (1) expendable and (2) non-expendable mold casting technologies (Figure 6a). Expendable mold castings are permanent or temporary molds produced by sand casting and plaster of Paris (POP) molds. Most bronze bell castings utilize expendable mold castings with temporary pattern casting such as lost wax casting (LWC) or semi-permanent mold castings such as rotational molding or pep-set casting (Figure 6b). In this section, we discuss the casting technologies related to bronze cast bells in particular.

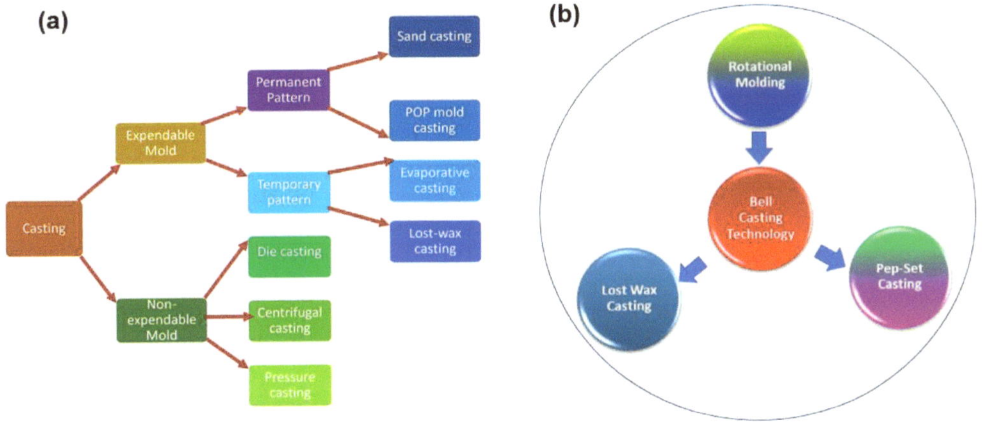

Figure 6. (**a**) Metal casting technologies and (**b**) popular bronze cast technologies for bells.

The traditional method of bronze casting using template molding expands the specified position of a mold during metal pouring. These template molds are considered improper because they do not take advantage of directional solidification [81]. The internal shrinkage pores inside the castings can disturb the sound vibrations in bells by introducing additional nodal points. Specifically, the casting technology of bronze bells can be divided into four methods, such as beeswax casting, rotational molding, polymer-set process, and LWC [82–85]. Among these, the rotational molding and beeswax casting methods are traditional casting techniques that have been used since ancient times. Beeswax remnants are not common in prehistoric European locations, with only a few exceptions to date [86,87]. The beeswax casting method was later used in Asia, especially in China and the Korean Peninsula, for the fabrication of bells, swords, and artifacts [88,89]. It is no exaggeration to say that almost all Buddhist bells in Korea were produced using the beeswax casting method from the unified Silla and Goryeo period [90].

4.1. Traditional Bronze Casting Methods for Bells

4.1.1. Importance of Beeswax in LWC

LWC or beeswax casting employs wax obtained from honey bees, an essential organic mixture of various compounds. The composition of beeswax differs quantitatively from different bee species [91]. The major components of beeswax include hydrocarbons, fatty acids, esters, and traces of a few unknown compounds [92–95]. Most recently, seven homologous series of beeswax have been used for identification, such as odd and even numbers of C- atoms (C17 to C35) (C22 to C34); odd numbers of monounsaturated fatty acids (C21 to C35) except HC29:1, HC31:1, and HC33:1; and monoesters (palmitate, (C34 to C50); oleate (C18:1–18, C18:1–20), and hydroxypalmitate with large chain alkanols (an isomer of C34 to C50)) [96].

Beeswax is believed to have specific importance in ancient technological, cultural, artistic, and symbolic areas as shown in Figure 7. The various applications of beeswax were summarized by Regert et al. in 2001 [97]. The last few years have shown new applications of beeswax as a glaze over pigment [98], a waterproof coating in ceramics [99,100], and an adhesive for a binding agent on Chinese turquoise-inlaid bronzes [101,102], in addition to its use as a sealing material for funereal homes [103]. With the development of lost wax metal castings, a significant amount of beeswax is required for the production od bronze castings using this method [104].

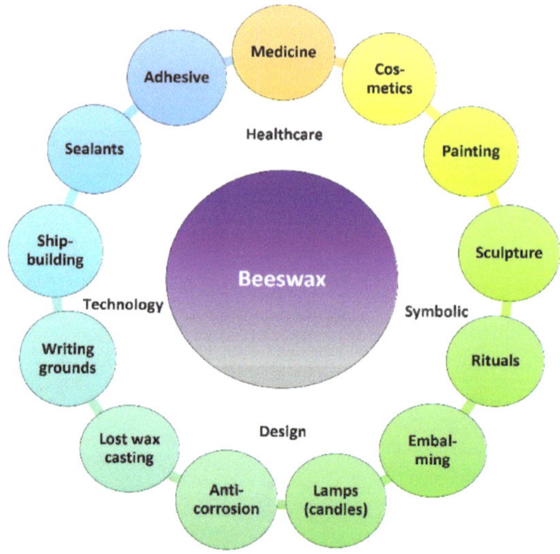

Figure 7. Application of beeswax in technological and artistic areas.

Beeswax has been used as a dental filling in Neolithic Slovenia [105]. Archaeological surveys have confirmed that the beeswax present in cooking pots is the remainder of beeswax castings [106]. Lamps or candles have played diverse technological, symbolic, and artistic roles in ancient times [107,108].

4.1.2. LWC Background

LWC is the most ancient casting technology for the production of cast artifacts, sculptures, bells, and swords. LWC is also known as investment or precision casting and has been widely known for centuries. The cast components have an excellent high surface finish, dimensional stability, and complex shapes (near-net-shape geometry) that are made possible when micromachining is not feasible or wasteful. Recent advancements in casting technologies have made it the most versatile casting method among the different casting technologies. According to Taylor [109], LWC originated in 5000 BCE when primitive men used it to fabricate tools for hunting animals.

Figure 8a–d shows the various cast objects used by primitive men to cast swords and pointed weapons using LWC. The assembled wax model was immersed in a clay mold followed by heating and squeezing of the wax out of the mold, leaving behind a hollow object that was filled with liquid metal. The solidified weapon heads were then detached and finished. Kotzin [110] mentioned the importance of this process in various jewelry, idols, and art castings for centuries. LWC has been found in various areas across the globe such as the treasures of the Pharaohs in Egypt, Inca tombs of South America, the ancient Etruscans, the Greeks in Europe, the Chinese Bronze Age, Indus Valley and Harappan civilization, etc. An example of a Harappan bronze sculpture, a creative dancing girl, is shown (Figure 8e). Later, this LWC was utilized to produce artifacts of Cu, Cu-Sn, and Au. The review by Kotzin states that LWC helped in promoting the cultural civilization of people across the globe. LWC offered solutions to various complex-shaped objects, undercut parts with a smooth finish, and fine details. In recent times, the use of LWC has continued to increase in the manufacturing of a variety of products.

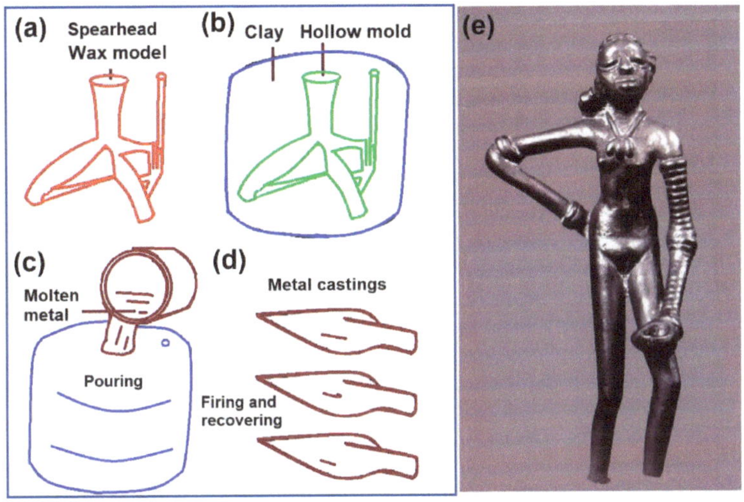

Figure 8. LWC of ancient spearheads. (**a**) Wax model assembly: (**b**) Placing of the wax model in the clay mold; (**c**) pouring of liquid metal into the clay mold after wax draining; and (**d**) detachment of the mold to obtain the solid product [109]. (**e**) Dancing girl sculpture.

4.1.3. Merits and Drawbacks of LWC

The various applications of LWC in the modern age include charger wheels, electronic gadgets, golf club heads, biomedical hip implants, and aerospace components for defense outlets according to Eddy et al. [111]. LWC has no metallurgical limitation on the products, including ferrous and non-ferrous alloys. No additional costly tools are involved in this process. However, LWC involves expensive manual labor to prepare the wax pattern and ceramic slurry.

The extensive investigations of LWCs carried out by Craig et al. [112] showed that the suitability of a wax pattern as a lost wax pattern depends on the following factors: (1) the lowest possible thermal expansion to match the desired dimensional accuracy; (2) a melting point that is lower than ambiance to prevent thermal distortions and cavitation issues; (3) high resistance to deformation at room temperature for easy handling; (4) high wettability and smoothness. (5) low viscosity to fill the thinnest sections of the mold during pouring; (6) easy detachment from the mold after casting; (7) ash free and should not leave any residue in the mold; and (8) environmentally safe. Other factors include the expenses, recyclability, availability, and toxicity when choosing a wax pattern. The efficiency of the lost wax can be achieved by employing additives, mixing it with different wax types, and optimizing the process parameters.

4.1.4. LWC for Bells

LWC using the beeswax method is the first ancient casting technique used in China around the 10th century BCE. It was introduced into Korea around the 3rd century BCE and was believed to be the origin of the modern LWC method. Beeswax casting is still a traditional method in Southeast Asia, Europe, and Africa. Graig et al. [113] studied the use of beeswax in dentistry at the University of Michigan and reported that the mechanical strength properties are especially important when considerable expansion occurs during investment casting for dental applications.

The beeswax casting method produces a model that is identical to the shape of the bell that is made of beeswax and overlaid with molding sand mixed with fine-grained clay and sand followed by drying [114]. After, the mold is heated to liquefy the wax inside, and liquid melt is poured into the space where the wax is dissolved. Beeswax is seldom mixed with rosin, wood oil, or beef oil to soften the wax so that the outer plates rotate well [41,45]. The process flow of the beeswax method consists of pattern engraving, wax pouring, wax pattern plate production, core production, wax layer on the core, wax pattern plate assembly, casting sand, drying, wax removal, firing, melting, and casting.

Kissi [115] evaluated the casting of hollow artifacts produced by Ghanaian traditional metalsmiths and suggested that the use of liquid wax to produce hollow wax frames in POP molds should be used to ensure direct duplication of the master without the creation of parting lines in the inner walls of the model.

Hossain et al. [116] studied the physicomechanical characteristics of paraffin and beeswax to simulate the rocking behavior for water jet drilling and concluded that natural beeswax could be a good substitute for reservoir rocks. Giuseppe et al. [117] studied the thermos-mechanical properties of beeswax-halloysite nanotube (HNT) composites and stated that a slight decrease in the beeswax crystallinity occurred after the HNT addition. Zhang et al. [118] investigated the thermal behavior of four insect waxes and obtained a melting point of 70.34 °C and melting enthalpy of 168.1 J/g for beeswax.

Previously, Dong-Joo et al. [119] studied the influence of the temperature and cooling speed on the mechanical characteristics of a pressure cast thermoplastic composite. The results showed that the crystallinity decreased with the increased cooling rate, with the slowly cooled specimen having a high fracture toughness.

The patterns are engraved on stone, clay, or wood such as talc with an intaglio, and then the pattern is re-sculpted with wax (Figure 9a,b). The preparation of 3D patterns such as dragon string or a sculpture is created using the beeswax form (Figure 9c–e). Larger bells are difficult to make using the beeswax casting method, and it is presumed that the rotary method and beeswax casting method coexisted at the time [120]. Unfortunately, the key technology of this traditional method was lost during Japanese colonial rule and the Korean War. The demand for bells increased after the Korean War due to the rapid spread of Protestants, and the first bell manufacturing company was established in Korea in 1954, which used the rotational molding method learned from the Japanese [121].

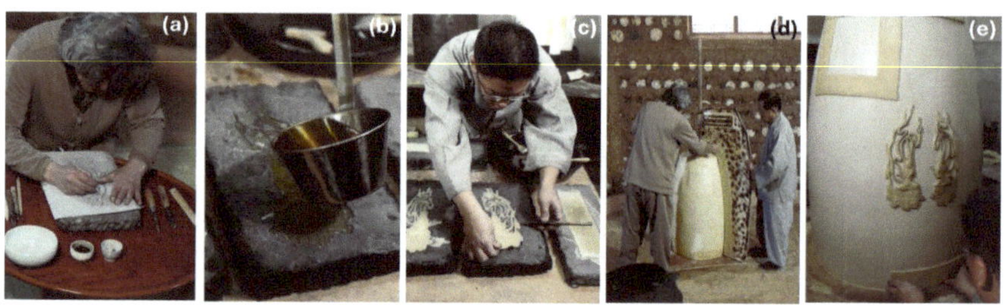

Figure 9. The fabrication process of a beeswax form [110,116]. (**a**) Pattern engraving, (**b**) beeswax pouring, (**c**) beeswax patterning, (**d**) plane beeswax form, and (**e**) beeswax bell form.

Modern LWC for the *pomjong* bell was developed in the last few decades and was patented in Korea in 2004 [122]. The biggest difference between LWC and conventional wax casting is whether the model of the bell is made only with beeswax or a mixture of beeswax and FRP. In LWC, the bell frame is made of FRP—only the pattern part is separately made of wax—and the wax pattern is inserted into the FRP frame (Figure 10a–b). Although it is a simple process, it is important to know how to insert beeswax into the FRP bell mold and separate the FRP bell mold so that the molding sand does not break after coating with the casting sand.

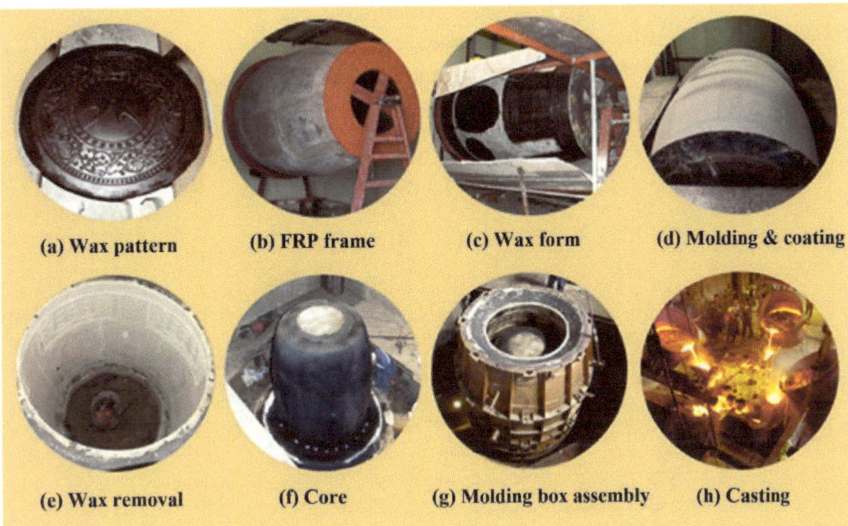

Figure 10. The LWC method used to produce a temple bell. (**a**) Wax pattern, (**b**) FRP frame, (**c**) wax form, (**d**) molding and coating, (**e**) wax removal, (**f**) core, (**g**) molding box assembly, and (**h**) casting [122].

When the model of the bell is completed using FRP and beeswax as shown in Figure 10c, unlike the traditional wax method, which uses clay and sand as the casting sand, in this method, ceramic materials such as zircon flour and chamotte sand (fire clay) are used as the casting sand, and colloidal silica is used as the binder. A slurry of colloidal silica and zircon flour is coated on the surface of the beeswax model and then chamotte sand is applied (Figure 10d). According to the size of the species, this process is repeated up to 30 times or more. After the casting sand coating is finished, the FRP and beeswax inside are removed and the outer shape is ceramicized through firing (Figure 10e,f). The finished outer frame is placed on the core, and a separately molded dragon string is placed on it to complete the form (Figure 10g,h).

Previous bell manufacturers have attempted to produce large and small bells using a variety of materials: cast iron, steel, Zn and Al alloys, glass, China clay, or pottery. However, the alloy consists of about 80 wt.% Cu and 20 wt.% Sn is still regarded as the basic material for the production of bells, the so-called 'bell bronze' [123,124]. Despite many trials of replacing the costly Sn with other elements, no alloy with similarly good acoustic properties has been achieved so far [36,125]. There were incidents where some large-scale projects failed to combine these properties. Some bells cracked after a short time of use, e.g., the Aleksejevskij bell, which worked for only one year, and the largest bell in the world, the Tsar Bell, also known as Tsarsky Kolokol, with a mass equal to 250×10^3 kg, has never stricken a note. Since 2000, the largest and heaviest ringing bell in the world has been the Bell of Good Luck at the Foquan Temple in Pingdingshan city, China, whose mass is 116×10^3 kg [126,127]. The shape of bells has changed over the centuries [128–130]; nevertheless, their construction was designed to have suitable strength and divine sound using properly selected parameters, i.e., wall thickness, bell diameter, and the composition of the alloy, which are decisive concerning the sound tone and timbre [131–133]. The optimal mechanical strength is easy to achieve by increasing the bell wall thickness, but the relationship between the shape of the bell and its sound is hard to grasp, especially because the bell sound consists of a series of merged tones and overtones, which give the so-called strike tone [134,135].

4.2. Rotational Molding Method

4.2.1. Rotomolding Process

Rotational molding is also termed rotomolding or rotational casting. It is the most popular processing method for molding. It was first developed in the early 20th century, but it became popular after the 1960s when Lyondell Basell replaced vinyl plastisol resins with polyolefin resins [136]. The rotomolding process consists of four stages: charging, heating, cooling, and de-molding, as shown in Figure 11 [137–141].

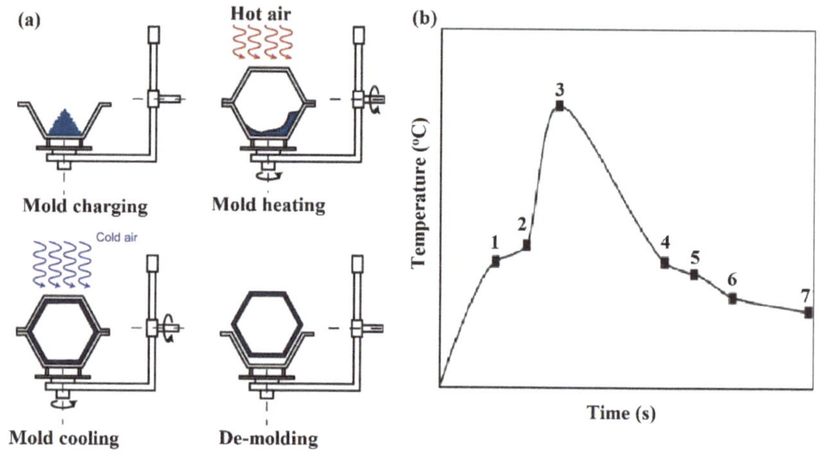

Figure 11. (**a**) Rotomolding process and (**b**) thermal cycle for rotomolding of a semi-crystalline polymer [141]. There are seven steps in the thermal cycle profile.

The various stages in rotomolding are illustrated in Figure 11a,b. The various steps are as follows. The first step is mold charging, where the cast material is poured into the mold. The cast material is in the form of a powder, with or without additives, to achieve a uniform heat distribution. After charging, the mold is heated with uniaxial or biaxial rotation. The rotation speed is optimized, which is relatively low to achieve a uniform distribution. The uniaxial to biaxial rotation ratio is usually maintained at 4:1 to avoid the adherence of melt over the mold surface. After heating, the mold is cooled with air or water spray and the solidification of the cast sample follows. After solidification, the final cast product is detached from the mold and recovered [141].

4.2.2. Materials for Rotomolding

The numerous materials and their respective properties used as rotomolded cast products are summarized in Table 4. Rotomolding has also been used to prepare automobile prototypes. Although this method has the disadvantages of poor strength and degradation over time, the process is now becoming more popular due to the production of stress-free plastic products. Due to these challenges, suitable surface functionalization is exercised by choosing a suitable resin and additive polymer. The optimum process parameters in rotomolding can give rise to promising plastic products for industries. Rotomolding products are expected to increase in the near future [124,125].

Table 4. Various polymers and their properties needed for rotomolding [142,143].

Polymer	Moldability	Impact Strength	Advantage	Limitations
Polyethylene	Excellent	Good	Low cost	Lower strength
Polypropylene	Good	Poor	Better stiffness	Low strength at cryo temperatures, costly than polyethylene
Polystyrene	Good	Poor	Minimum shrinkage	Poor strength
Polyether ether ketone	Moderate	Poor	Rigid, fire safety	Expensive compared to polyethylene
Polyvinyl chloride	Good	Good	Easily paintable	costly than polyethylene, lower stiffness than polyethylene
Acrylonitrile butadiene styrene	Moderate	Good	Rigid and paintable	costly than polyethylene
Ethylene butyl acetate	Good	Good	Stretchable	Powder form
Epoxy	Moderate	Poor	Thick and rigid, high impact strength	Highly expensive
Fluropolymers	Good	Good	Chemical resistant	toxic
Nylon	Good	Good	Heat resistance, better impact resistance	Expensive
Polycarbonate	Good	Good	Transparent, tough	Harder moldability compared to polyethylene
Polyurethane	Good	Good	Wear resistant	Higher cost than polyethylene

Rotational molding of temple bells is associated with the production of a core and an outer bell shape using separate rotating plates for the core and the external shape, their fixation on the central axis, and attachment of the casting sand while the rotating plate is rotating (Figure 12). To make the cross-section of the core, the outer shape should be cut exactly in half. The size must be decided by considering the shrinkage rate of the casting, and the difference between the diameters of the two rotating plates determines the thickness and shape of the bell. The core is fabricated using stacked bricks with a size slightly smaller than the actual core to be manufactured or by reinforcing bars to form the core (Figure 12a). A rotating plate is attached to the center for sand casting while rotating it (Figure 12b). Green sand is used as the casting sand mixed with clay as a binder. When the shape of the core is established, the surface of the core is smoothed, and then a graphite coating with water is painted and left to dry (Figure 12c,d).

Figure 12. The fabrication process of the bell core using rotational molding. (a) Building core bricks, (b) rotational molding, (c) mold wash coating, and (d) core drying [11].

When the core is completed, beeswax is applied while rotating according to the thickness of the bell, and then the surface is finely trimmed to finish. To improve the fluidity, beeswax is mixed with animal oil. After, the melted beeswax is pasted on a pre-patterned plate and inserted onto the surface of the wax model. The dragon strings are formed directly, with beeswax attached to the upper part and cast sand over the beeswax. To prevent cracking and improve the strength, broken earthenware fragments are also attached to the outer surface as reinforcements as the outer surface dries naturally. After complete drying, the outside of the mold is heated to melt the wax inside, leaving behind the bell core and outer part. The productivity of this method is low due to several complicated workflows. Raw sand is used as casting sand, and clay or bentonite is used as the binder. The proper mixing ratio of raw sand and bentonite is about 100:4 and is kneaded with a little water [144]. The quality of the bell core can be improved by choosing high-quality zircon sand or other ceramics with excellent fire resistance [145,146].

The relationship between the moisture content of the molding sand (the breathability) and the strength of the molding sand was investigated in previous study [144]. The breathability was optimal at a 2~3% moisture content, and the strength of the mold increased up to a 10~20% moisture content, and it tended to decrease when it exceeded the moisture content. Considering the air permeability, the authors suggested that mold strength and workability was optimal at a moisture content of 6~8% [144].

4.2.3. Major Applications of Rotomolding

Rotomolding processes are cost-effective compared to other casting methods due to their lightweight, flexible, and corrosion resistance properties, etc. [147–149]. However, there are certain limitations of rotomolding such as its low strength and stability [150–154]. Certain additives have been used to improve the rotocast strength [155–160]. Therefore, rotomolding is often used to fabricate hollow, multilayered seam-free products. This technique has attracted enough attention in the past few decades and produces stress-free products with heat [161–164]. Rotomolding molds are less expensive than other cast molds and material wastage is minimized [165,166]. Rotomolding has been commonly used in tanks, medical instruments, fuel storage tanks, toys, traffic signs, toys, furniture, toolbox, etc. [167,168]. Other miscellaneous applications include light laundry mats, kayaks, vehicle crash bars, oxygen mask lids, vending and display items, aquarium accessories, drug dispensers, contact lenses, etc. [169]. As early as the 1970s, cross-linkable and modified polyethylene-grade polymers were present in the rotomolding market. These new polymers again fueled new market areas, especially the production of large tanks.

4.3. *Pep-Set Casting Method*

4.3.1. Pep-Set System

Pep-set is a highly reactive process used to make sand cores and molds in a foundry using the no-bake process. This process requires a pep-set system, which needs a relatively long time for polymerization and simultaneous fast curing and provides higher productivity [170]. The cast product strength is very high and there is no need for the further addition of a binder with the pep-set system, which in turn minimizes contamination and the emission of impurities during molding. The standard concentration of a pep-set system is less than 0.6% per part. The curing speed depends on the ambient temperature and can be controlled by the quantity and/or quality of catalyst used in the ratio (1:0.5–5.0% per part).

The pep-set system is a potential casting process, where the binders rely on the polymerization reaction of phenolic resin with an isocyanate additive. The process involves a three-part binder system (binder, hardener, and liquid catalyst). The hardening proceeds without any formation of by-products such H_2O or HCHO [170].

4.3.2. Background of Pep-Set Casting

The pep-set casting method and the LWC method are modern casting techniques that began to be used in the 1980s. The pep-set process was developed mostly in Europe and later introduced into Korea in the 1980s [171]. Pep-set binders are primarily a product of polymerized phenolic resin and an isocyanate component. The process involves the three steps of a binder, hardener, and liquid catalyst [172]. The pep-set method is more productive than the conventional rotary method but also results in a better quality of the product. The pep-set process, one of the self-hardening mold casting methods, is a modern casting method developed in Europe and introduced into Korea in the 1980s through Japan. It was used for the casting of automobile engines and precision machine parts but started to be used to case sculptures such as bells and statues in 1986. Even though there was resistance in the main industry at the time, it was an opportunity to raise the quality of Korean bells. More than 10 years later, in 1997, the traditional method of beeswax casting was restored by Won Kwang-Sik, which is an important cultural asset [173].

A limited number of studies exists on the casting of Buddhist temple bells. A few notable works were summarized by Y.H. Yeom in around 1991–1995 [11,149]. There are a lack of studies that have been carried out on the use of the pep-set method for temple bells. In 2006, a new beeswax casting method was used for large Buddhist bells for the first time in the world [174,175]. However, due to the low yield, the beeswax casting method fell into hibernation at the industrial level, but it was later improved and is being used today.

4.3.3. Molding in Pep-Set

This method requires a high initial cost due to the various sizes of the molds and their usage. The mold used in the pep-set casting method is manufactured by a method that is completely different from the method used in the rotary or other casting methods. First, the rotary plate is made with POP, which is the same shape as the bell, and then a pattern engraved on POP is inserted to make a bell model out of POP (Figure 13a,b). FRP (fiber-reinforced plastic) is again applied to the POP model, and the process of hardening is repeated to make a 凹-shaped FRP model by removing the plaster inside when the appropriate thickness is reached. If FRP is applied to the inside of this 凹-shaped FRP model and separated again, a 凸-shaped FRP model is made, which is used as a mold (Figure 13c). At this moment, the mold is divided into three to four parts so that it can be separated from the formwork after molding.

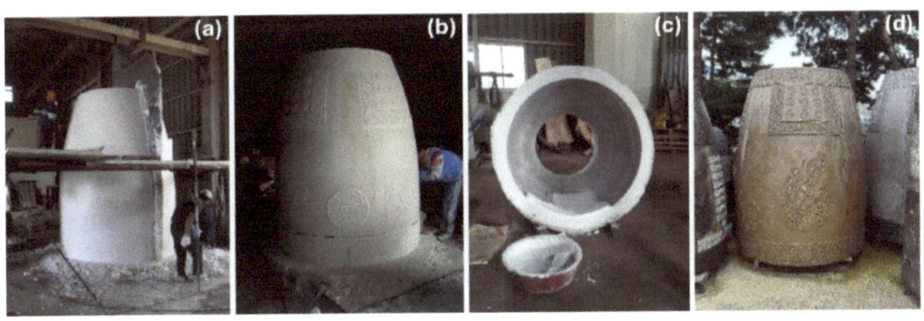

Figure 13. The fabrication process of a bell form. (**a**) A POP bell form, (**b**) an engraved plaster bell form, (**c**) an intaglio FRP bell form, and (**d**) a completed bell form.

The molding operation is carried out by putting a mold that is large enough to be filled with the molding sand on the outside of the vertical mold made of FRP, mixing resin (pep-set) and molding sand with a kneader, and filling the space between the mold and mold walls.

Artificial silica sand is mainly used with casting sand, mixed in a ratio of 50:50. The smaller the particle size of the molding sand, the more delicate the surface obtained. Since it is impossible to obtain a sound casting, it is important to select a particle size that is suitable for the casting. When the molding sand is hardened, the FRP mold inside is disassembled and removed, and a $ZnCO_3$-based coating agent is applied to the mold surface engraved with the pattern (Figure 13d).

According to historical records, casting technology dates to 5000 BCE and originated in the Middle East and later entered Asian countries such as China, Korea, Japan, and India. The earliest records of casting technology are preserved in the writings of the monk Theophilus Presbyter in his book *Schedula Diversarum Atrium* in 12th century BCE [176]. According to Theophilus Presbyter, the operational sequence of bell-making consists of several steps, such as building and shaping the clay core with a horizontal lathe, removing the lath spindle, and closing the hole with plastic clay to support the U-bent iron staple hanging on the clapper. After that, the upper part of the mold is added to the core, and four pole guides are used as a sink. The casting pit is prepared with Roman tiles plastered with clay. Heat is provided from both the upper and lower part of the square outer furnace. After firing, the pit is quickly emptied, and the mold is slowly lifted out and detached to recover the bell and the iron hoops [177].

In the late 1950s, a detailed operational sequence for casting neck collars of the Early and Middle Bronze Age was created by Hans Drescher [178]. Later, in the 1980s, Rønne added an important knowledge base regarding stamping and spiral techniques [179,180]. In summary, these developments led to the assumption that these decorated artifacts (such as neck collars, belt discs, and tutuli) were crafted via the wax model into their final forms [178]. Secondary decorative components such as rims and geometrical elements were either stamped [180] or added to the wax.

With the advancement of modern casting technology, computational tools have enabled foundrymen to bridge the gap between design and manufacturing. New computational approaches should be used to identify defects (hot tear, shrinkage pore, cold shut, etc.), determine the casting time, guide the morphology, and optimize the entire process.

5. Advanced Rapid Prototyping Techniques in Investment Casting

As already discussed, conventional wax casting technology is expensive for low-volume manufacturing such as customized or prototype component casting. As such, the process is highly time-consuming and spans weeks to months depending on the machine shop's capacity and schedule. Design errors and iterations add up and impact the final manufacturing cost. In these situations, rapid prototyping (RP) techniques are receiving increasing attraction for casting objects with freedom from design constraints. LWC has been integrated with RP techniques to improve the yield and flexibility of manufacturing in various applications [181,182].

Advanced computer-aided manufacturing (CAD) design has helped to achieve the best RP techniques. The CAD model of a 3D object is sliced into several sequential layers that can be joined to consequent layers in a layerwise fashion. The different RP techniques' process flow is shown in Figure 14a. The diversity in the materials and processes used for binding the layers in sequence constitutes the different RP processes. The existing RP techniques are classified into four major classes: liquid, powder, sheet, and gas-based platforms (Figure 14b). The first RP technique, stereolithography (SLA), was discovered in the late 1980s. After, several improvements of the RP techniques were performed. According to Liu et al. [177], other commercial RP techniques include fused deposition modeling (FDM), laminated object manufacturing (LOM), selective laser melting (SLM), and 3D printing.

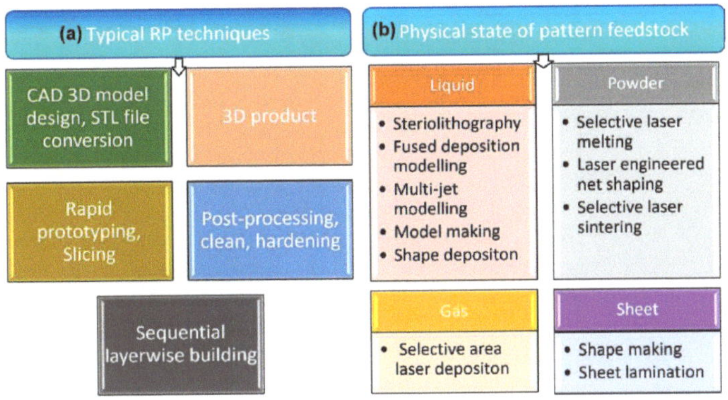

Figure 14. (**a**) Typical RP process flow [150]. (**b**) Classification of RP techniques based on physical state of feedstock [182].

According to these technological advances, this classification will allow 3D manufacturers to choose a suitable RP technique with increased accuracy, performance, and durability. Cheah et al. [181] reviewed the diverse applications of RP techniques to several LWC processes ranging from jewelry casting, sports goods, biomedical implants, injection molded parts, and die casting to the automotive/aircraft industries. The integration of RP techniques with LWC has potential merit for rapid and cost-effective production of high-precision castings.

Although RP techniques are highly productive and flexible, the application of these techniques to temple bells is new due to the expenses incurred over the raw materials. The printing of bronze bells requires the integration of wax 3D printing and LWC technologies. Wax 3D printing uses the SLA technique to create the wax pattern from a wax-like resin. For additional support to stand the wax pattern, support structures are also printed together with the wax model. The support structures are removed manually after the printing process followed by cleaning of the wax pattern for casting.

Initially, one or more wax sprues can be joined to the pattern. Next, these sprues can be attached to a wax 'tree' together with other wax patterns. This wax tree is then placed in a flask and covered with a fine POP. As soon as the POP solidifies, a mold is formed for bronze casting. The POP mold is then fired or oven heated to burn all the wax. The liquid bronze is then poured to fill the cavities left by the wax followed by cooling, solidification, and breaking off the mold to obtain the cast model.

The process of employing RP techniques together with the fabrication of wax patterns is referred to as rapid investment casting (RIC). The advantages of RIC include its cost effectiveness, high manufacturing ability, freedom in the design of parts that were initially difficult or impossible to make via machining, feasibility of design iterations or tool modification, and ability to facilitate parametric optimization effectively. However, RIC patterns are not economical and are limited to mass production due to the expensive RP materials involved. For a high production volume, rapid tooling (RT) can effectively manufacture tens to millions of wax patterns in an economical way.

6. Conclusions

In this paper, the history, materials, and casting methods used to produce bells were overviewed. Primitive bells were manufactured using casting and bell-making technology such that even a complex bell material composition was well-designed. Korea's bell-making technology has been steadily developing, breaking away from the traditional method due to the development of new bell materials and processes. Owing to a lack of metallurgical knowledge of compositions in earlier bells, the alloys used in the Middle Ages were significantly different from the current bell metals. A study of earlier bells through metallography revealed substantial differences in the microstructure and compositions through modified melting and casting approaches. During the post-independence era, Korea steadily innovated different rotational methods, pep-set casting, and LWC methods, becoming a world leader in cast bell technology. Future research should be carried out with other low-cost wax and additives to improve the castability of bells. Nowadays, the use of pure elements and progressed melting practices has decreased contamination levels due to the addition of P and S, which makes bells more robust and less sensitive to cracking when hit by a hammer or wooden block. However, limited attempts have been carried out to cast a temple bell to replace conventional metal casting techniques. Future research should be directed towards RP techniques to fabricate wax patterns directly from the CAD file by layerwise deposition of wax droplets. Though these RP techniques are more advanced, conventional LWC techniques are still being used today for different metal casting and temple bell fabrication because of the obstacles of the high cost of materials and experimentation so that small- and medium-sized industries are unable to procure them. Therefore, further amendments of the current methodologies that reduce tooling costs are desired. Technological developments of RP techniques are expected to bring the cost down and lead to a better casting quality, temple bells, and acoustics, resulting in a longer service life to mankind.

Author Contributions: Conceptualization, C.-S.W. and J.P.J.; methodology, C.-S.W.; formal analysis, K.-S.W.; investigation, K.-S.W.; resources, A.S.; writing—original draft preparation, C.-S.W. and A.S.; writing—review and editing, C.-S.W. and K.-S.W.; project administration, A.S.; funding acquisition, A.S. All authors have read and agreed to the published version of the manuscript.

Funding: This research received no external funding.

Data Availability Statement: Not applicable.

Conflicts of Interest: The authors declare no conflict of interest.

References

1. Cebulak, M.; Kowalski, J.S.; Wojciechowski, W. Bell founding—The past and the present. *Arch. Foundary Eng.* **2010**, *10*, 167–170.
2. DeSapio, V.; Fan, Q.; Fravel, D.; Houck, G.; Lenchitz, H.; Mapes, C.; Sykes, H.; Thomsen, C.; Treat, A.; Tsuji, K.; et al. *Foundry Products—United States International Trade Commission Report. Investigation No. 332-USITC Publication May 2005*; U.S. International Trade Commission: Washington, DC, USA, 2005.
3. Legge, K.A.; Petrolito, J. Designing idiophones with tuned overtones. *Acoust. Aust.* **2007**, *35*, 47–50.
4. Isogawa, S. Casting sites of bronze bell and iron kettle in ancient and medieval Japan. *ISIJ Int.* **2014**, *54*, 1123–1130. [CrossRef]
5. Doney, L. Chapter 4—Bronze temple bells from the Tibetan imperial period: Buddhist material culture in context. In *Beyond Boundaries*; Willis, M., Schaik, S.V., Doney, L., Lucas, D.B., Elizabeth, A.C., Eds.; Walter de Gruyter GmbH: Berlin, Germany; Boston, MA, USA, 2020; Volume 2.
6. The Beginning of the Country's History, Prehistoric Times—Gojoseon, The Korean Cultural Center New York (KCCNY), 460 Park Avenue 6th Floor, New York, NY 10022. Available online: https://www.koreanculture.org/korea-information-history (accessed on 26 January 2022).
7. *When Nazis Took All the Bells*; The National Bell Festival, Inc. and Bells.org.: Washington, DC, USA, 2022. Available online: https://www.bells.org/blog/when-nazis-took-all-bells (accessed on 26 January 2022).
8. Bells and Clocks Since 1842, The Verdin Company, 444 Reading Road, Cincinnati, OH 45202. 2022. Available online: https://www.verdin.com/bells/history-of-bells/ (accessed on 26 January 2022).
9. Xi'an Municipal Documents Repository. *Xi'an City Archives Announces Historical Materials of the Anti-Japanese War. A Report*; 31 August 2005; Xi'an Municipal Documents Repository: Xi'an, China, 2005.
10. Zoom Korea, Meet the Master behind Korea's Most Iconic Bell. *Korea JungAng Daily in Association with The New York Times*. 9 December 2021. Available online: https://koreajoongangdaily.joins.com/2021/12/09/culture/features/korean-bell-korean-temple-bell-Won-Kwangsik/20211209154758585.html (accessed on 26 January 2022).
11. Yum, Y.H. *Korean Bell*; Seoul National University Press: Seoul, Korea, 1991; pp. 3–43.
12. Nathaniel, S., Jr. Bell (Yongzhong) 5th–3rd century B.C, China. In *A Treasury of Archaeological Bells*; Hastings House Publishers: New York, NY, USA, 1978; Available online: https://www.metmuseum.org/art/collection/search/61053 (accessed on 1 February 2022).
13. Kanzeon-ji Temple Bonsho Bell, Japan Heritage Dazaifu, Dazaifu City Japan Heritage Utilization Council. 2016. Available online: https://www.dazaifu-japan-heritage.jp/en/bunkazai/detail.php?cId=326 (accessed on 1 February 2022).
14. Bronze Bell of Sangwonsa Temple. In *Korea JungAng Daily in Association with The New York Times*; 7 July 2013. Available online: https://koreajoongangdaily.joins.com/news/article/article.aspx?aid=2974158 (accessed on 1 February 2022).

15. Catherine, S.J. Standing Waves and Musical Instruments, Connexions Module: m12413. pp. 1–10. Available online: http://cnx.org/content/m12413/latest/ (accessed on 1 February 2022).
16. Jinpo, T.; Kobayashi, H.; Shigeo, T.; Tokihito, N. Papermaking Parts for Casting Production. Japanese Patent JP4002200B2, 31 October 2007.
17. Cartwright, M. The Bronze Bells of Ancient Korea, World History Encyclopedia. Last Modified 18 October 2016. Available online: https://www.worldhistory.org/article/961/the-bronze-bells-of-ancient-korea/ (accessed on 1 February 2022).
18. The Age-Old Chinese Bell Culture, China.org.cn. Available online: http://www.china.org.cn/features/fbich/2003--11/12/content_1078450.htm (accessed on 1 February 2022).
19. The Story of Culture and Arts—Dongjong, Bronze Bells of Korea. Available online: http://contents.history.go.kr/mobile/tz/view.do?levelId=tz_b19 (accessed on 1 February 2022).
20. Deep Echo of the Divine Bell; the New Beginning of the Silla Great Bell, Gyeonju City Council. 2017. Available online: https://www.gyeongju.go.kr/ (accessed on 1 February 2022).
21. Kim, S.; Lee, J.H.; Byeon, J.H. Vibration and Sound of Silla Great Bell. *J. Acoust. Soc. Kor.* **2017**, *36*, 186–193. (In Korean) [CrossRef]
22. Park, S.H.; Lee, M.; Hahn, N.; Sung, K.M. Acoustic characteristics and timbre preferences of Korean bells. *J. Korean Acoust. Soc.* **2011**, *30*, 273–280. [CrossRef]
23. Jincheon Bell Museum, Jincheon-Gun, Korea. Available online: https://www.jincheon.go.kr/site/eng/sub.do?menukey=2687 (accessed on 1 February 2022).
24. Jang, I.N.; Cho, N.C. A study on archaeological production sites of South Korean bronze bells through slag and soil analysis. *Archaeol. Anthropol. Sci.* **2022**, *14*, 3. [CrossRef]
25. Na, D.K. Korean Bells—Design and Characteristics. In Proceedings of the Ninth International Conference on the Beginnings of the Use of Metals and Alloys, Busan, Korea, 16–19 October 2017; The Korean Institute of Metals and Materials: Seoul, Korea, 2018; Volume 1, pp. 1–11.
26. Steve46814. Korean Buddhist 'Emille' Bell. World History Encyclopedia. Last Modified 5 October 2016. Available online: https://www.worldhistory.org/image/5838/korean-buddhist-emille-bell/ (accessed on 1 February 2022).
27. Dae, C.C. Sacred Spirit of Emille Bell, The Korea Times, 14 January 2013. Available online: https://www.koreatimes.co.kr/www/opinion/2018/06/162_128814.html (accessed on 20 March 2022).
28. Gyeongju Hwangnyongsa Temple Site, Current Status—Historic Site No. 6 (Designated on 21 January 1963). Available online: http://www.gyeongju.go.kr/tour (accessed on 20 March 2022).
29. Lee, J.; Kim, C. A study on formative style of Silla bell. *J. Korea Multimed. Soc.* **2013**, *16*, 1214–1229. [CrossRef]
30. Landers, J. A Rare Collection of Bronze Age Chinese Bells Tells a Story of Ancient Innovation. *Smithsonian Magazine*, 5 October 2017. Available online: https://www.smithsonianmag.com/ (accessed on 20 March 2022).
31. *Iron Bell of Jeondeungsa Temple*; Cultural Heritage Administration: Daejeon, Korea, 1963. Available online: https://english.cha.go.kr/html/HtmlPage.do?pg=/contact/contactsEng.jsp&mn=EN_05_03 (accessed on 20 March 2022).
32. Fletcher, N.H. The nonlinear physics of musical instruments. *Rep. Prog. Phys.* **1999**, *62*, 723–764. [CrossRef]
33. Schilling, M. *Bells. Shape, Sound and Ornament*; Verlag der Kunst: Dresden, Germany, 1988.
34. Audy, J.; Audy, K. Analysis of bell materials: Tin bronzes. *China Foundry* **2009**, *58*, 77–81.
35. Bartocha, D.; Baron, C.; Suchon, J. The influence of solidification rate on high-tin bronze microstructure. *Arch. Foundry Eng.* **2019**, *19*, 89–97.
36. Cekus, D.; Kwiaton, P.; Nadolski, M.; Sokół, K. Quality assessment of a manufactured bell using a 3D scanning process. *Sensors* **2020**, *20*, 7057. [CrossRef] [PubMed]
37. Strafford, K.N.; Newell, R.; Audy, K.; Audy, J. Analysis of bell material from the Middle Ages to the Recent Time. *Endeavour* **1996**, *20*, 22–27. [CrossRef]
38. Cheng, C.F.; Schwitter, C.M. Nickel in Ancient Bronzes. *Am. J. Archaeol.* **1957**, *61*, 351–365. [CrossRef]
39. Rosenfield, A.R. The crack in the liberty bell. *Int. J. Fract.* **1976**, *12*, 791–797. [CrossRef]
40. Rainio, R. Classifying iron age bells, pellet bells and bell pendants. In *Studies in Music Archeology*; German Archeological Institute: Berlin, Germany, 2008; Volume VI. (In German)
41. Yun, Y.H.; Cho, N.C.; Doh, J.M. Experimental research of lost-wax casting: Gilt-Bronze Pensive Bodhisattva. *J. Conserv. Sci.* **2020**, *36*, 123–142. [CrossRef]
42. Vináš, J.; Vrabel, M.; Greš, M.; Brezina, J.; Sabadka, D.; Fedorko, G.; Molnár, V. Restoration of worn movable bridge props with use of bronze claddings. *Materials* **2018**, *11*, 459. [CrossRef]
43. Audy, K. The Metallurgy of Ancient Artefacts. Ph.D Thesis, University of South Australia, Adelaide, Australia, 1999.
44. Audy, J. Bells Evaluation from Buchner Workshop from Casting Point of View. Bachelor's/Master's Eng Thesis, The Technical University in Kosice City, Kosice, Czechoslovakia, June 1988.
45. Audy, J.; Cech, J.; Beno, J. Composition and microstructure of medieval bells. *Pract. Metallogr.* **1992**, *29*, 74–84. [CrossRef]
46. Audy, K.; Audy, J. Analysis of bell-making process from the middle ages to recent times. *Pract. Metallogr.* **2006**, *43*, 271–291. [CrossRef]
47. Buchner, A. *Workshop Diary and Recipes (1884–1975)*; Item Number HU 1384/193/84; A Report on Collection of Slovak Technical Museum (STM): Kosice City, Slovakia, 1984.

48. Sarudyova, M. *Work of A. Buchner—The Bell-Maker and Art Black-Smith (1884–1975)*; Brochure of Permanent Museum Exhibition; A Report on Collection of Slovak Technical Museum (STM): Kosice City, Slovakia, 1987.
49. Grigerova, T. *Metallurgy of Casting Alloys of Nonferrous Materials*; Technical University (VST): Kosice City, Slovakia, 1986.
50. Vilcko, J.; Slovak, S. *Casting Technology*; SNTL: Bratislava, Slovakia, 1987.
51. American Society for Metals. *Casting Design Handbook*; McGraw Hill Inc.: Columbus, OH, USA, 1958.
52. American Foundrymen's Society. *Pattern-Makers' Manual*; McGraw Hill Inc.: Columbus, OH, USA, 1953.
53. Schubert, R.; D' Egidio, S.M. The surface composition of copper with indoor exposures ranging from 3 to 47 years. *Corr. Sci.* **1990**, *30*, 999–1008. [CrossRef]
54. Borland, D.W. The Structure of Materials. In *The University of Melbourne Handbook*; University of Melbourne: Melbourne, Australia, 1989; pp. 24–25.
55. Heine, R.W.; Loper, C.R.; Rosenthal, P.C. *Principles of Metal Casting*; McGraw Hill Inc.: Columbus, OH, USA, 1967.
56. Zeuner, F. *Dating the Past*; Methuen & Co. Ltd.: London, UK, 1970.
57. Shim, J.-H.; Oh, C.-S.; Lee, B.-J.; Lee, D.N.Z. Thermodynamic assessment of the Cu-Sn system. *Metallkde* **1996**, *87*, 205–212. [CrossRef]
58. Pillai, R.M.; Pillai, S.G.K.; Damodaran, A.D. Shaping of High Bronze in Ancient Kerala and the Metallurgy Behind. *Met. Mater. Process.* **2002**, *13*, 291–300.
59. Manu, K.; Abhinav, J.; Shankar, K.V. Morphological and Mechanical Behaviour of Cu–Sn Alloys—A review. *Met. Metal Int.* **2021**, *27*, 1915–1946.
60. Paul, C.; Sellamuthu, R. The effect of Sn content on the properties of surface refined Cu-Sn bronze alloys. *Procedia Eng.* **2014**, *97*, 1341–1347. [CrossRef]
61. Nadolski, M. The evaluation of mechanical properties of high-tin bronzes. *Arch. Foundry Eng.* **2017**, *17*, 127–130. [CrossRef]
62. Cekus, D.; Nadolski, M. Assessment of the change in the tin concentration in bronzes to the basic components of the sound of bells. *Arch. Metall. Mater.* **2021**, *66*, 531–535.
63. Debuta, V.; Carvalho, M.; Figueiredo, E.; Antunes, J.; Silva, R. The sound of bronze: Virtual resurrection of a broken medieval bell. *J. Cult. Herit.* **2016**, *19*, 544–554. [CrossRef]
64. Slamet, S.; Suyitno, S.; Kusumaningtyas, I.; Miasa, I.M. Effect of high-tin bronze composition on physical, mechanical, and acoustic properties of Gamelan materials. *Arch. Foundry Eng.* **2021**, *21*, 137–145.
65. Abdulhussaien kadhim, H.; Abed, I.J. Investigation wear behaviour of tin bronze alloy prepared by different casting techniques. *IOP Conf. Ser. Mater. Sci. Eng.* **2021**, *1094*, 012126. [CrossRef]
66. Quaranta, M.; Catelli, E.; Prati, S.; Sciutto, G.; Mazzeo, R. Chinese archaeological artefacts: Microstructure and corrosion behaviour of high-leaded bronzes. *J. Cult. Herit.* **2014**, *15*, 283–291. [CrossRef]
67. Vazdirvanidis, A.; Pantazopoulos, G. Metallographic study of great anthony historical bronze bells of apostle Andrew Skete in Mount Athos, Greece. *Metallogr. Microstruct. Anal.* **2017**, *6*, 340–351. [CrossRef]
68. Sarkar, P.K.; Kakoty, S.K. Effect of quenching parameters on mechanical properties of bell metal. *Mater. Today Proc.* **2021**, *44*, 4179–4183. [CrossRef]
69. Srinivasan, S. Iron Age beta (23% tin) bronze: Peninsular Indian bowls of Adichanallur, Nilgiris, and Boregaon. *Mater. Manuf. Process.* **2016**, *32*, 807–812. [CrossRef]
70. Liu, J.; Zheng, W.; Yin, H. Effect of heat treatment temperature on the spinning structure and properties of a Cu–Sn alloy. *Microsc. Microanal.* **2020**, *26*, 29–35. [CrossRef] [PubMed]
71. Pillai, R.M.; Pillai, S.G.K.; Damodaran, A.D. The lost-wax casting of icons, utensils, bells, and other items in South India. *JOM* **2002**, *54*, 12–16. [CrossRef]
72. Park, J.S.; Park, C.W.; Lee, K.J. Implication of peritectic composition in historical high-tin bronze metallurgy. *Mater. Char.* **2009**, *60*, 1268–1275. [CrossRef]
73. *ASM Handbook, Volume 3: Phase Diagrams*; ASM International: Materials Park, OH, USA, 1992.
74. *ASM Specialty Handbook, Copper and Copper Alloys*; ASM International: Materials Park, OH, USA, 2001.
75. Higgins, R.A. *Engineering Metallurgy*, 6th ed.; Arnold: London, UK, 1993.
76. Pantazopoulos, G.; Preka-Alexandri, K.; Varoufakis, G. Metallurgical investigation of bronze urns found at Corfu Island, Greece. *Rev. Min. Geotech. Metall. Eng.* **2010**, *20*, 123–139. (In Greek)
77. Cekus, D.; Kwiaton, P.; Nadolski, M.; Sokół, K. Effect of material-technological properties on the reconstruction of utility parameters of bells. *Eng. Sci. Technol. Int. J.* **2021**, *24*, 1042–1048. [CrossRef]
78. Pantazopoulos, G.; Vazdirvanidis, A. Characterization of the microstructural aspects of machinable α-β phase brass. *Microsc. Anal.* **2008**, *22*, 13–16.
79. Rzadkosz, S.; Garbacz-Klempka, A.; Kozana, J.; Piękoś, M.; Czekaj, E.; Perek-Nowak, M. Research of aluminium influence on tin bronzes. *Arch. Foundry Eng.* **2015**, *15*, 93–98.
80. Campbell, J. Perspective Chapter: A Personal Overview of Casting Processes. In *Casting Processes and Modelling of Metallic Materials*; Abdallah, Z., Aldoumani, N., Eds.; IntechOpen: London, UK, 2020.
81. Nadolski, M.; Cekus, D. The influence of casting parameters on the sound of a bell. In Proceedings of the 26th International Conference on Metallurgy and Materials, Brno, Czech Republic, 24–26 May 2017; pp. 1893–1898.
82. Lee, W. The Casting Method of a Temple Bell Using Beeswax. Korean Patent KR100952444B1, 12 April 2010.

83. Greco, A.; Maffezzoli, A.; Vlachopoulos, J. Simulation of Heat Transfer During Rotational Molding. *Adv. Polym. Technol.* **2003**, *22*, 271–279. [CrossRef]
84. Holtzer, M.; Bobrowski, A.; Dro ˙zy ´nski, D.; Makselon, J.; Isendorf, B. Investigations of Properties of Moulding Sands with Resins Applied in the Alphaset Technology. *Arch. Found. Eng.* **2013**, *13*, 31–37.
85. Bell, K. Lost Wax: An Exploration of Bronze Sculpture in Senegal. In *Independent Study Project (ISP) Collection*; University of Denver: Denver, CO, USA, 2009; Volume 735.
86. Heron, C.; Nemcek, N.; Bonfield, K.M.; Dixon, D.; Ottaway, B.S. The chemistry of Neolithic beeswax. *Naturwissenschaften* **1994**, *81*, 266–269. [CrossRef]
87. Frade, J.C.; Monge Soares, A.M.; Candeias, A.; Ribeiro, M.I.M.; Nunes da Ponte, T.; Serra, M.; Porfírio, E. Beeswax and propolis as sealants of funerary chambers during the Middle Bronze Age in the South-Western Iberian Peninsula. In Proceedings of the 39th International Symposium for Archaeometry, Leuven, Belgium, 28 May–1 June 2012; Scott, R.B., Braekmans, D., Carremans, M., Degryse, P., Eds.; Centre for Archaeological Science Leuven: Leuven, Belgium, 2014; pp. 141–145.
88. Hung, W. A sampan shan chariot ornament and the Xiangrui design in Western Han art. *Arch. Asian Art.* **1984**, *37*, 38–59.
89. Kim, H.Y. Korea's Movable Metal Type Opens the Era of Modern Civilization. *Koreana* **2005**, *19*, 60–65.
90. Shin, S. *Artistic Achievements and International Exchange of Metalwork of the Unified Silla Period: Gilt-bronze Candlesticks with Inlaid Crystal Ornaments*; Art History Research No. 290/291; National Treasure No. 174: Seoul, South Korea, 2016; Volume 9, pp. 43–66.
91. Tulloch, P. Composition of some natural waxes. *Cosmet. Perfumery* **1974**, *89*, 53–54.
92. Jiménez, J.J.; Bernal, J.L.; del Nozal, M.J.; Martín, M.T.; Bernal, J. Sample preparation methods for beeswax characterization by gas chromatography with flame ionization detection. *J. Chromatogr. A* **2006**, *1129*, 262–272. [CrossRef]
93. Aichholz, R.; Lorbeer, E. Investigation of combwax of honeybees with high temperature gas chromatography and high-temperature gas chromatography– chemical ionization mass spectrometry. I. High-temperature gas chromatography. *J. Chromatogr. A* **1999**, *855*, 601–615. [CrossRef]
94. Evershed, R.P.; Vaughan, S.J.; Dudd, S.N.; Soles, J.S. Fuel for thought? Beeswax in lamps and conical cups from late Minoan Crete. *Antiquity* **1997**, *71*, 979–985. [CrossRef]
95. Tulloch, P.; Hoffman, L.L. Canadian beeswax: Analytical values and composition of hydrocarbons, free acids and long chain esters. *J. Am. Oil Chem. Soc.* **1972**, *49*, 696–699. [CrossRef]
96. Maia, M.; Nunes, F.M. Authentification of beeswax (*Apis mellifera*) by hightemperature gas chromatography and chemometric analysis. *Food Chem.* **2013**, *136*, 961–968. [CrossRef]
97. Regert, M.; Colinart, S.; Degrand, L.; Decavallas, O. Chemical alteration and use of beeswax through time: Accelerated ageing tests and analysis of archaeological samples from various environmental contexts. *Archaeometry* **2001**, *43*, 549–569. [CrossRef]
98. Daniels, V. Analyses of copper- and beeswax-containing green paint on Egyptian antiquities. *Stud. Conserv.* **2007**, *52*, 13–18. [CrossRef]
99. Garnier, N.; Cren-Olivé, C.; Rolando, C.; Regert, M. Characterization of archaeological beeswax by electron ionization and electrospray ionization mass spectrometry. *Anal. Chem.* **2002**, *74*, 4868–4877. [CrossRef] [PubMed]
100. Regert, M.; Vacher, C.S.; Moulherat, C.; Decavallas, O. Adhesive production and pottery function during the Iron Age at the site of Grand Aunay. *Archaeometry* **2003**, *45*, 101–120. [CrossRef]
101. Luo, W.; Li, T.; Wang, C.; Huang, F. Discovery of beeswax as binding agent on a 6th-century BC Chinese turquoise-inlaid bronze sword. *J. Archaeol. Sci.* **2012**, *39*, 1227–1237. [CrossRef]
102. Regert, M.; Rolando, C. Identification of archaeological adhesives using direct inlet electron ionization mass spectrometry. *Anal. Chem.* **2002**, *74*, 965–975. [CrossRef]
103. Amir, A.; Gadot, Y.; Weitzel, J.; Finkelstein, I.; Neumann, R.; Bezzel, H.; Covello-Paran, K.; Sergi, O. Heated beeswax usage in mortuary practices: The case of Ḥorvat Tevet (Jezreel Valley, Israel) ca. 1000 BCE. *J. Archaeol. Sci. Rep.* **2021**, *36*, 102904. [CrossRef]
104. Davey, C.J. The early history of lost-wax casting. In *Metallurgy and Civilisation: Eurasia and Beyond*; Mei, J., Rehren, T., Eds.; Archetype: London, UK, 2009; pp. 147–154.
105. Bernardini, F.; Tuniz, C.; Coppa, A.; Mancini, L.; Dreossi, D.; Eichert, D.; Turco, G.; Biasotto, M.; Terrasi, F.; de Cesare, N.; et al. Beeswax as dental filling on a Neolithic human tooth. *PLoS ONE* **2012**, *7*, e44904. [CrossRef]
106. Kimpe, K.; Jacobs, P.A.; Waelkens, M. Mass spectrometric methods prove the use of beeswax and ruminant fat in late Roman cooking pots. *J. Chromatogr. A* **2002**, *968*, 151–160. [CrossRef]
107. Zografou, A. Magic lamps, luminous dreams: Lamp in PGM recipes. In *Light and Darkness in Ancient Greek Myth and Religion*; Christopoulos, M., Karakantza, E.D., Levaniouk, O., Eds.; Lexington Books: Plymouth, UK, 2010; pp. 276–294.
108. Ma, S.P. Study of Lamp in Han Dynasty. Ph.D. Thesis, Fudan University, Shanghai, China, 2012.
109. Taylor, P.R. An illustrated history of lost wax casting. In Proceedings of the 17th Annual BICTA Conference, Stratford-upon-Avon, UK, 4–7 September 1983.
110. Kotzin, E.L. *Metal Casting and Moulding Processes*; American Foundrymen Society Inc.: Schaumburg, IL, USA, 1981.
111. Eddy, W.P.; Barbero, R.J.; Dieters, W.I.; Esarey, B.J.; Frey, L.; Gros, J.R.; Lyman, T. (Eds.) *Investment Casting*; American Society for Metals: Materials Park, OH, USA, 1974; pp. 237–261.
112. Craig, R.G.; Eick, J.D.; Peyton, F.A. Properties of natural waxes used in dentistry. *J. Dental Res.* **1965**, *44*, 1308–1316. [CrossRef]
113. Craig, R.G.; Eick, J.D.; Peyton, F.A. Strength properties of waxes at various temperatures and their practical application. *J. Dental Res.* **1967**, *46*, 300–305. [CrossRef] [PubMed]

114. Won, K.S. The Beeswax Casting Method for a Big Size Temple Bell. Korean Patent KR101487449B1, 3 February 2015.
115. Kissi, B.S. Evaluation of beeswax and core materials for lost wax casting. Ph.D. Thesis, Faculty of Art, College of Art and Social Sciences, KNUST, Kumasi, Ghana, 2011.
116. Hossain, E.M.; Ketata, C.; Islam, R. Experimental study of physical and mechanical properties of natural and synthetic waxes using uniaxial compressive strength test. In Proceedings of the Third International Conference on Modeling, Simulation and Applied Optimization, Sharjah, United Arab Emirates, 20–22 January 2009.
117. Giuseppe, C.; Giuseppe, L.; Milioto, S.; Parisi, F.; Sparacino, V. Thermal and dynamic mechanical properties of beeswax-halloysite nano composites for consolidating waterlogged archaeological woods. *Polym. Degrad. Stab.* **2015**, *120*, 220–225.
118. Zhang, H.; Feng, Y.; Li, K.; Zhang, W. Thermal analysis of four insect waxes based on differential scanning calorimetry (DSC). *Procedia Eng.* **2011**, *18*, 101–106.
119. Dong-Joo, L.; Ick-Jae, S. Effects of vacuum, mold temperature and cooling rate on mechanical properties of press consolidated glass fiber/PET composite. *Compos. Part A Appl. Sci. Manuf.* **2002**, *33*, 1107–1114.
120. Park, M.Y. An experimental study on the restoration of beeswax casting method. *J. Stud. Bibliograph.* **2006**, *33*, 79–106.
121. Sohn, J.W.; Park, J.H.; Choi, S.B. Vibration characteristics of high speed rotary bell cup. *Trans. Korean Soc. Noise Vibr. Eng.* **2015**, *25*, 771–778. [CrossRef]
122. Won, K.S. Lost Wax Casting Method of the Temple Bell. Korean Patent KR-10–0587775B1, 14 June 2006.
123. Hossain, M.A.; Khan, S.A.R.; Kaiser, M.S.; Nur, M.A. Experimental Analysis on the Physical Manners of Thermally Affected Bell Metal. In Proceedings of the International Conference on Mechanical, Industrial and Energy Engineering 2020, Khulna, Bangladesh, 19–21 December 2020.
124. Mukherjee, D. Investigating the technological aspects of bell metal with special reference to West Bengal Heritage. *J. Multidisciplinary Stud. Archaeol.* **2018**, *6*, 1004–1012.
125. Lee, J.M.; Kim, S.H.; Lee, S.J.; Jeong, J.D.; Choi, H.G. A study on the vibration characteristics of a large size Korean bell. *J. Sound Vibr.* **2002**, *257*, 779–790. [CrossRef]
126. Lisovskii, V.A.; Lisovskaya, O.B.; Kochetkova, L.P.; Favstov, Y.K. Sparingly Alloy. Bell bronzes with elevated parameters of mechanical properties. *Metal Sci. Heat Treat.* **2007**, *49*, 232–235. [CrossRef]
127. Nevworldwonders. Pingdingshan and Spring Temple. Available online: http://www.nevworldwonders.com/2012/04/day-214-pingdingshan-and-spring-temple.html (accessed on 20 March 2022).
128. Gierlotka, S. *Bells, History, Technique of Production, and Driving Mechanisms*; Edycja—Książki Naukowe i Specjalistyczne: Katowice, Poland, 2013; p. 17, ISBN 978-83-935389-9-7. (In Polish)
129. Piaskowski, J. *Technology of Ancient Artistic Castings*; Foundry Institute: Krakow, Poland, 1981. (In Polish)
130. Merechal, K. *Bell Casting Technology*; Przegląd Odlewnictwa: Krakow, Poland, 1974; pp. 269–273. (In Polish)
131. Hua, J. The sound of chime bells of 2400 years ago. *Endeavour* **1993**, *17*, 32–37.
132. Saraiva, A.S.; Figueiredo, E.; Águas, H.; Silva, R.J.C. Characterisation of archaeological high-tin bronze corrosion structures. *Stud. Conserv.* **2022**, *67*, 222–236. [CrossRef]
133. Audy, J.; Audy, K. Effects of microstructure and chemical composition on strength and impact toughness of tin bronzes. *MM Sci. J.* **2009**, *6*, 124–129. [CrossRef]
134. Orr, S.A. In situ acoustic analysis of two twentieth-century heritage carillons. *Herit. Sci.* **2021**, *9*, 16. [CrossRef]
135. Swadowski, J. Some problems concerning the acoustic analysis of bells. In *Bellfounding in Poland*; District Museum in Toruń: Toruń, Poland, 2003; p. 119. (In Polish)
136. *A Guide to Rotational Molding*; LyondellBasell Industries: Houston, TX, USA, 2000.
137. Sarrabi, S.; Lacrampe, M.-F.; Krawczak, P. Phosphorous Antioxidants against Polypropylene Thermal Degradation during Rotational Molding-kinetic Modeling. *J. Appl. Polym. Sci.* **2015**, *132*, 41285. [CrossRef]
138. Kontopoulou, M.; Vlachopoulos, J. Bubble dissolution in molten polymers and its role in rotational molding. *Polym. Eng. Sci.* **1999**, *39*, 1189–1198. [CrossRef]
139. Shan, T.K.; Chen, C.T.; Huang, J.D. Design of a plastic auto's door based on the rotational molding process. *Key Eng. Mater.* **2016**, *717*, 79–85. [CrossRef]
140. Löhner, M.; Drummer, D. Experimental studies on the bonding strength and fracture behavior of incompatible materials bonded by mechanical adhesion in multilayer rotational molding. *J. Polym.* **2016**, *2016*, 5768453. [CrossRef]
141. Ogila, K.O.; Shao, M.; Yang, W.; Tan, J. Rotational molding: A review of the models and materials. *Express Polym. Lett.* **2017**, *11*, 778–798. [CrossRef]
142. *Rotational Moulding Technology Handbook*; Engineers India Research Institute, EIRI Consultants and Engineers: New Delhi, India, 2009.
143. Gupta, M.; Ramkumar, P.L.; Sangani, V. An approach toward augmenting materials, additives, processability and parameterization in rotational molding: A review. *Mater. Manuf. Process* **2020**, *35*, 1539–1556. [CrossRef]
144. Yeom, Y.H. *Internationalization of Korean Traditional Technology-Bronze Age*; National Science Foundation: Daejeon, Korea, 1995.
145. Lee, W.H.; Wu, Y.F.; Ding, Y.C.; Cheng, T.W. Fabrication of ceramic moulds using recycled shell powder and sand with geopolymer technology in investment casting. *Appl. Sci.* **2020**, *10*, 4577. [CrossRef]
146. Carey, P.R.; Swartzlander, M. *Sand Binder Systems. Technical Paper, Part-2*; Ask Chemicals: Hilden, Germany, 2022; pp. 1–9.
147. Callister, W.D. *Fundamentals of Materials Science and Engineering*; John Wiley and Sons: New York, NY, USA, 2001.

148. Balasubramanian, A. *Classification of Materials*; University of Mysore: Mysore, India, 2017. [CrossRef]
149. Drouet, C.; Leriche, A.; Hampshire, S.; Kashani, M.; Stamboulis, A.; Iafisco, M.; Tampieri, A. Types of ceramics: Material class. In *Advances in Ceramic Biomaterials*; Elsevier: Amsterdam, The Netherlands, 2017; pp. 21–82.
150. Bagherpour, E.; Pardis, N.; Reihanian, M.; Ebrahimi, R. An overview on severe plastic deformation: Research status, techniques classification, microstructure evolution, and applications. *Int. J. Adv. Manuf. Technol.* **2019**, *100*, 1647–1694. [CrossRef]
151. Ramkumar, P.L.; Kulkarni, D.M.; Abhijit, V.V.R.; Cherukumudi, A. Investigation of melt flow index and impact strength of foamed LLDPE for rotational moulding process. *Procedia Mater. Sci.* **2014**, *6*, 361–367. [CrossRef]
152. Geyer, R.; Jambeck, J.R.; Law, K.L. Production, use, and fate of all plastics ever made. *Sci. Adv.* **2017**, *3*, e1700782. [CrossRef] [PubMed]
153. Liu, H.; Wang, F.; Shi, M.; Tian, W. Mechanical behavior of polyurethane polymer materials under triaxial cyclic loading: A particle flow code approach. *J. Wuhan Univ. Technol. Sci. Ed.* **2018**, *33*, 980–986. [CrossRef]
154. Jia, Z.; Xia, X.; Wang, X.; Wang, T.; Xu, G.; Liu, B.; Zhou, J.; Li, F. All-conjugated amphiphilic diblock copolymers for improving morphology and thermal stability of polymer/nanocrystals hybrid solar cells. *Front. Mater. Sci.* **2018**, *12*, 225–238. [CrossRef]
155. Crawford, R.J. *Plastics Engineering*; Pergamon Press: Oxford, UK, 1987.
156. Liu, G.; Park, C.B.; Lefas, J.A. Production of low-density LLDPE foams in rotational molding. *Polym. Eng. Sci.* **1998**, *38*, 1997–2009. [CrossRef]
157. Wang, B.; Panigrahi, S.; Tabil, L.; Crerar, W. Pre-treatment of flax fibers for use in rotationally molded biocomposites. *J. Reinf. Plast. Compos.* **2007**, *26*, 447–463. [CrossRef]
158. Pal, R.; Narasimha Murthy, H.N.; Sreejith, M.; Vishnu Mahesh, K.R.; Krishna, M.; Sharma, S.C. Effect of laminate thickness on moisture diffusion of polymer matrix composites in artificial seawater ageing. *Front. Mater. Sci.* **2012**, *6*, 225–235. [CrossRef]
159. Bart, J.C.J. *Additives in Polymers: Industrial Analysis and Applications*; John Wiley & Sons Ltd.: West Sussex, UK, 2005.
160. Alshabanat, M. Morphological, thermal, and biodegradation properties of LLDPE/treated date palm waste composite buried in a soil environment. *J. Saudi Chem. Soc.* **2019**, *23*, 355–364. [CrossRef]
161. Nugent, P. Rotational Molding. In *Applied Plastics Engineering Handbook*; Elsevier: Amsterdam, The Netherlands, 2011; pp. 311–332.
162. Ramkumar, P.L.; Kulkarni, D.M. Impact strength investigation of foamed linear low density polyethylene in rotational moulding process. *Int. J. Mater. Eng. Innov.* **2016**, *7*, 159. [CrossRef]
163. Liu, S.-J.; Fu, K.-H. Effect of enhancing fins on the heating/cooling efficiency of rotational molding and the molded product qualities. *Polym. Test.* **2008**, *27*, 209–220. [CrossRef]
164. Torres, F.G.; Aragon, C.L. Final Product Testing of Rotational Moulded Natural Fibre-reinforced Polyethylene. *Polym. Test.* **2006**, *25*, 568–577. [CrossRef]
165. Crawford, R.J.; Spence, A.G.; Cramez, M.C.; Oliveira, M.J. Mould pressure control in rotational moulding. *Proc. Inst. Mech. Eng. Part B J. Eng. Manuf.* **2004**, *218*, 1683–1693. [CrossRef]
166. Spence, A.G.; Crawford, R.J. Removal of pinholes and bubbles from rotationally moulded products. *Proc. Inst. Mech. Eng. Part B J. Eng. Manuf.* **1996**, *210*, 521–533. [CrossRef]
167. Gogos, G. Bubble removal in rotational molding. *Polym. Eng. Sci.* **2004**, *44*, 388–394. [CrossRef]
168. Qin, L.; Ding, Y.-M.; Zhu, G.-C.; Yu, H.-C.; Yang, W.-M. Heat flow analysis and efficiency optimization of rotational molding equipment for large plastic products. *Int. Polym. Process.* **2015**, *30*, 194–201. [CrossRef]
169. Kunzler, W.F.; Coombs, W.F. Rotational Molding of Contact Lenses. U.S. Patent US4555372A, 26 November 1985.
170. De Donno, G. Pep Set, an Efficient and Environmentally Friendly Binder System. Posted on 8 November 2016. Available online: https://www.metalworkingworldmagazine.com/7239-2/ (accessed on 20 March 2022).
171. Choi, C.; Kim, C. Technological trends for special casting systems and environmental improvement in foundries. *J. Korean Soc. Cast. Eng.* **2017**, *37*, 2.
172. Zhao, X.; Ning, Z.; Li, Z.; Zou, W.; Li, B.; Huang, Y.; Cao, F.; Sun, J. Evolved gas analysis of PEP-SET sand by TG and FTIR. J. *Anal. Appl. Pyrol.* **2017**, *127*, 490–495. [CrossRef]
173. Kyung-Jun, R.A.; Kim, H.Y.; Won, K.S. Restoration of "JikJi" movable metal type and the related issues. *J. Hist. Sci. Korea* **2006**, *28*, 139–160.
174. Kienlin, T.L. *Traditions and Transformations: Approaches to Eneolithic (Copper Age) and Bronze Age metalworking and Society in Eastern Central Europe and the Carpathian Basin*; BAR Int Series; Archaeopress: Oxford, UK, 2010.
175. Baron, J.; Miazga, B.; Ntaflos, T.; Puziewicz, J.; Szumny, A. Beeswax remnants, phase and major element chemical composition of the bronze age mould from Gaj Oławski (SW Poland). *Archaeol. Anthropol. Sci.* **2016**, *8*, 187–196. [CrossRef]
176. History of Casting, Bernier Cast Metals, Inc. 2018. Available online: https://bernierinc.com/history-of-casting (accessed on 20 March 2022).
177. Vidale, M.; Melucco Vaccaro, A.; Salvatore, M.R.; Micheli, M.; Balista, C. From Theophilus to C. S. Smith: Discovery of an eleventh century AD bell-casting mold from venosa (Southern Italy). *Mat. Res. Soc. Symp. Proc.* **1992**, *267*, 757–779. [CrossRef]
178. Drescher, H. Eine technische Betrachtung bronzezeitlicher Halskragen. *Offa* **1953**, *12*, 67–72.
179. Rønne, P. Early Bronze Age Spiral Ornament–The technical Background. *J. Dan. Archaeol.* **1989**, *8*, 126–143. [CrossRef]
180. Rønne, P. *Forsogarkaeologi og Bronzealderens Ornamentik*; Historical-Archaeological Research Centre, Experimental Archaeology, Studies in Technology and Culture: Lejre, Denmark, 1991.

181. Cheah, C.M.; Chua, C.K.; Lee, C.W.; Feng, C.; Totong, K. Rapid prototyping and tooling techniques: A review of applications for rapid investment casting. *Int. J. Adv. Manuf. Technol.* **2005**, *25*, 308–320. [CrossRef]
182. Liu, Q.; Sui, G.; Leu, M.C. Experimental study on the ice pattern fabrication for the investment casting by rapid freeze prototyping. *Comput. Ind.* **2002**, *48*, 181–197. [CrossRef]

metals

Article

Phase Prediction, Microstructure and Mechanical Properties of Fe–Mn–Ni–Cr–Al–Si High Entropy Alloys

Essam R. I. Mahmoud [1,*], Awaluddin Shaharoun [1], Mohamed A. Gepreel [2] and Saad Ebied [3]

1 Department of Mechanical Engineering, Islamic University of Madinah, Madinah 42351, Saudi Arabia; prof.awaluddin@gmail.com
2 Materials Science and Engineering Department, Egypt-Japan University of Science and Technology, New Borg El-Arab 16448, Egypt; mohamed.gepreel@ejust.edu.eg
3 Department of Production Engineering and Mechanical Design, Faculty of Engineering, Tanta University, Tanta 31527, Egypt; saad_ebied@f-eng.tanta.edu.eg
* Correspondence: essamibrahim2@yahoo.com; Tel.: +966-543876061

Abstract: The selection of high-entropy alloys (HEAs), which are relatively lightweight and have unique mechanical properties, remains a substantial challenge. In this study, six new HEAs were designed from the relatively low-cost Fe–Mn–Ni–Cr–Al–Si system using Thermo-Calc software, and then manufactured using a casting process. The effects of the atomic ratio of the alloying elements on the microstructures and mechanical properties of these alloys in the as-cast condition were systematically investigated. Brittle body-centered cubic BCC/B2 and silicide phases were found in relatively large amounts in the form of dendritic structure within large equiaxed grains with fine needle-shaped phases in the $Fe_{30}Mn_{15}Ni_{20}Cr_{15}Al_{10}Si_{10}$ and $Fe_{35}Mn_{15}Ni_{20}Cr_{15}Al_{10}Si_5$ alloys, in addition to the face-centered cubic (FCC) phase. When the contents of Mn and Ni were increased in the $Fe_{35}Mn_{25}Ni_{15}Cr_{15}Al_5Si_5$ and $Fe_{35}Mn_{20}Ni_{20}Cr_{15}Al_5Si_5$ alloys, the amounts of brittle phases were reduced; however, the ductile FCC phase is not significant. The FCC phase amount, which appeared as a honeycombed structure, was more than enough when the Si content was decreased to 3%. Broad relationships between the chemical composition of the alloys, especially the Si content, and the hardness and compression properties' measurements were established. As the Si content decreased, both the hardness and compression properties of the resulting alloy also decreased. The experimental observation of the six HEAs matched the equilibrium phases predicted by the Thermo-Calc calculations.

Keywords: high-entropy alloy; thermo-calc calculation; casting; microstructure analysis; hardness measurements

Citation: Mahmoud, E.R.I.; Shaharoun, A.; Gepreel, M.A.; Ebied, S. Phase Prediction, Microstructure and Mechanical Properties of Fe–Mn–Ni–Cr–Al–Si High Entropy Alloys. *Metals* **2022**, *12*, 1164. https://doi.org/10.3390/met12071164

Academic Editors: Carlos Alexandre Dos Santos and Eleani Maria Da Costa

Received: 8 June 2022
Accepted: 4 July 2022
Published: 8 July 2022

1. Introduction

High-entropy alloys (HEAs) have attracted many researchers' attention since 2004, after Jien-Wei Yeh and Brian Cantor [1,2] achieved good results from them in their work. High-entropy alloys (HEAs) can be defined as multi-metallic materials, containing four or more basic alloying elements in equal atomic percentages (at.%) or near-equiatomic proportions [3,4]. The atomic portion of each element is often in excess of 5 at.%. This design philosophy was initially aimed at stabilizing the massive solid solutions of a single phase over high configurational entropy. These newly designed alloys have exceptional mechanical properties, which may vary entirely from their basic elements [5–8]. HEAs are a favored approach for the production of high-performance alloys with improved mechanical toughness and strength, higher thermal stability, enhanced oxidation, soft magnetic properties, and corrosion resistance [9–12]. In addition, several HEAs have shown high resistance to irradiation, exhibiting lower irradiation-stimulated segregation and reduced density of dislocation loops, compared to ordinary alloys [13,14], and possessing self-healing properties [15,16]. These unique properties position them as the first choice for structural applications, especially in the improvement of power plants' efficiency, which need to employ novel materials that can withstand extreme working conditions, particularly for the new generation of nuclear fission. On the other hand, other HEAs show both high wear and corrosion resistance, accepted mechanical properties, and high biocompatibility, in addition to the low cost, which make these HEAs favorable candidates for biomedical implantation materials [17–20]. Furthermore, a metal film of HEAs with accepted toughness, electrical resistivity, and high fatigue performance could also be a promising candidate for flexible electronic devices [21]. Another HEA shows excellent thorough properties including accepted mechanical properties, excellent antifouling abilities, outstanding resistance to wear and corrosion for marine applications [22]. Moreover, the mechanical properties of high-entropy alloys (HEAs) can be improved significantly by thermo-mechanical processing, which gives these alloys unique properties, making them vastly suitable for different applications as described by Wani et al. [23]. For example, the recent study on the AlCoCrFeNi2.1 high-entropy alloy with eutectic structure [24,25], tensile strength > 1000 MPa, and ductility of 10% was obtained by rolling the ingot to a reduction ratio of up to 90% and then annealing at a temperature of 1200 °C. Further examples of these multi-component alloys are $Fe_{40}Mn_{40}Co_{10}Cr_{10}$ HEAs, produced by Deng et al. [26], and the same alloy with Ni addition, $Fe_{40}Mn_{27}Ni_{26}Co_5Cr_2$, studied by Yao et al. [27]. They elucidated that these types of alloys have superior mechanical properties analogous to some types of tool steels. Another study suggested inserting Al into these alloys in the state of Co to improve the mechanical properties and to achieve cost reduction [28,29]. On the other hand, Li et al. improved the hardness and corrosion resistance of the $FeCoNiCrCu_{0.5}$ alloy by adding 1.5 at.% Al. In another study, N.D. Stepanov et al. [30] studied the effect of an Al fraction on the microstructure and mechanical properties of the Fe–Mn–Cr–Ni–Al non-equiatomic high-entropy alloys system with various Al percentages (x = 0–14 at.%). They found that an Al addition to a BCC-prone $Fe_{40}Mn_{25}Cr_{20}Ni_{15}$ alloy is useful for the improvement of the studied alloys with the mixture of bcc matrix and the inserted B2 precipitation showing promising mechanical properties. Cui et al. [31] showed that the addition of Al 0.77 at.% to the FeCrNiAlx HEA established it as a promising material for application in aerospace fields, due to its high specific strength and accepted ductility at high temperatures. Moreover, Kishore et al. [32] investigated the effect of the addition of Mn to an equi-atomic CoCrNi alloy. They concluded that a Mn addition enhanced the hardening ability because of the steep increase in hardness (~650 Hv in $Co_{33}Ni_{33}Cr_{19}Mn_{15}$ alloy vs. ~610 Hv in CoCrNi alloy). In addition, Oh et al. [33] stated that the addition of Mn increases the fraction of the BCC phase and, as a result, the nanohardness in the HEA AlCoCrFeN also increases. In another high-entropy alloy, $Fe_{40}Co_{40}Ni_{10}M_{10}$ (M = Al, Mn) [34], Al addition was useful for obtaining a complete BCC structure formation using the casting process. On the other hand, Mn produces dual phase development with high mechanical properties as a result of grain refinement. Meng et al. [35] investigated the effect of Cr on the mechanical properties of $Fe_{30}Ni_{20}Mn_{35}Al_{15}$. They elucidated that the ductility at room temperature increased with increasing the Cr content to 6 at.%, because the addition of Cr caused the complete deactivation of the environmental embrittlement. Gu et al. [36] studied the effect of Si addition to the Al0.3 CoCrFeNi high-entropy alloy. They clarified that adding Si could increase the hardness and strength of alloys. They concluded that Si addition increases the development of a new (Al, Ni, Si) rich phase with a B2 structure (BCC1 phase), and a $(Fe, Cr)_3Si$ phase with L21 structure (BCC2 phase). Guo et al. [37] explained that the addition of Si to $TaMo_{0.5}NbZrTi_{1.5}Al_{0.1}$ enables significant grain refinement and development of intergranular silicides after annealing. Furthermore, they mentioned that the Si addition improves the mechanical properties alongside increasing the deformability of the alloy by facilitating the development of fine silicides and the extra solute effects. Xu et al. [38] showed that the Si addition could manipulate the microstructure of VNbTiTaSix and, in turn, improve the mechanical properties when the Si addition reaches 10%. In the coating domain, Liu et al. [39] investigated the effect of adding Si to the AlCoCrFeNiSix coatings. They found that the enhancement of microhardness was controlled by the influence of dislocation strengthening, rather than fine structure strengthening and solution strengthening. Additionally, the wear resistance was enhanced due to the formation of different oxides including SiO_2 and SiO. On the other hand, Ma et al. [40] concluded that at high temperatures adding Si improves the phase stability of the CoCr2FeNb0.5Ni coating by reducing the interplanar distance of the crystal planes. In addition, the formation of the (Cr, Si)Ox amorphous oxide layer can inhibit oxygen from diffusing inwards, and hence increase the high temperature oxidation resistance. Huang et al. [41] explained that the Si addition improved both the microhardness and wear resistance of the FeCoCrNiSix alloys. The improvement in the microhardness can be related to the formation of BCC phase because of the addition of Si. Hou et al. [42] studied the effect of adding Si to the metastable $Fe_{50}Mn_{30}Co_{10}Cr_{10}$ and found that significant improvement in toughness occurred in the SLMed metastable high-entropy alloy. This improvement can be related to the numerous deformation mechanisms included because the addition of Si. Guo et al. [43] found that adding Ni improved the mechanical properties of the $AlCoCrFeTi_{0.5}$ high-entropy alloys by achieving dual phase structure. Qiu et al. [44] studied the effect of Fe fraction in four FexCoNiCu HEAs and found that, by increasing the Fe fraction, the crystal structure of the alloys progressively altered from FCC to a mixture of BCC and FCC phases. In addition, the deformation mechanism can be manipulated via controlling the chemical composition of the HEAs alloys. For example, Huang et al. [45] studied the effect of the Ta composition on the deformation mechanism of the TiZrHfTaX (x = 1, 0.8, 0.6, 0.5) alloys. They found that, by reducing the percentage of Ta, both TWIP and TRIP strengthening mechanisms are included. Furthermore, Xiaoyi et al. [46] studied the Mo and Nb composition effect on the mechanical properties of the $AlCrFe_2Ni_2(MoNb)x$ alloys and deduced that Mo and Nb content in the range 0.1–0.7 increased yield strength in compression (from 878 MPa to 1549 MPa) though the plastic strain from 43.7% to 8.6%. The design of new high-performance HEAs has been attracting much attention from researchers recently. However, HEAs have intrinsic properties of sluggish diffusion. Therefore, the equilibrium phase formation may be kinetically very hard to achieve. Therefore, there are many empirical methodologies to predict phase formation in HEAs based on the aforementioned research. Among these methodologies are thermodynamic parameters, and recently machine learning, as discussed elsewhere.

In this work, we investigate different non-equiatomic FeAlNiCrMnSi HEA compositions. The new alloys are close in composition to high-Mn stainless steel. Fe, Ni, and Cr elements are the main elements of stainless steel. In addition, Mn is used as an austenite stabilizer. Al and Si elements are used to improve the mechanical and chemical properties as well as to decrease the density. Various characterization methods were applied to characterize the new alloys' microstructure and phase composition. The phase formation in HEAs is controlled by different factors. The thermodynamic parameters, such as mixed *entropy* (ΔS_{mix}), mixing enthalpy (ΔH_{mix}), Gibbs free energy (ΔG_{mix}), valence electron concentration (VEC), and others have direct relation to the phase formation. Empirically, VEC can provide an idea about the formed phases. For example, single FCC and BCC solid solution phases form when VEC ≥ 8 and VEC < 6.87, respectively, while a mixture of FCC and BCC solid solution phases are expected if $6.87 \leq$ VEC < 8, as reported in [47]. However, in this work, we aim to connect the equilibrium phase formation in this type of low-cost Fe–Cr–Mn–Ni–Al–Si high-entropy alloy system, predicted by Thermo-Calc TCHEA software, with the formed equilibrium/nonequilibrium phases during real solidification process without further treatment process.

2. Materials and Methods

Six new FeMnNiCrAlSi high-entropy alloys (HEAs) were designed using a platform of Thermo-Calc software (Version 2022a, THERMOCALC TCHEA6, Stockholm, Sweden) equipped with a HEA database (TCHEA2021). The effect of the contents of the different elements of the alloys in the as-cast condition on the phase constituents, microstructure, and hardness of the resulting materials was studied. The new HEAs were prepared from the following elements of Fe, Mn, Ni, Cr, Al, and Si according to the chemical composition in Table 1 using an electric arc furnace (ARCAST 200, Maine, ME, USA) under a high-purity argon atmosphere. First, the ingots were produced using high-purity elemental Fe (99.99%) shot, Mn (99.95%) flake, Ni (99.95%) granule, Cr (99.95%) granule, Al (99.96%) wire, and Si (99.99%) granule. To ensure full melting and the homogeneity of the ingot, it was melted 4 times by using an electromagnetic stirrer and flipping each time. Before the microstructural analysis, metallographic procedures, which include grinding, polishing, and etching, were applied to the samples. The microstructure investigation was carried out using an optical microscope after electrochemical etching with 10% Oxalic acid. X-ray diffraction (Model-6100, Shimadzu, Kyoto, Japan) was used in the scanning range of $20° \leq 2\theta \leq 80°$ intervals with a step size of 0.05 deg. The scan rate of 1 deg./s was used to investigate the crystal structure of the six alloys in the as-cast condition. A Shimadzu microhardness tester (Shimadzu, Kyoto, Japan) was used to measure the microhardness with a testing load 5 N, and the time of indentation was 30 s. An average of 5 indents was taken for each condition. The compression tests were carried out on the alloys using the (AGS-X, Shimadzu) universal testing machine (Shimadzu, Kyoto, Japan) at a strain rate of 10^{-3} s^{-1} to an engineering strain of 0.4 on cylindrical specimens Ø5 mm × 7 mm in size.

Table 1. The chemical composition of FeMnNiCrAlSi HEA alloys (at.%).

Alloy	Fe	Mn	Ni	Cr	Al	Si
1	30	15	20	15	10	10
2	35	15	20	15	10	5
3	35	25	15	15	5	5
4	35	20	20	15	5	5
5	35	20	22	15	5	3
6	36	20	20	16	5	3

3. Results and Discussion

3.1. Phase Constitution

The Thermo-Calc software phase prediction of the alloy $Fe_{30}Mn_{15}Ni_{20}Cr_{15}Al_{10}Si_{10}$ (Alloy 1) shows the formation of the brittle BCC (B2) phase at first to solidify phase in addition to silicides, such as MnNiSi and CrSi intermetallic compounds phases, as shown in Figure 1. This is due to the high percentage of Si, which promotes the formation of the brittle BCC (B2) phase. At the same time, the high content of Al (10 at.%) plays a significant role as a BCC stabilizer. This was confirmed by the XRD analysis shown in Figure 2, which clearly shows BCC (B2) is the main phase beside some unknown peaks, which is most probably due to silicides and/or sigma phase formation during solidification.

When the Si content was decreased to 5 at.% and the Fe was increased to 35 at.% in Alloy 2 ($Fe_{35}Mn_{15}Ni_{20}Cr_{15}Al_{10}Si_5$), the FCC phase beside the BCC phases were detected as predicted to appear during solidification in the Thermo-Calc chart shown in Figure 3. This is also confirmed with the XRD peaks in Figure 2. These two alloys (Alloy 1 and 2) were very brittle and broke during the processing of the cast ingot. To increase the ratio of the ductile FCC phase, the content of Mn, which is an FCC stabilizer, was increased from 15 at.% to 25 at.% in Alloy 3 ($Fe_{35}Mn_{25}Ni_{15}Cr_{15}Al_5Si_5$) at the expense of Al and Ni. The phase prediction of Alloy 3 is shown in Figure 4. There was a reasonable amount of the FCC phase appearing during the solidification process at a temperature close to the solidus line. However, the brittle B2 structure is still the main phase in Alloy 3, as well as formation of silicides at high temperature close to the melting point. The fourth trial was to increase the FCC fraction by increasing the Ni content, a strong FCC stabilizer, to 20 at.% as in Alloy 4 ($Fe_{35}Mn_{20}Ni_{20}Cr_{15}Al_5Si_5$) at the expense of the Mn content. The Thermo-Calc calculation of this Alloy 4, Figure 5, predicts the formation of the BCC phases—first solid to form; however, the ductile FCC phase is the main phase at the end of the solidification process. Further reductions in the Si content to 3 at.% while increasing the Ni content in Alloy 5 ($Fe_{35}Mn_{20}Ni_{22}Cr_{15}Al_5Si_3$) led to a delay in the formation of the silicides at high temperatures; the results are shown in Figure 6. The alloy contained mainly the FCC phase, and a small amount of B2 phase was expected while the silicides precipitation was delayed at lower temperatures. According to Figure 6, Alloy 5 would have full FCC phases at high temperatures before sigma; whereas silicides precipitate at lower temperatures. Alloy 5 is expected to have good deformability, and it showed high cold workability compared to Alloys 1–4, as will be explained in a different work.

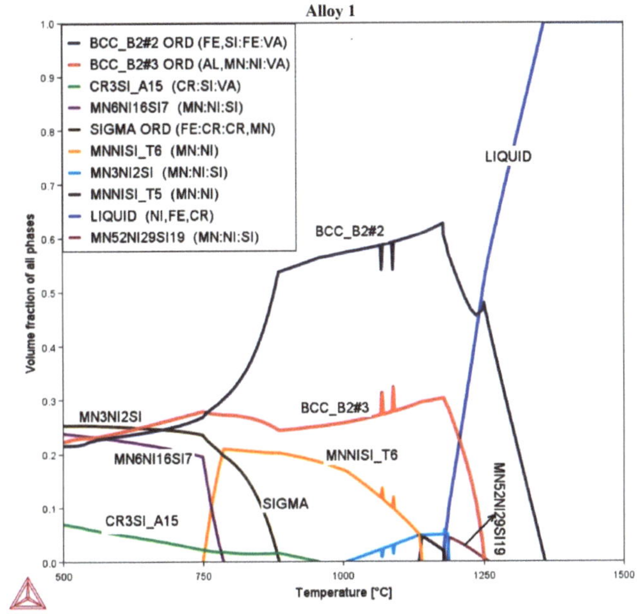

Figure 1. Phase diagram of Alloy 1 ($Fe_{30}Mn_{15}Ni_{20}Cr_{15}Al_{10}Si_{10}$) HEA.

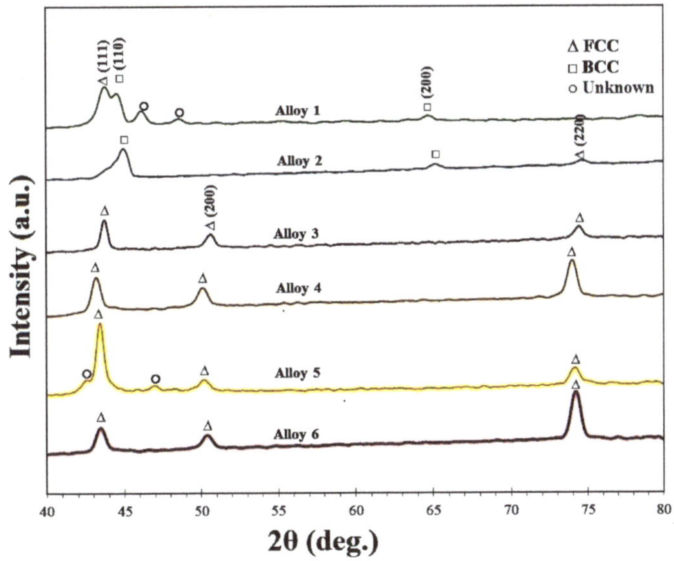

Figure 2. XRD patterns of the as-cast six HEAs.

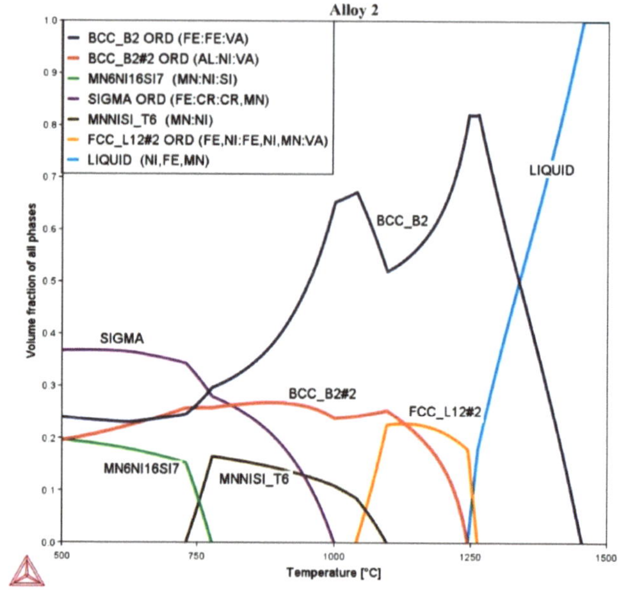

Figure 3. Phase diagram of Alloy 2 ($Fe_{35}Mn_{15}Ni_{20}Cr_{15}Al_{10}Si_5$) HEA.

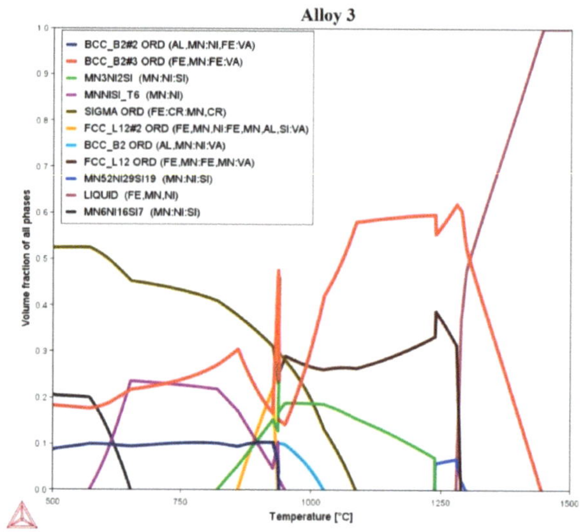

Figure 4. Phase diagram of Alloy 3 ($Fe_{35}Mn_{25}Ni_{15}Cr_{15}Al_5Si_5$) HEA.

Alloy 6 ($Fe_{36}Mn_{20}Ni_{20}Cr_{16}Al_5Si_3$) represents the last trial of this study, as shown in Figure 7. The aim with Alloy 6 was to increase the strength by increasing the BCC phase fraction by a slight increase in the Cr content to be 16 at.% and reducing Ni back to 20 at.%, while maintaining less silicide formation by keeping Si content as its lower content of 3%. In Alloy 6, the FCC phase became the main constituent, and a slightly higher amount of B2 phase was observed compared to Alloy 5, and no silicides present at high temperature. Alloy 6 was also deformable and the results will be presented elsewhere. The phase identification of the 6 alloys in the study was carried out by XRD. Figure 2 shows a high degree of consensus with the phase constitution of the alloys predicted by Thermo-Calc in Figures 1 and 3–7, as explained below.

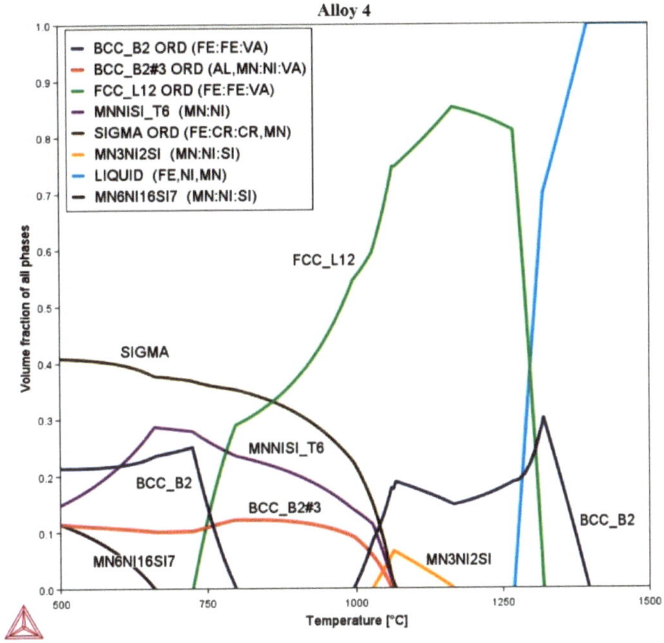

Figure 5. Phase diagram of Alloy 4 ($Fe_{35}Mn_{20}Ni_{20}Cr_{15}Al_5Si_5$) HEA.

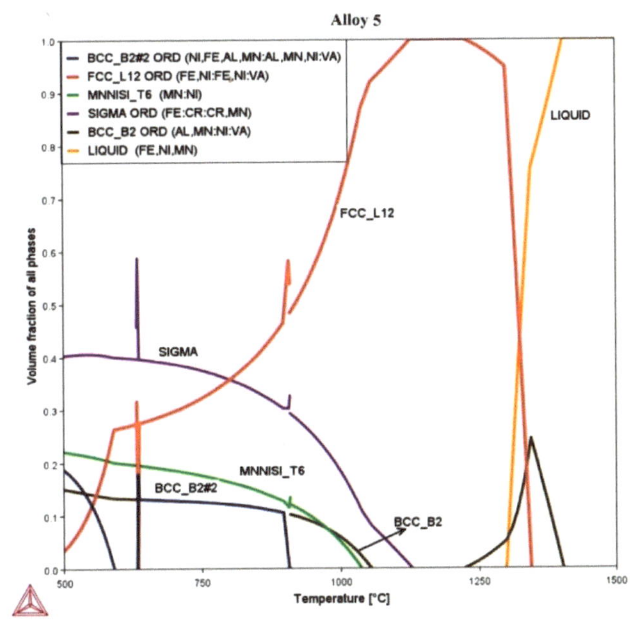

Figure 6. Phase diagram of Alloy 5 ($Fe_{35}Mn_{20}Ni_{22}Cr_{15}Al_5Si_3$) HEAs.

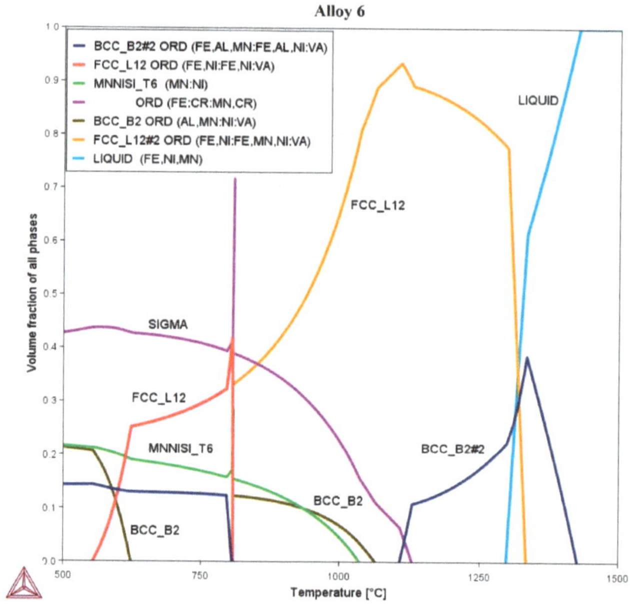

Figure 7. Phase diagram of Alloy 6 ($Fe_{36}Mn_{20}Ni_{20}Cr_{16}Al_5Si_3$) HEAs.

3.2. Microstructure Characterization

Figure 8 shows the microstructure of HEAs (Alloys 1–3). In Alloy 1, dendritic structure within large equiaxed grains with fine, needle-shaped phases was the matrix of this alloy along with the white BCC phase. This is due to the high percentage of Si, which promotes the formation of this kind of needle-like silicide structures. According to the XRD results and the Thermo-Calc prediction, Alloy 1 consists mainly of BCC/B2 and silicide phases, which is a very hard and brittle mixture. The dark, round, fine phase may be a different type of silicide formed by precipitation in solid state at high temperatures, according to XRD results and the Thermo-Calc prediction in Figure 1.

According to many published works, the high content of Al (10 at.%) can form aluminides (Si does the same), which act as heterogenous nucleation sites during solidification leading to the formation of equiaxed grain microstructures [48]. For Alloy 2 ($Fe_{35}Mn_{15}Ni_{20}Cr_{15}Al_{10}Si_5$), BCC/B2 grains were observed together with very fine precipitates, as shown in Figure 8b. According to XRD in Figure 2, BCC/B2 is the main phase, and fine FCC and B2 phases may be formed during cooling, matching the Thermo-Calc prediction. Generally, the higher amount of Al (in Alloys 1 and 2) enhanced the formation of BCC phases, as reported by Singh et al. [49]. A semi- honeycomb BCC structure with fine arms (dendritic structure) was observed in Alloy 3 ($Fe_{35}Mn_{25}Ni_{15}Cr_{15}Al_5Si_5$) to be distributed in a matrix of FCC phase, as shown in Figure 8c. Fine, dark, minor phases were observed which may be related to silicide formation during cooling in the solid state as in Alloy 1 and 2. For the structure of Alloy 4 ($Fe_{35}Mn_{20}Ni_{20}Cr_{15}Al_5Si_5$), dendritic FCC grains were observed as shown in Figure 9a. Small amounts of the darker brittle BCC phase were detected in the inter-dendrite zone. However, the XRD results reported that the BCC phase would be a minor phase while an unknown phase was quite detectable. In the microstructure of Alloy 4 in Figure 9a, there are two inter-dendrite phases: one is in gray color and the other is in black. These unknown phases may be silicides formed by precipitation in the solid state during cooling and/or the BCC/B2 phases. Further studies are needed to clarify this issue.

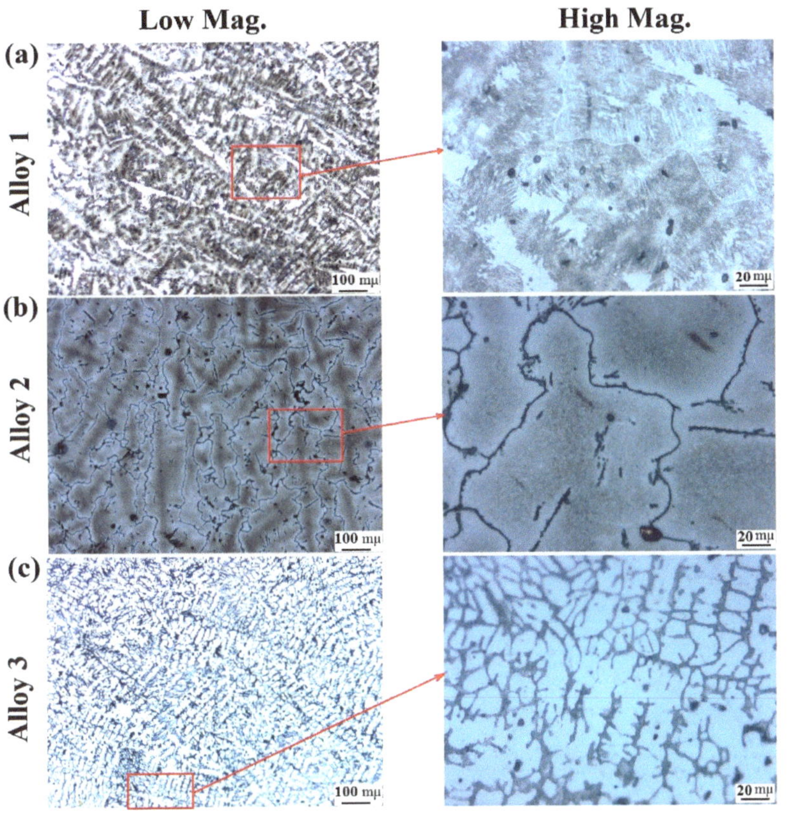

Figure 8. Microstructural images of as-cast HEAs (**a**) Alloy 1, (**b**) Alloy 2, and (**c**) Alloy 3. Low and high magnification are presented in left and right, respectively.

The structure of Alloy 5 ($Fe_{35}Mn_{20}Ni_{22}Cr_{15}Al_5Si_3$) consisted of a large columnar and dendritic grains of mainly the FCC phase, as shown in Figure 9b. The high FCC phase constitution in the alloy is due to high Ni and Mn content and less Al and Si content. Minor, aligned, black, droplet-shaped phases were precipitated in the inter-dendritic zones. These minor, black, droplet-shaped phases were most probably formed during cooling in the solid state and may be silicides or Al-Ni-rich phases [48], even though only the FCC phase was detected by XRD in Alloy 5. In accordance with this, Alloy 5 contains the biggest amount of the FCC phase among the six studied alloys, as predicted by Thermo-Calc where the FCC phase becomes the only solid phase at high temperature range. Therefore, Alloy 5 would be softer than others and shows high deformability.

The structure of Alloy 6, with a slightly higher amount of Cr while keeping low content of Al and Si (Alloy 6 with $Fe_{36}Mn_{20}Ni_{20}Cr_{16}Al_5Si_3$ system), showed a honeycomb dendritic structure consisting mainly of the FCC phase, as shown in Figure 9c. Higher precipitates than in Alloy 5 appeared on the boundary of the grains and concentrated on the triple point of grain boundaries in Alloy 6. These are most often the BCC/B2 phases as required from the design and proven by the XRD results. Almost all microstructural observations of the present six alloys were in good agreement with the Thermo-Calc calculations predictions and the XRD patterns.

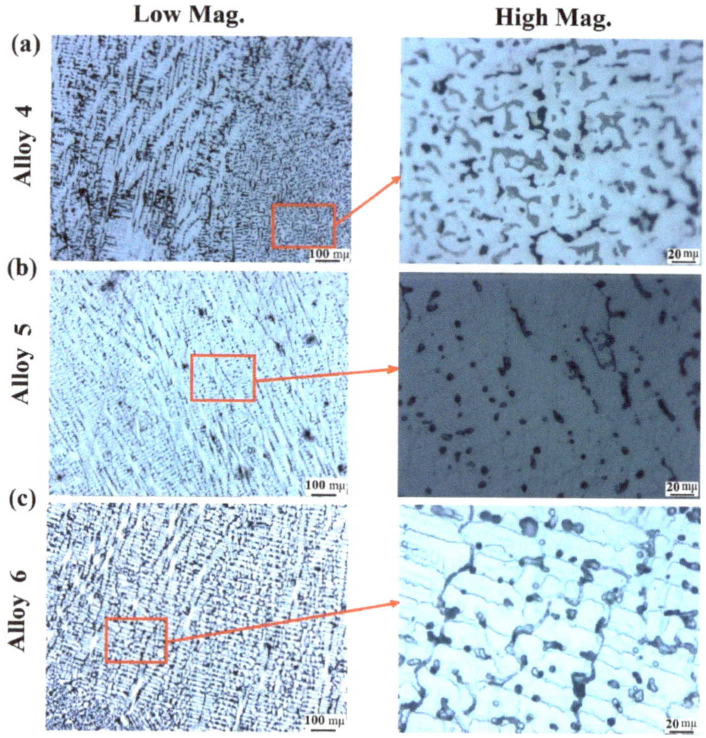

Figure 9. Microstructural images of as-cast HEAs (**a**) Alloy 4, (**b**) Alloy 5, and (**c**) Alloy 6. Low and high magnification are presented in left and right, respectively.

3.3. Mechanical Properties

The microhardness of the six HEAs under investigation was measured using a Vickers microhardness tester, and the results are shown in Figure 10. It is clear that the hardness is noticeably affected by the chemical composition of the alloy. It can be seen that the key factor in the increase in hardness is the Si and Al fractions. Firstly, hardness decreased with decreasing both Si and Al fractions. This clearly appeared in the remarkably high hardness values of Alloy 1 and Alloy 2, which have high Al and Si contents. Al and Si stabilize the hard BCC phases. Additionally, they restrict the formation of the ductile FCC phase. Moreover, hard silicide intermetallics can easily form, which can harden these higher Si content alloys. Even when the Si content was reduced to half the amount in Alloy 2, the other strong BCC phase stabilizer (Al) increases the hardness to almost 500 HV. When both of them (Al and Si) were reduced to 5 at.% in Alloy 3 and Alloy 4, the hardness values were reduced to about 350 HV and 250 HV, respectively. Reduction in Al and Si provide the opportunity for the ductile FCC phase to form and reduce the hardness. The difference in hardness between Alloy 3 and Alloy 4 comes from the amount of formed FCC phase during solidification. Alloy 4 has a higher Ni content (20 at.%), which is a strong FCC phase stabilizer. When the Si content was reduced further to 3 at.% in Alloy 5 and Alloy 6, the hard silicides almost disappeared and the amount of the ductile FCC phase became the main constituent of the structure alongside a small amount of the brittle BCC (B2) phase. This results in a reduction in hardness to less than 200 HV, and provides a chance for it to be easily cold-deformed.

For more clarification of the mechanical properties, compression tests of the six alloys in their as-cast condition were performed and the results are shown in Figure 11. Firstly, it is important to record here that the casting defects such as pores, segregation, and uneven residual stresses can negatively impact the mechanical properties [50]. Alloy 1 was damaged during the test due to its higher brittleness. This was expected due to the ultra-high content of Al (10 at.%) and Si (10 at.%). The ultimate compression strength increases with higher Al and moderate Si content (Alloy 2) due to the formation of higher volume fractions of BCC phases. It reached as high as 1700 MPa. The compressive ductility often decreased when the strength increased. The ductility is very sensitive to the number of brittle phases, such as the BCC phase and other intermetallics. Therefore, the ductility of Alloy 1 was very low to the point of being broken during the initial stages of the test, while Alloy 2 shows ductility of only 10%. By reducing the amount of Al and Si by half, the ultimate strength reduced to almost half, about 900 MPa, and the ductility was increased to about 17% in Alloy 4. Unexpected results were recorded for Alloy 5 and 6, with higher values of compressive ultimate strength of about 1000 MPa. The formation of FCC phase with higher volume fraction increased the ductility to 25%. At the same time, the honeycomb dendritic structure increased the strength. In addition, the B2 precipitates that appeared in the triple point of grain boundaries act as obstacles against fracture during the test and increased the strength to 950 MPa for Alloy 6.

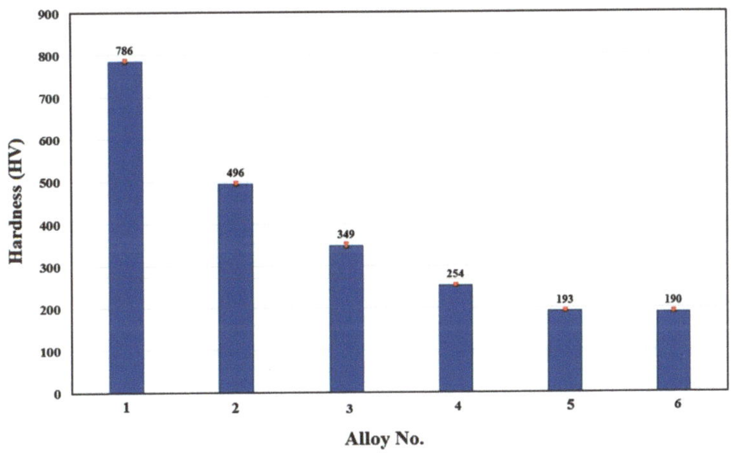

Figure 10. Microhardness of the as-cast six Fe–Mn–Ni–Cr–Al–Si HEAs.

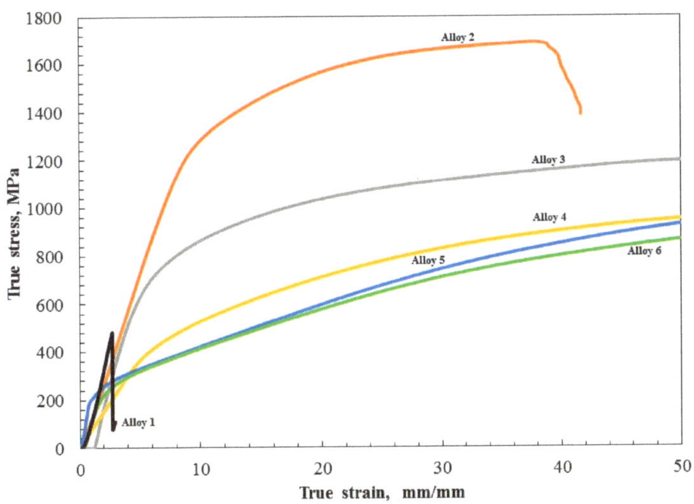

Figure 11. Typical true stress–strain curves of compression for the as-cast six Fe–Mn–Ni–Cr–Al–Si HEAs.

4. Conclusions

In this study, the effect of changing the fraction of alloying elements on the microstructures, and the mechanical properties of these alloys in the as-cast condition, were systematically investigated. The Thermo-Calc software was very efficient in predicting the phases in the as-cast condition, as confirmed by the experimental results. The first solid phases predicted to form during cooling from melt state are in all likelihood the present phases at room temperature. The major results of this study can be summarized as follows:

1. Alloys with high Si fraction (Alloy 1 and Alloy 2) have mainly brittle silicides and BCC/B2 phases, and showed high hardness with values more than 750 Hv.
2. The Si addition has a crucial role in the compression properties of the Fe–Mn–Ni–Cr–Al–Si HEAs system; as the Si addition increased, the compression strength increased and the ductility decreased.
3. Decreasing the Si and Al content, while increasing the Mn and Ni contents to work as FCC stabilizing elements in Alloys 3, 5 and 6, caused the amounts of brittle phases were to be significantly reduced, and the ductile FCC phase was maintained as the main phase at room temperature. This resulted in hardness as low as 190 Hv.
4. The microstructure and hardness can be easily manipulated in easy deformable Fe–Cr–Mn–Ni–Al–Si alloys by changing the alloying elements' content while keeping Al at 5 at.% and Si at 3 at.%.
5. There was a good consensus between experimental observations in the six high-entropy alloys and the equilibrium phase diagram predicted by the Thermo-Calc calculations.

Author Contributions: Conceptualization, E.R.I.M.; methodology, E.R.I.M. and A.S.; validation, S.E. and M.A.G.; formal Analysis, M.A.G. and E.R.I.M.; investigation, E.R.I.M.; data curation, S.E.; writing—original draft preparation, E.R.I.M. and S.E.; writing—review and editing, M.A.G.; supervision, M.A.G.; project administration, S.E.; funding acquisition, E.R.I.M. All authors have read and agreed to the published version of the manuscript.

Funding: Scientific Research Deanship, Islamic University of Madinah, Tamayyuz—2 grant number 714.

Institutional Review Board Statement: Not applicable.

Informed Consent Statement: Not applicable.

Data Availability Statement: Not applicable.

Acknowledgments: The authors would like to express their appreciation for the support provided by the Scientific Research Deanship, Islamic University of Madinah, with Tamayyuz—2 grant number 714.

Conflicts of Interest: The authors declare that they have no known competing financial interests or personal relationships that could have appeared to influence the work reported in this paper.

References

1. Yeh, J.W.; Chen, S.K.; Lin, S.J.; Gan, J.Y.; Chin, T.S.; Shun, T.T.; Tsau, C.H.; Chang, S.Y. Nanostructured high-entropy alloys with multiple principal elements: Novel alloy design concepts and outcomes. *Adv. Eng. Mater.* **2004**, *6*, 299–303. [CrossRef]
2. Cantor, B.; Chang, I.T.H.; Knight, P.; Vincent, A.J.B. Microstructural development in equiatomic multicomponent alloys. *Mater. Sci. Eng. A* **2004**, *375–377*, 213–218. [CrossRef]
3. Li, Z.; Ludwig, A.; Savan, A.; Springer, H.; Raabe, D. Combinatorial metallurgical synthesis and processing of high-entropy alloys. *J. Mater. Res.* **2018**, *33*, 3156–3169. [CrossRef]
4. Otto, F.; Dlouhý, A.; Somsen, C.; Bei, H.; Eggeler, G.; George, E.P. The influences of temperature and microstructure on the tensile properties of a CoCrFeMnNi high-entropy alloy. *Acta Mater.* **2013**, *61*, 5743–5755. [CrossRef]
5. Miracle, D.B.; Senkov, O.N. A critical review of high entropy alloys and related concepts. *Acta Mater.* **2017**, *122*, 448–511. [CrossRef]
6. Bian, B.B.; Guo, N.; Yang, H.J.; Guo, R.P.; Yang, L.; Wu, Y.C.; Qiao, J.W. A novel cobalt-free FeMnCrNi medium-entropy alloy with exceptional yield strength and ductility at cryogenic temperature. *J. Alloys Compd.* **2020**, *827*, 153981. [CrossRef]
7. Liu, D.; Jin, X.; Guo, N.; Liaw, P.K.; Qiao, J.W. Non-equiatomic FeMnCrNiAl high-entropy alloys with heterogeneous structures for strength and ductility combination. *Mater. Sci. Eng. A* **2021**, *818*, 141386. [CrossRef]

8. Elkatatny, S.; Gepreel, M.A.H.; Hamada, A.; Nakamura, K.; Yamanaka, K.; Chiba, A. Effect of Al content and cold rolling on the microstructure and mechanical properties of $Al_5Cr_{12}Fe_{35}Mn_{28}Ni_{20}$ high-entropy alloy. *Mater. Sci. Eng. A* **2019**, *759*, 380–390. [CrossRef]
9. Mishra, R.K.; Sahay, P.P.; Shahi, R.R. Alloying, magnetic and corrosion behavior of AlCrFeMnNiTi high entropy alloy. *J. Mater. Sci.* **2019**, *54*, 4433–4443. [CrossRef]
10. Soare, V.; Mitrica, D.; Constantin, I.; Popescu, G.; Csaki, I.; Tarcolea, M.; Carcea, I. The Mechanical and Corrosion Behaviors of As-cast and Re-melted AlCrCuFeMnNi Multi-Component High-Entropy Alloy. *Metall. Mater. Trans. A Phys. Metall. Mater. Sci.* **2015**, *46*, 1468–1473. [CrossRef]
11. Pang, J.; Zhang, H.; Zhang, L.; Zhu, Z.; Fu, H.; Li, H.; Wang, A.; Li, Z.; Zhang, H. A ductile $Nb_{40}Ti_{25}Al_{15}V_{10}Ta_5Hf_3W_2$ refractory high entropy alloy with high specific strength for high-temperature applications. *Mater. Sci. Eng. A* **2022**, *831*, 142290. [CrossRef]
12. Chen, C.; Zhang, H.; Fan, Y.; Zhang, W.; Wei, R.; Wang, T.; Zhang, T.; Li, F. A novel ultrafine-grained high entropy alloy with excellent combination of mechanical and soft magnetic properties. *J. Magn. Magn. Mater.* **2020**, *502*, 166513. [CrossRef]
13. Jin, K.; Lu, C.; Wang, L.M.; Qu, J.; Weber, W.J.; Zhang, Y.; Bei, H. Effects of compositional complexity on the ion-irradiation induced swelling and hardening in Ni-containing equiatomic alloys. *Scr. Mater.* **2016**, *119*, 65–70. [CrossRef]
14. Kumar, N.A.P.K.; Li, C.; Leonard, K.J.; Bei, H.; Zinkle, S.J. Microstructural stability and mechanical behavior of FeNiMnCr high entropy alloy under ion irradiation. *Acta Mater.* **2016**, *113*, 230–244. [CrossRef]
15. Egami, T.; Guo, W.; Rack, P.D.; Nagase, T. Irradiation resistance of multicomponent alloys. *Metall. Mater. Trans. A Phys. Metall. Mater. Sci.* **2014**, *45*, 180–183. [CrossRef]
16. Kareer, A.; Waite, J.C.; Li, B.; Couet, A.; Armstrong, D.E.J.; Wilkinson, A.J. Short communication: 'Low activation, refractory, high entropy alloys for nuclear applications'. *J. Nucl. Mater.* **2019**, *526*, 151744. [CrossRef]
17. Gouda, M.K.; Salman, S.A.; Ebied, S.; Ashmawy, A.M.; Gepreel, M.A.H.; Chiba, A. Biocompatibility and corrosion resistance of low-cost Ti–14Mn–Zr alloys. *J. Mater. Res.* **2021**, *36*, 4883–4893. [CrossRef]
18. Wang, W.; Yang, K.; Wang, Q.; Dai, P.; Fang, H.; Wu, F.; Guo, Q.; Liaw, P.K.; Hua, N. Novel Ti-Zr-Hf-Nb-Fe refractory high-entropy alloys for potential biomedical applications. *J. Alloys Compd.* **2022**, *906*, 164383. [CrossRef]
19. Schönecker, S.; Li, X.; Wei, D.; Nozaki, S.; Kato, H.; Vitos, L.; Li, X. Harnessing elastic anisotropy to achieve low-modulus refractory high-entropy alloys for biomedical applications. *Mater. Des.* **2022**, *215*, 110430. [CrossRef]
20. Yang, W.; Pang, S.; Liu, Y.; Wang, Q.; Liaw, P.K.; Zhang, T. Design and properties of novel Ti–Zr–Hf–Nb–Ta high-entropy alloys for biomedical applications. *Intermetallics* **2022**, *141*, 107421. [CrossRef]
21. Strozi, R.B.; Leiva, D.R.; Huot, J.; Botta, W.J.; Zepon, G. An approach to design single BCC Mg-containing high entropy alloys for hydrogen storage applications. *Int. J. Hydrogen Energy* **2021**, *46*, 25555–25561. [CrossRef]
22. Yu, Y.; Xu, N.; Zhu, S.; Qiao, Z.; Zhang, J.; Yang, J.; Liu, W. A novel Cu-doped high entropy alloy with excellent comprehensive performances for marine application. *J. Mater. Sci. Technol.* **2021**, *69*, 48–59. [CrossRef]
23. Wani, I.S.; Sathiaraj, G.D.; Ahmed, M.Z.; Reddy, S.R.; Bhattacharjee, P.P. Evolution of microstructure and texture during thermo-mechanical processing of a two phase $Al_{0.5}$CoCrFeMnNi high entropy alloy. *Mater. Charact.* **2016**, *118*, 417–424. [CrossRef]
24. Wani, I.S.; Bhattacharjee, T.; Sheikh, S.; Bhattacharjee, P.P.; Guo, S.; Tsuji, N. Tailoring nanostructures and mechanical properties of AlCoCrFe$Ni_{2.1}$ eutectic high entropy alloy using thermo-mechanical processing. *Mater. Sci. Eng. A* **2016**, *675*, 99–109. [CrossRef]
25. Bhattacharjee, T.; Wani, I.S.; Sheikh, S.; Clark, I.T.; Okawa, T.; Guo, S.; Bhattacharjee, P.P.; Tsuji, N. Simultaneous Strength-Ductility Enhancement of a Nano-Lamellar AlCoCrFe$Ni_{2.1}$ Eutectic High Entropy Alloy by Cryo-Rolling and Annealing. *Sci. Rep.* **2018**, *8*, 3276. [CrossRef]
26. Deng, Y.; Tasan, C.C.; Pradeep, K.G.; Springer, H.; Kostka, A.; Raabe, D. Design of a twinning-induced plasticity high entropy alloy. *Acta Mater.* **2015**, *94*, 124–133. [CrossRef]
27. Yao, M.J.; Pradeep, K.G.; Tasan, C.C.; Raabe, D. A novel, single phase, non-equiatomic FeMnNiCoCr high-entropy alloy with exceptional phase stability and tensile ductility. *Scr. Mater.* **2014**, *72–73*, 5–8. [CrossRef]
28. Baker, I.; Meng, F. Lamellar coarsening in $Fe_{28}Ni_{18}Mn_{33}Al_{21}$ and its influence on room temperature tensile behavior. *Acta Mater.* **2015**, *95*, 124–131. [CrossRef]
29. Liao, Y.; Baker, I. On the room-temperature deformation mechanisms of lamellar-structured $Fe_{30}Ni_{20}Mn_{35}Al_{15}$. *Mater. Sci. Eng. A* **2011**, *528*, 3998–4008. [CrossRef]
30. Stepanov, N.D.; Shaysultanov, D.G.; Chernichenko, R.S.; Tikhonovsky, M.A.; Zherebtsov, S.V. Effect of Al on structure and mechanical properties of Fe–Mn–Cr–Ni–Al non-equiatomic high entropy alloys with high Fe content. *J. Alloys Compd.* **2019**, *770*, 194–203. [CrossRef]
31. Cui, P.; Liu, Y.; Zhou, F.; Lai, Z.; Zhu, J. Enhancing high temperature mechanical properties via modulating B2 phase with Al contents in FeCrNiAlx(x = 0.63,0.71,0.77) high entropy alloys. *J. Alloys Compd.* **2022**, *903*, 163883. [CrossRef]
32. Kishore, K.; Chandan, A.K.; Hung, P.T.; Kumar, S.; Kawasaki, M.; Gubicza, J. On the enhanced hardening ability and plasticity mechanisms in a novel Mn-added CoCrNi medium entropy alloy during high-pressure torsion. *J. Alloys Compd.* **2022**, *904*, 163941. [CrossRef]
33. Oh, M.C.; Sharma, A.; Lee, H.; Ahn, B. Phase separation and mechanical behavior of AlCoCrFeNi-X (X = Cu, Mn, Ti) high entropy alloys processed via powder metallurgy. *Intermetallics* **2021**, *139*, 107369. [CrossRef]

34. Chen, C.; Wang, W.; Li, H.; Wei, P.; Zhang, W.; Fan, Y.; Chen, J.; Yuan, S.; Wei, R.; Zhang, T.; et al. Effects of Al and Mn on microstructure, magnetic and mechanical properties of $Fe_{40}Co_{40}Ni_{10}M_{10}$ (M=Al, Mn) medium entropy alloys. *J. Alloys Compd.* **2022**, *890*, 161779. [CrossRef]
35. Meng, F.; Qiu, J.; Baker, I. The effects of chromium on the microstructure and tensile behavior of $Fe_{30}Ni_{20}Mn_{35}Al_{15}$. *Mater. Sci. Eng. A* **2013**, *586*, 45–52. [CrossRef]
36. Gu, X.; Zhuang, Y.; Jia, P. Evolution of the microstructure and mechanical properties of as-cast $Al_{0.3}$CoCrFeNi high entropy alloys by adding Si content. *Mater. Sci. Eng. A* **2022**, *840*, 142983. [CrossRef]
37. Guo, Y.; He, J.; Li, Z.; Jia, L.; Wu, X.; Liu, C. Strengthening and dynamic recrystallization mediated by Si-alloying in a refractory high entropy alloy. *Mater. Sci. Eng. A* **2022**, *832*, 142480. [CrossRef]
38. Xu, Z.Q.; Ma, Z.L.; Tan, Y.; Wang, M.; Zhao, Y.; Cheng, X.W. Effects of Si additions on microstructures and mechanical properties of VNbTiTaSix refractory high-entropy alloys. *J. Alloys Compd.* **2022**, *900*, 163517. [CrossRef]
39. Liu, H.; Sun, S.; Zhang, T.; Zhang, G.; Yang, H.; Hao, J. Effect of Si addition on microstructure and wear behavior of AlCoCrFeNi high-entropy alloy coatings prepared by laser cladding. *Surf. Coat. Technol.* **2021**, *405*, 126522. [CrossRef]
40. Ma, M.; Han, A.; Zhang, Z.; Lian, Y.; Zhao, C.; Zhang, J. The role of Si on microstructure and high-temperature oxidation of CoCr2FeNb0.5Ni high-entropy alloy coating. *Corros. Sci.* **2021**, *185*, 109417. [CrossRef]
41. Huang, L.; Wang, X.; Jia, F.; Zhao, X.; Huang, B.; Ma, J.; Wang, C. Effect of Si element on phase transformation and mechanical properties for FeCoCrNiSix high entropy alloys. *Mater. Lett.* **2021**, *282*, 128809. [CrossRef]
42. Hou, Y.; Liu, T.; He, D.; Li, Z.; Chen, L.; Su, H.; Fu, P.; Dai, P.; Huang, W. Sustaining strength-ductility synergy of SLM $Fe_{50}Mn_{30}Co_{10}Cr_{10}$ metastable high-entropy alloy by Si addition. *Intermetallics* **2022**, *145*, 107565. [CrossRef]
43. Guo, R.; Pan, J.; Liu, L. Achieving dual-phase structure and improved mechanical properties in AlCoCrFe$Ti_{0.5}$ high-entropy alloys by addition of Ni. *Mater. Sci. Eng. A* **2022**, *831*, 142194. [CrossRef]
44. Qiu, H.; Zhu, H.; Zhang, J.; Xie, Z. Effect of Fe content upon the microstructures and mechanical properties of FexCoNiCu high entropy alloys. *Mater. Sci. Eng. A* **2020**, *769*, 138514. [CrossRef]
45. Huang, Y.; Gao, J.; Wang, S.; Guan, D.; Xu, Y.; Hu, X.; Rainforth, W.M.; Zhu, Q.; Todd, I. Influence of tantalum composition on mechanical behavior and deformation mechanisms of TiZrHfTax high entropy alloys. *J. Alloys Compd.* **2022**, *903*, 163796. [CrossRef]
46. Li, X.; Gao, X.; Lu, H.; Shi, C.; Guo, N.; Yin, F. Microstructure evolution and mechanical properties of $AlCrFe_2Ni_2(MoNb)_x$ high entropy alloys. *J. Mater. Res. Technol.* **2022**, *17*, 865–875. [CrossRef]
47. Wu, Y.; Zhang, Y. Design and High-Throughput Screening of High Entropy Alloys. In *Advances in High-Entropy Alloys-Materials Research, Exotic Properties and Applications*; IntechOpen: London, UK, 2021; pp. 1–16. [CrossRef]
48. Kurz, W.; Bezençon, C.; Gäumann, M. Columnar to equiaxed transition in solidification processing. *Sci. Technol. Adv. Mater.* **2001**, *2*, 185–191. [CrossRef]
49. Singh, S.; Wanderka, N.; Murty, B.S.; Glatzel, U.; Banhart, J. Decomposition in multi-component AlCoCrCuFeNi high-entropy alloy. *Acta Mater.* **2011**, *59*, 182–190. [CrossRef]
50. Ghaderi, A.; Moghanni, H.; Dehghani, K. Microstructural Evolution and Mechanical Properties of $Al_{0.5}$CoCrFeNi High-Entropy Alloy after Cold Rolling and Annealing Treatments. *J. Mater. Eng. Perform.* **2021**, *30*, 7817–7825. [CrossRef]

MDPI AG
Grosspeteranlage 5
4052 Basel
Switzerland
Tel.: +41 61 683 77 34

Metals Editorial Office
E-mail: metals@mdpi.com
www.mdpi.com/journal/metals

www.ingramcontent.com/pod-product-compliance
Lightning Source LLC
Chambersburg PA
CBHW041351210926
51850CB00013B/3046